CRASH COURSE
Endocrine and Reproductive Systems
SECOND EDITION

Series editor
Daniel Horton-Szar
BSc (Hons), MB BS (Hons)
GP Registrar
Northgate Medical Practice
Canterbury
Kent

Faculty advisor
Dr Susan Whiten
MA PhD
Senior Lecturer
School of Medicine
University of St Andrews
Fife

Endocrine and Reproductive Systems

SECOND EDITION

Stephan Sanders
BMedSci (Hons)
Nottingham Medical School
University of Nottingham

First Edition Author
Madeleine Debuse

Edinburgh • London • New York • Oxford • Philadelphia • St Louis • Sydney • Toronto 2002

MOSBY
An imprint of Elsevier Science Limited

Commissioning Editor   Alex Stibbe
Project Manager   Colin Arthur
Project Development Manager   Ruth Swan
Designer   Andy Chapman
Cover Illustrations   Kevin Faerber
Illustration Management   Mick Ruddy

© 2003, Elsevier Science Limited. All rights reserved.

The right of S Sanders to be identified as author of this work has been asserted by him in accordance with the Copyright, Designs and Patents Act 1988

No part of this publication may be reproduced, stored in a retrieval system, or transmitted in any form or by any means, electronic, mechanical, photocopying, recording or otherwise, without either the prior permission of the publishers or a licence permitting restricted copying in the United Kingdom issued by the Copyright Licensing Agency, 90 Tottenham Court Road, London W1T 4LP. Permissions may be sought directly from Elsevier's Health Sciences Rights Department in Philadelphia, USA: phone: (+1) 215 238 7869, fax: (+1) 215 238 2239, e-mail: healthpermissions@elsevier.com. You may also complete your request on-line via the Elsevier Science homepage (http://www.elsevier.com), by selecting 'Customer Support' and then 'Obtaining *Permissions*'.

First edition 1998
Second edition 2003
Reprinted 2003

ISBN 0723432457

British Library Cataloguing in Publication Data
A catalogue record for this book is available from the British Library

Library of Congress Cataloging in Publication Data
A catalog record for this book is available from the Library of Congress

Notice
Medical knowledge is constantly changing. Standard safety precautions must be followed, but as new research and clinical experience broaden our knowledge, changes in treatment and drug therapy may become necessary or appropriate. Readers are advised to check the most current product information provided by the manufacturer of each drug to be administered to verify the recommended dose, the method and duration of administration, and contraindications. It is the responsibility of the practitioner, relying on experience and knowledge of the patient, to determine dosages and the best treatment for each individual patient. Neither the Publisher nor the author assumes any liability for any injury and/or damage to persons or property arising from this publication.

www.elsevierhealth.com

Typeset by Kolam, Pondicherry, India
Printed in Spain by Graphycems

Preface

Without the endocrine system the world would be populated by very short and infertile people with an incredibly low life expectancy. It is clearly an advantage to have such a system, although this can be forgotten when you need to learn how it works. It is easy to get bogged down by the details of individual hormones whilst forgetting how subtly they work together to regulate so many vital processes.

This book aims to present the endocrine and reproductive systems in a clear and concise manner. It focuses on information that is useful in clinical medicine and also on subjects that frequently come up in examinations.

The first page of each chapter that deals with a hormone system contains a summary of the important points, including a diagram of how that hormone system works. I hope these will act as a quick revision tool for those vital moments of panic knowledge acquisition before an exam.

It is important to note that this book will not teach you to spell the word 'epididymis'. If you find that you can spell this word you may want to investigate a career outside of medicine.

'Hic sunt dracones.'

Stephan Sanders

Nowadays, instead of thinking of yourself as a medical student, you may be encouraged to consider yourself as a 'doctor in training', entering a professional environment from the very first days of your Medical School education. You are certainly entering a profession that will change beyond our recognition within your professional lifetime. Medicine continues to be an enormous challenge: that is why it is so fascinating and why it can bring such enormous satisfaction.

You probably have impossibly high expectations of becoming a brilliant, empathetic clinician, happily juggling a sophisticated social life with your research and teaching responsibilities. The reality is that there is scarcely time to be successful in even one of these areas! Modern evidence-based medical practice depends on advancing scientific knowledge, but we all have to compromise between breadth and depth of knowledge in any field. Medical educators in all disciplines are struggling to define 'core' knowledge in order to rationalize the medical curriculum. Core knowledge is the body of information that forms the basis for understanding both wider issues and future developments. The philosophy of the Crash Course books is based on the fact that successful students have a very good idea of what that core is. They are close to the information, quick to see the difficult areas that require explanation, and bring a lively approach to the subject.

Preface

You will find a wealth of well-organized information in this book, reviewed in a way that will facilitate your understanding. I particularly recommend that you use the self-assessment section: it is a very powerful means of learning. I wish you every success in your study of Endocrinology and in your chosen profession.

Dr Susan Whiten
Faculty Advisor

In the six years since the First Editions were published, there have been many changes in medicine, and in the way it is taught. These Second Editions have been largely rewritten to take these changes into account, and keep Crash Course up to date for the twenty-first century. New material has been added to include recent research and all pharmacological and disease management information has been updated in line with current best practice. We have listened to feedback from hundreds of students who have been using Crash Course and have improved the structure and layout of the books accordingly: pathology material has been closely integrated with the relevant basic medical science; there are more multiple-choice questions and the clarity of text and figures is better than ever.

The principles on which we developed the series remain the same, however. Medicine is a huge subject, and the last thing a student needs when exams are looming is to waste time assembling information from different sources, and wading through pages of irrelevant detail. As before, Crash Course brings you all the information you need, in compact, manageable volumes that integrate basic medical science with clinical practice. We still tread the fine line between producing clear, concise text and providing enough detail for those aiming at distinction. The series is still written by medical students with recent exam experience, and checked for accuracy by senior faculty members from across the UK.

I wish you the best of luck in your future careers!

Dr Dan Horton-Szar
Series Editor (Basic Medical Sciences)

Acknowledgements

I would like to thank Dan Horton-Szar and Susie Whiten for their excellent guidance and counsel. Thank you also to Alex Stibbe and Ruth Swan for their 'gentle encouragement' with deadlines and for making my writing into a real book. I must also thank caffeine, in its many varied and wonderful forms, for adding hours to days that at first glance appeared to have too few.

Thanks also to Kevin Hanretty from the Glasgow Nuffield Hospital for his advice on the clinical chapters and to Barry Kelly from the Royal Victoria Hospital, Belfast, for his help with the Investigations and Imaging chapter.

On a more personal level, my thanks to: Steve Holden, Phil Webster and Doug Forrester for their friendship and for making life entertaining, Peter and Deborah Sanders for their support and DNA and Piers Sanders for being that little bit less tough than me (officially true now—it's a published fact). Finally, thank you to Imogen Hart for being the most amazing person I have ever met and her revelations in geography.

Figure acknowledgements
Figure 14.12 Adapted from KL Moore, TVN Persaud, The Developing Human, Clinically Oriented Embryology, 5th edition, by permission of WB Saunders.
Figure 14.13 Adapted from D Llewellyn-Jones, Fundamentals of Obstetrics and Gynaecology, 6th edition, 1994, by permission of Suzanne Abraham and Mosby.
Figures 17.7, 17.9, 17.12A, 17.15A Reproduced with permission from R Grainger and D Allison, eds, Diagnostic Radiology: A Textbook of Medical Imaging, 4th edition, Churchill Livingstone.
Figures 17.8, 17.10A and B, 17.11A, 17.14 Reproduced with permission from D Sutton, Textbook of Radiology and Imaging, 6th edition, Churchill Livingstone.
Figure 17.13 Reproduced with permission from CRW Edwards et al, eds, Davidson's Principles and Practice of Medicine, 17th edition, Churchill Livingstone.
Figures 17.15B, C and D Reproduced with permission from IPC Murray & PJ Ell, eds, Nuclear Medicine, 2nd edition, Churchill Livingstone.

Dedication

To Imogen

Contents

Preface v
Acknowledgements vii
Dedication viii

Part I: Basic Medical Science 1

1. **Overview of the Endocrine System** 3
   The role of the endocrine system 3
   Hormones and endocrine secretion 3
   Organization of the endocrine system 4
   Hormone types and secretion 7
   Paracrine hormones 9
   Hormone receptors 10
   Relationship of the nervous and endocrine systems 12

2. **The Hypothalamus and the Pituitary Gland** 15
   Anatomy 15
   Development 17
   Microstructure 18
   Hormones 20
   Disorders of the hypothalamus 23
   Disorders of the anterior pituitary 23
   Disorders of the posterior pituitary 27

3. **The Thyroid Gland** 29
   Anatomy 30
   Microstructure 31
   Development 31
   Hormones 32
   Disorders of the thyroid gland 34

4. **The Adrenal Glands** 41
   Anatomy 42
   Development 42
   Microstructure 42
   Hormones of the adrenal cortex 45
   Disorders of the adrenal cortex 48
   Hormones of the adrenal medulla 53
   Disorders of the adrenal medulla 54

5. **The Pancreas** 57
   Location and anatomy 58
   Microstructure 59
   Development 59
   Hormones 59
   Endocrine control of glucose homeostasis 62
   Disorders 63

6. **Up and Coming Hormones** 73
   Endocrine role of the gastrointestinal tract 73
   Pineal gland 76
   Endocrine role of adipose tissue 78

7. **Endocrine Control of Fluid Balance** 81
   Fluid balance 82
   Hormones involved in fluid balance 83
   Natriuretic factors 85
   Disorders of fluid balance 85

8. **Endocrine Control of Calcium Homeostasis** 87
   The role of calcium 87
   Mechanisms involved in calcium homeostasis 87
   Hormones involved in calcium homeostasis 89
   Disorders of calcium regulation 92

9. **Endocrine Control of Growth** 97
   Direct control of growth 97
   Indirect control of growth 98
   Determination of height 100
   Disorders of growth 100

10. **Endocrine Disorders of Neoplastic Origin** 103
    Multiple endocrine neoplasia syndromes 103
    Ectopic hormone syndromes 103

11. **Development of the Reproductive System** 107
    Embryological development of gender . .107
    Development of the breast 112
    Postnatal development 112
    Puberty 112

12. **The Female Reproductive System** 115
    Organization 116
    Oogenesis 121
    Hormones 122
    The menstrual cycle 124

ix

Contents

Disorders of the ovaries and fallopian
tubes 127
Disorders of the endometrium and
myometrium 131
Menstrual disorders and the menopause 135
Disorders of the cervix 138
Disorders of the vagina and vulva 140
Disorders of the female breast 144

13. The Male Reproductive System 149
Organization 149
Spermatogenesis .154
Hormones 156
Disorders of the testes and epididymis . . .157
Disorders of the prostate 160
Disorders of the penis 162
Disorders of the male breast 165

14. The Process of Reproduction 167
Sexual intercourse and dysfunction 167
Fertilization and contraception 170
Infertility 176
Therapeutic abortion 178
The placenta 178
Reproductive hormones in pregnancy . .181
Maternal adaptations to pregnancy 183
Parturition and labour 187
Disorders of labour 189

Lactation 190
Disorders of pregnancy and the placenta 192

Part II: Clinical Assessment 197

**15. Common Presentations of Endocrine and
Reproductive Disease 199**

16. History and Examination 209
History 209
Communication skills 211
Examination 214

17. Investigations and Imaging 227
Investigating hormones 227
Other investigations 232
Imaging of the endocrine and reproductive
systems 234

Part III: Self-assessment 241

Multiple-choice Questions 243
Short-answer Questions 253
Essay Questions .254
MCQ Answers 255
SAQ Answers 264

Index 267

ENDOCRINOLOGY

1. Overview of the Endocrine System — 3
2. The Hypothalamus and the Pituitary Gland — 15
3. The Thyroid Gland — 29
4. The Adrenal Glands — 41
5. The Pancreas — 57
6. Up and Coming Hormones — 73
7. Endocrine Control of Fluid Balance — 81
8. Endocrine Control of Calcium Homeostasis — 87
9. Endocrine Control of Growth — 97
10. Endocrine Disorders of Neoplastic Origin — 103
11. Development of the Reproductive System — 107
12. The Female Reproductive System — 115
13. The Male Reproductive System — 149
14. The Process of Reproduction — 167

1. Overview of the Endocrine System

The role of the endocrine system

The endocrine system allows cells to communicate using chemical messengers called hormones. This communication is essential for the maintenance of homeostasis (Greek for 'staying the same'). Homeostasis is an ongoing process that minimizes change from the ideal physiological conditions, creating a suitable environment for life. As a result, hormones are important components of all major body systems; you cannot escape them.

The endocrine system also regulates long-term changes in the body, including:
- Growth.
- Sexual development.
- Pregnancy.

After reading this chapter you should be able to:
- Explain what is meant by the term 'hormone'.
- Picture the general organization of the endocrine system.
- Understand how hormone secretion is controlled.
- Describe the synthesis of the main types of hormones.
- Understand how these hormones act through their cellular receptors.
- Discuss the integration and role of the endocrine and nervous systems.

Important words:
Hormone: A chemical signal transported in the blood that is secreted by endocrine cells
Endocrine tissue: A group of cells that secrete hormones
Target cell: A cell that responds to a specific hormone
Receptor: A protein in target cells that detects hormones
Second messenger: A chemical that transmits the hormone message from the receptor to the effector
Effector: A protein regulated by a hormone that brings about the cellular effects

Hormones and endocrine secretion

Hormones
Classical definition
Classically a hormone is described as a chemical substance that is secreted by specialized endocrine cells directly into the blood to exert an effect on distant target cells. This process is endocrine secretion.

The word 'endocrine' means 'internal secretion' while 'hormone' is derived from the Greek verb '*hormao*' meaning 'I excite'.

Modern definition
Recent research has revealed many locally acting chemical substances that have challenged the classical definition of hormones. Four modes of delivery are recognized (Fig. 1.1). They are:
- Endocrine—chemicals that act on distant cells via the blood stream, e.g. thyroxine.
- Paracrine—chemicals that act on the surrounding cells without entering the blood, e.g. gut hormones.
- Autocrine—chemicals that act on the cell they are secreted from, e.g. nitric oxide.
- Neurocrine—signals between neurons, e.g. neurotransmitters.

Different textbooks suggest different explanations of the term 'hormone' ranging from the classical definition to a definition that encompasses all chemical signals external to cells. If clinicians talk about hormones, they generally mean chemical signals that pass through the blood (endocrine delivery).

Types of hormone
Three classes of hormone are secreted into the blood; the characteristics of these are explained later in the chapter:

Overview of the Endocrine System

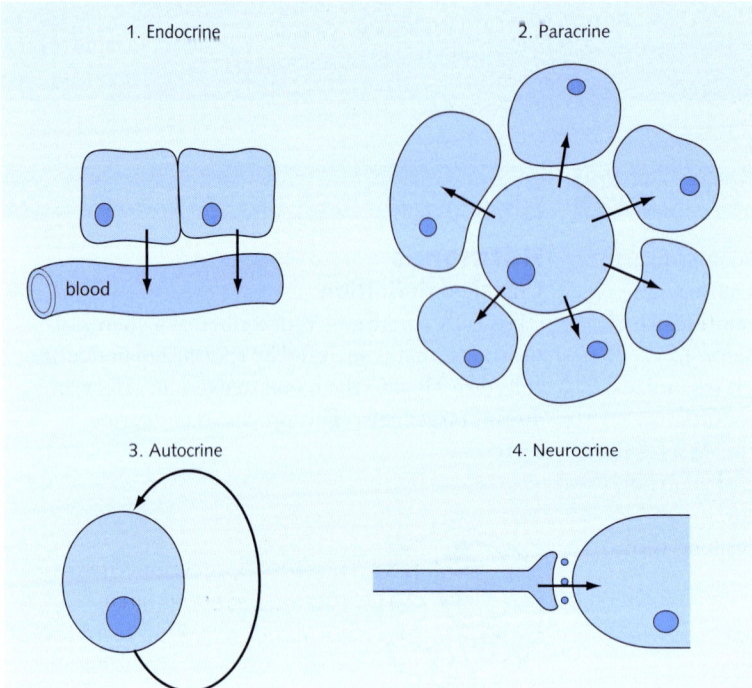

Fig. 1.1 The routes by which chemical signals are delivered to cells.

- Polypeptides (also called proteins).
- Steroids.
- Modified amino acids.

Endocrine tissues
Definition
An endocrine tissue is simply one that secretes a hormone. These tissues respond to signals that either stimulate or inhibit the release of the specific hormone.

Arrangement of endocrine tissues
Endocrine tissues contain cells that secrete hormones; these cells can be arranged in three patterns:
- As an endocrine organ devoted to hormone synthesis, e.g. the thyroid gland.
- As clusters of cells within an organ, e.g. the islets of Langerhans in the pancreas.
- Individual cells scattered diffusely throughout an organ, e.g. the gastrointestinal (GI) tract.

Endocrine organs
The term 'endocrine organ' originally referred to organs in which specialized endocrine cells formed a significant component. These 'traditional' endocrine organs are shown in Fig. 1.2 along with the hormone they secrete.

However, we now know that almost all organs contain some endocrine tissue, for example:
- Adipose tissue, secretes leptin.
- Lungs, secrete 5-hydroxytryptamine (5-HT; serotonin).
- Heart, secretes atrial natriuretic factor (ANF).

Organization of the endocrine system

The regulation and control of many major hormones follows a similar pattern that starts in the brain and ends with a hormone being secreted. Understanding this pattern is the key to understanding how the endocrine system works. There are three steps, each of which involves the secretion of a hormone that stimulates the next step (Fig. 1.3). The control of hormones released by the thyroid gland will be used to illustrate this pathway throughout.

The main components
Hypothalamus
The endocrine system is coordinated by the hypothalamus. This is a part of the brain that acts as a bridge between the nervous system and endocrine system, translating neural messages into chemical (hormonal) signals. It initiates the secretion of

Organization of the Endocrine System

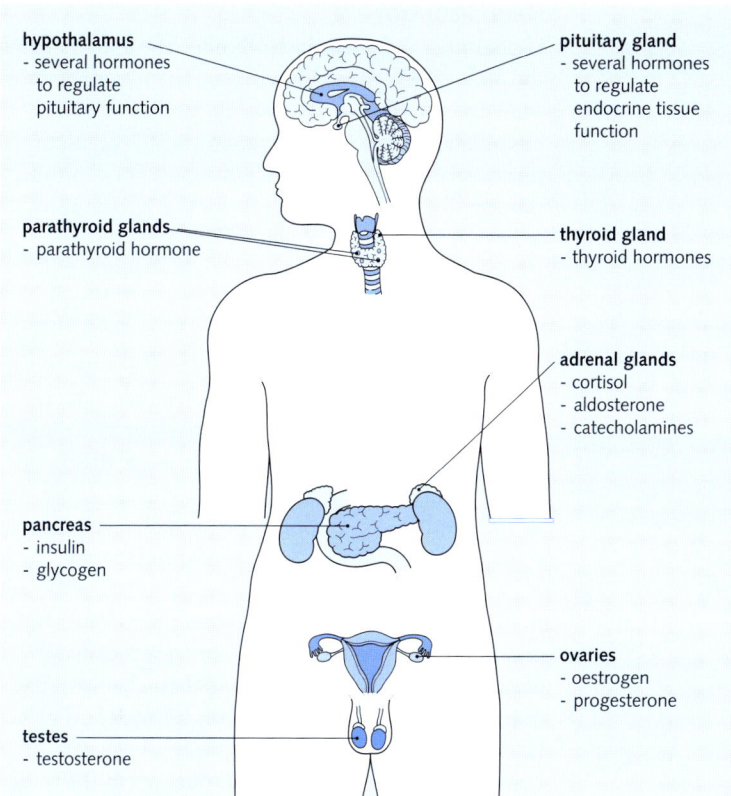

Fig. 1.2 The location of major endocrine organs and the hormones secreted by them.

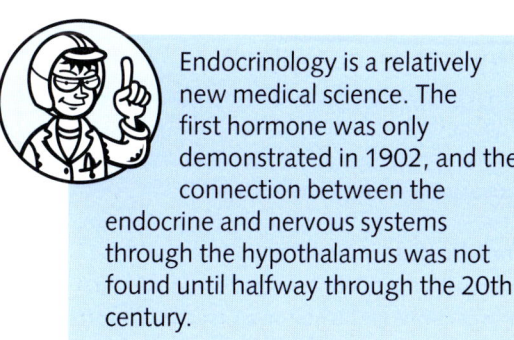

Endocrinology is a relatively new medical science. The first hormone was only demonstrated in 1902, and the connection between the endocrine and nervous systems through the hypothalamus was not found until halfway through the 20th century.

hormones by controlling the function of the pituitary gland via 'releasing hormones'. These hormones do not act directly on peripheral endocrine tissues. Hormones secreted from the hypothalamus are released in pulsatile manner often with a circadian rhythm (regular changes through a 24 h cycle). Thyrotrophin-releasing hormone (TRH) is secreted into the blood by the hypothalamus; this initiates the hormone cascade resulting in the release of thyroid hormones.

Pituitary gland

The pituitary gland is found at the base of the brain beneath the hypothalamus. It releases hormones into the blood in response to signals from the hypothalamus. The hormones from the pituitary gland regulate the function of peripheral endocrine tissues throughout the body. TRH from the hypothalamus acts on the pituitary gland to cause the release of thyroid stimulating hormone (TSH) into the blood stream.

Peripheral endocrine tissues

The hormones secreted by the pituitary gland act on peripheral endocrine tissues. These tissues respond by increasing or decreasing secretion of specific hormones into the blood. It is the hormones secreted by these peripheral tissues that affect the state of the body by acting on target cells. TSH from the pituitary gland stimulates the thyroid gland to release thyroid hormones into the blood.

Target cells

Different hormones act on different, specific cells. The cells that respond to a specific hormone are called its target cells; they can be found anywhere in

5

Overview of the Endocrine System

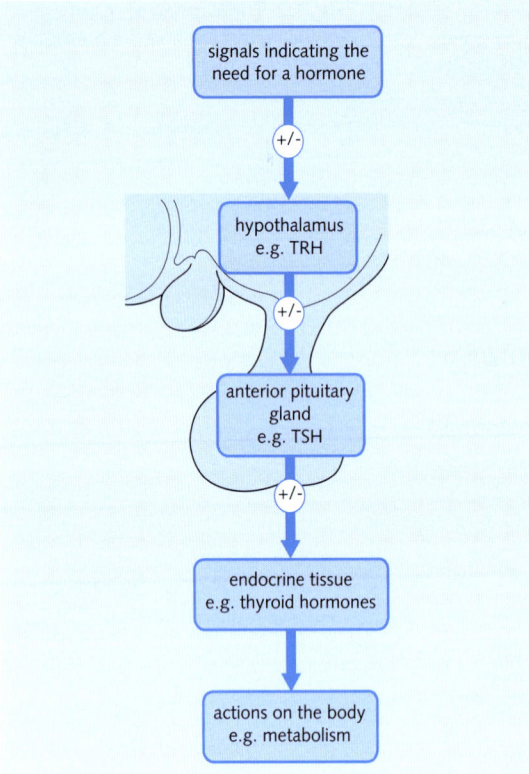

Fig. 1.3 The organization of the endocrine system.

Neural control
Higher neural centres can influence the activity of the endocrine system by acting on the hypothalamus. They can increase or decrease the secretion of hypothalamic releasing hormones, which regulate the secretion of pituitary gland hormones. For example, stress or fear will inhibit reproductive hormone secretion, and cold external temperatures will stimulate TRH.

Hormonal feedback
An almost universal feature of endocrine system regulation is feedback from the hormones that are released. The hormones can feed back by two means:
- Directly—e.g. the hormone thyroxine affects the hypothalamus and pituitary.
- Indirectly—e.g. through chemical changes caused by the hormones, such as glucose deficiency.

Feedback is usually inhibitory, thus a hormone can inhibit its own production; this process is called negative feedback. It is an essential mechanism that prevents excess secretion of many hormones. The level at which the feedback acts varies between hormones, however, many hormones act at the level of the hypothalamus and pituitary gland. For example, thyroid hormones feed back to the anterior pituitary where they inhibit the release of TSH (Fig. 1.4).

Why is it so complex?
At first glance the endocrine system seems incredibly complex for no obvious reason. Many students wish that the system had fewer hormones and organs, but there are a number of advantages.

Amplification
As described, endocrine signals begin in the hypothalamus and result in a cascade of hormones

the body. All target cells have receptors to detect the specific hormone, but the effect of the hormone can vary between cells. Thyroid hormones from the thyroid gland act on almost every cell in the body to increase the rate of metabolism through receptors on the cell surface.

Control of hormone secretion
Overall control
Endocrine tissues are regulated by signals from a variety of neural and systemic sources. These signals are processed by cells to determine the rate of hormone secretion. The strength and importance of the signals varies so that hormone secretion fits the needs of the body.

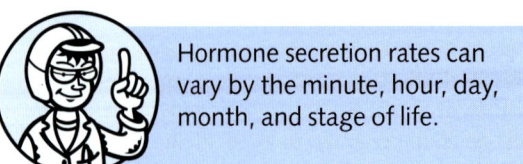

Hormone secretion rates can vary by the minute, hour, day, month, and stage of life.

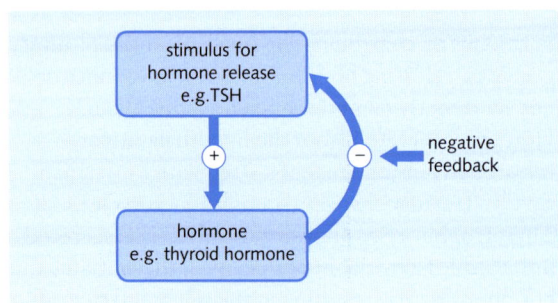

Fig. 1.4 Negative feedback.

Hormone Types and Secretion

from different endocrine glands. There is only a small population of cells in the hypothalamus that secrete each hormone; for example, about 2000 neurons secrete gonadotrophin releasing hormone (GnRH). Because of the small number of cells involved, they are able to respond to important but small neural signals, but they cannot secrete large amounts of hormone.

The very small quantities of hormone secreted directly into the blood stream by the hypothalamus can be detected by the closely related pituitary gland. This gland is able to secrete a greater quantity of hormone than the hypothalamus, but it is still too small to secrete enough for the whole body.

In response to hormones from the pituitary gland the peripheral endocrine tissues secrete hormones in large quantities that can act throughout the body. In this manner the signal of a small number of neurons in the hypothalamus is amplified in three stages to affect the entire body.

Control

The endocrine system regulates all major body processes that are essential for life including:
- Metabolic rate.
- Nutrient levels.
- Cardiac output and blood pressure.
- Reproduction.

Since they are so important, these processes must be controlled very tightly. The complex interactions of the endocrine system allow for many sites of control in order to prevent excessive or deficient hormone release, and to maintain homeostasis.

Hormone types and secretion

This section describes the properties and synthesis of the three classes of hormone (Fig. 1.5).

Polypeptide hormones

As their name suggests, polypeptide hormones are proteins that act as hormones. The size of the polypeptide varies widely from 3 to 200 amino acid residues; they cannot pass through cell membranes due to their size and water-soluble nature. Protein hormones are the most numerous type (often a safe bet in an exam). Accordingly they are secreted by many glands, including:
- Hypothalamus—TRH, GnRH, growth hormone releasing hormone (GHRH), etc.
- Pituitary—TSH, follicle stimulating hormone (FSH), luteinizing hormone (LH), oxytocin, etc.
- Pancreas and GI tract—insulin, glucagon, cholecystokinin (CCK), etc.

Synthesis

Polypeptide hormones are synthesized in the same manner as any other protein. DNA in the nucleus is transcribed to mRNA and translated into the protein by ribosomes. The protein is then processed by the Golgi apparatus and stored in secretory granules. Many hormones undergo changes in the Golgi apparatus or secretory granules including:
- Cleavage reactions to free a smaller polypeptide hormone from the larger prohormone.
- Addition of carbohydrate groups to form glycoproteins.

Fig. 1.5 Comparison of the different types of hormone.

Comparison of different types of hormone

	Polypeptides	Modified amino acids	Steroids
Size	Medium–large	Very small	Small
Ability to cross cell membrane	✗	✓	✓
Receptor type	Cell-surface	Cell-surface or intracellular	Intracellular
Soluble in:	Water	Water	Fat
Action	Protein activation	Protein activation or synthesis	Protein synthesis
Transport in the blood	Dissolved in the plasma	Dissolved in the plasma or bound to plasma proteins	Bound to plasma proteins

Overview of the Endocrine System

Secretion
The secretory granules are released by exocytosis, in which the membrane of the granule fuses with the membrane of the cell causing the contents to be ejected. This process is triggered by calcium entering the cell. Polypeptide hormone release is controlled mainly by regulating secretion rather than synthesis.

Polypeptide secreting cells
Polypeptide secreting cells all have a similar histological appearance (Fig. 1.6):
- Large, prominent nuclei.
- Small amount of cytoplasm.
- Prominent Golgi apparatus.
- Abundant rough endoplasmic reticulum (RER).
- Large numbers of secretory granules.
- Surrounded by fenestrated blood sinusoids.

Steroid hormones
Steroids are small, fat-soluble molecules that can pass through cell membranes but must circulate bound to plasma proteins since they are insoluble in the blood. They are secreted by:
- Adrenal cortex—cortisol and aldosterone.
- Ovaries—oestrogen and progesterone.
- Placenta—oestrogen and progesterone.
- Testes—testosterone.

Synthesis
Steroids are derived from cholesterol by a series of reactions in the mitochondria and smooth endoplasmic reticulum (SER). Cholesterol is acquired from the diet or synthesized within the cells; it is stored within lipid droplets seen in the cytoplasm of steroid cells. All steroids have the same basic structure formed by four rings of carbon (Fig. 1.7), but individual hormones differ in the following ways:
- Side chains attached to these rings.
- Bonds within the rings (double or single).

The exact sequence of reactions to synthesize each hormone varies, since there are many different enzyme pathways. However, the vast majority of steroid hormones share two common steps.

Step 1
Cholesterol is converted into pregnenolone by the desmolase enzyme found within the mitochondria of steroid-producing cells. Desmolase removes six carbon atoms from the cholesterol side chain of ring D. This reaction is the rate-limiting step in steroid synthesis.

Step 2
Pregnenolone is converted to progesterone by enzymes found in the mitochondria and cytoplasm. This reaction involves:
- Isomerization—the double bond moves from ring B to ring A.
- Oxidation—the hydroxyl group (OH) of ring A becomes a keto group (O).

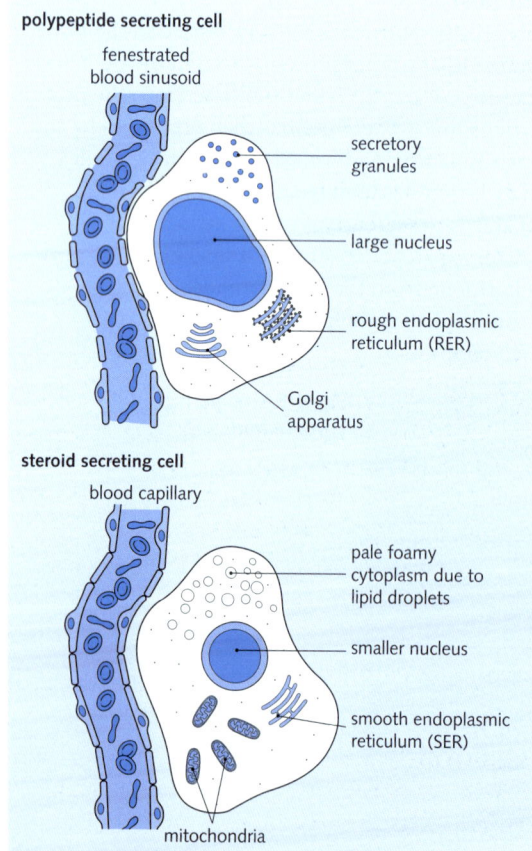

Fig. 1.6 Appearance of a polypeptide secreting cell and steroid secreting cell.

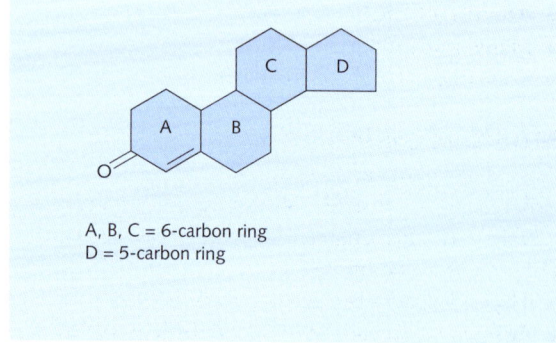

A, B, C = 6-carbon ring
D = 5-carbon ring

Fig. 1.7 Basic structure of a steroid hormone.

Paracrine Hormones

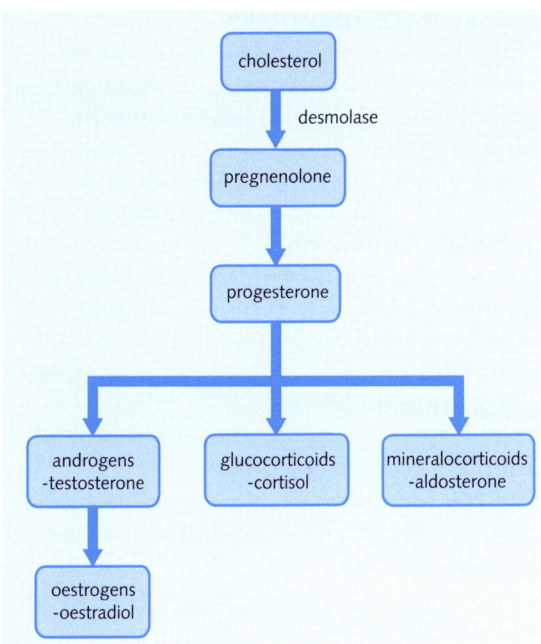

Fig. 1.8 Steroid synthesis. The initial stages are the same for all steroid hormones.

Further steps are very variable but the general pattern is shown in Fig. 1.8.

Secretion
The steroid hormone is released immediately so the rate of release is determined by the rate of synthesis, especially the synthesis of pregnenolone.

Steroid secreting cells
Steroid secreting cells also have a similar histological appearance to each other (see Fig. 1.6):
- Small, rounded nuclei.
- Large amount of cytoplasm.
- Large numbers of lipid droplets (foamy appearance).
- Abundant SER.
- Many mitochondria.
- Surrounded by blood capillaries.

Modified amino acids
Several hormones are formed by altering the structure of amino acids, producing small, water-soluble hormones that can cross cell membranes. They are secreted by the:
- Thyroid gland—thyroid hormones.
- Adrenal medulla—catecholamines (noradrenaline and adrenaline).
- Hypothalamus—dopamine.
- Pineal gland—melatonin.

Synthesis
These hormones are synthesized from two amino acids:
- Tyrosine—precursor of thyroid hormones, dopamine, and catecholamines.
- Tryptophan—precursor of melatonin and 5-HT.

The reactions to modify these amino acids vary significantly between hormones so the synthesis is described in the individual chapters. The hormones are stored in secretory granules except thyroid hormones, which uniquely are stored in follicles.

Secretion
The granules are released by exocytosis in the same way as polypeptide hormones. The rate of release is regulated mainly by secretion.

Modified amino acid secreting cells
The cells that secrete modified amino acid hormones vary more than the cells secreting polypeptide or steroid hormones, however the following features are often found:
- Large nuclei.
- Many mitochondria.
- Abundant RER.
- Prominent Golgi apparatus.
- Large numbers of secretory granules.
- Surrounded by blood capillaries.

 A single hormone may have multiple actions, equally multiple hormones may have the same action. This is demonstrated by insulin and the regulation of blood glucose respectively.

Paracrine hormones

Eicosanoids
Although eicosanoids are not always considered as hormones and they do not form one of the main classes of hormones, they are important in many physiological processes. They are, therefore, included in most endocrine courses. They are small,

Overview of the Endocrine System

lipid-soluble molecules that act in a paracrine (local) manner. They are derived from a phospholipid found in the cell membrane called arachidonic acid which is broken down by the enzyme phospholipase A_2. There are two pathways, which synthesize different groups of eicosanoids:

- Cyclooxygenase pathway—forms prostaglandins and thromboxanes.
- Lipoxygenase pathway—forms leukotrienes.

Eicosanoids are released immediately and readily cross cell membranes. Their action varies between cells and the specific eicosanoid molecule that is formed.

Aspirin and NSAIDs produce their anti-inflammatory action by blocking the cyclooxygenase pathway to prevent prostaglandin synthesis.

Hormone receptors

Target cells possess unique receptors that bind specific hormones; without these receptors the hormones can have no effect. The number of receptors per cell can be increased or decreased to alter the strength of the hormone's effect. Receptors are found in two locations:

- Cell-surface receptors—for polypeptides and catecholamines; they activate or inhibit enzymes, which may affect protein synthesis.
- Intracellular receptors—for steroids and thyroid hormones; they stimulate or inhibit protein synthesis directly.

The response to a hormone varies between target cells so that the same hormone can have different actions on different tissues. This variation is partly due to different receptor types but also the response to receptor stimulation.

Hormones that act via cell-surface receptors can respond faster than those stimulating intracellular receptors because the activation of pre-existing enzymes takes less time than synthesizing new proteins. This explains why catecholamines released for the 'fight-or-flight' response use cell-surface receptors even though they can cross cell membranes.

Cell-surface receptors

Cell-surface receptors are necessary for polypeptide hormones, which cannot cross the cell membrane, and catecholamines. The receptor must transmit the external signal into the cell where it can have an effect, therefore cell-surface receptors are glycoproteins that cross the cell membrane to create extracellular and intracellular domains. When the hormone binds to the receptor it triggers a cascade of changes within the cell that alter protein activity. There are two types of cell-surface receptor involved in the endocrine system:

- G-protein coupled receptors.
- Tyrosine kinase receptors.

G-protein coupled receptors

G-protein coupled receptors are extremely common throughout the endocrine system. They consist of two main elements:

- Receptor.
- G-protein.

The receptor is a glycoprotein with a hormone binding site on the extracellular surface and a G-protein binding site on the intracellular surface. When the hormone binds the receptor changes shape affecting the attached G-protein.

The G-protein is an enzyme that can break down guanosine triphosphate (GTP), hence the name. It is made of two functional subunits:

- α-subunit—bound to guanosine diphosphate (GDP) in the resting state.
- βγ-complex—bound to the α-subunit if GDP is present.

When the hormone binds, causing the receptor to change shape, the α-subunit exchanges the GDP for GTP. The G-protein splits into the two subunits described above, both of which leave the receptor and bind to effector proteins also found on the inside of the cell membrane.

These effector proteins often include the enzyme adenylate cyclase that synthesizes cyclic AMP (cAMP) from ATP. cAMP acts as a second messenger: a chemical signal that can enter the cell to activate or inhibit enzymes to bring about the effects of the hormone. The activated enzymes are often kinases, which add phosphate to other proteins, thereby activating them.

When the hormone signal stops, the α-subunit breaks down GTP into GDP and inorganic phosphate. The βγ-complex rejoins the α-subunit to form the G-protein, and this once again binds to the receptor. The effector proteins are no longer stimulated and cAMP is no longer produced. Fig. 1.9 shows the action of a G-protein receptor.

Other G-protein coupled receptors can use different effectors or second messengers including:
- Inhibition of adenylate cyclase.
- Stimulation of inositol triphosphate.
- Activation of ion channels.

Tyrosine kinase receptors

Insulin and insulin-like growth factors act through tyrosine kinase receptors. These receptors are glycoproteins with kinase activity (ability to add phosphate groups) that is triggered by the binding of the hormone. The receptors phosphorylate each other by adding phosphate to tyrosine residues within their structure. This attracts other proteins that also become phosphorylated. The mechanism of these receptors is not known beyond this point, however a second messenger may well be involved. This mechanism is shown in Fig. 1.10.

Intracellular receptors

Hormones that readily cross the cell membrane, especially steroids, use intracellular receptors. The receptors stimulate protein synthesis directly, so they are also called transcription factors.

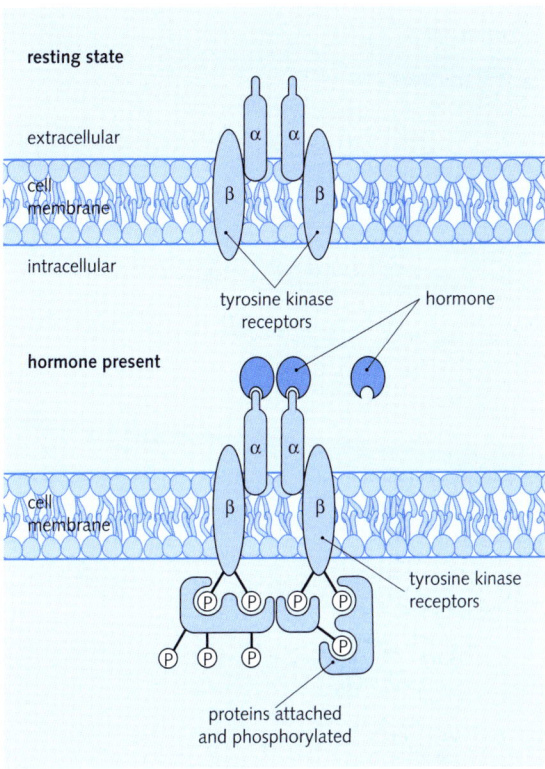

Fig. 1.10 Mechanism of action of a tyrosine kinase receptor.

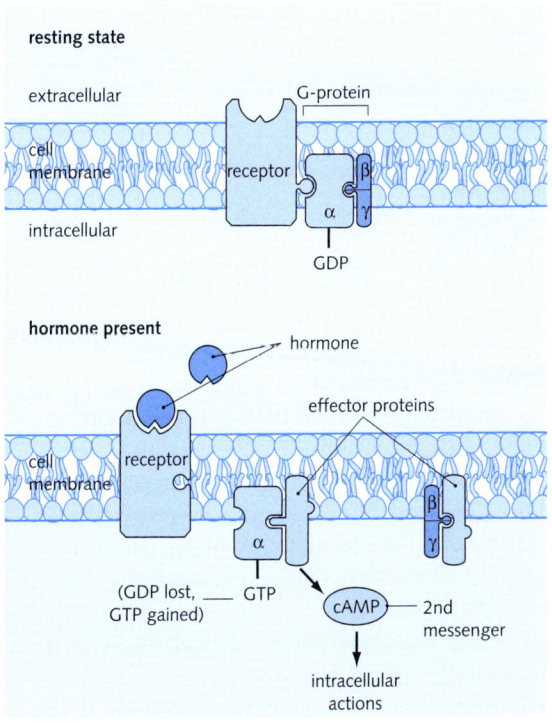

Fig. 1.9 Mechanism of action of a G-protein receptor.

The hormone binds to the receptor in the cytoplasm causing a change in shape that activates the receptor. The hormone and receptor enter the nucleus together, where they bind to specific sections of DNA called hormone response elements. This binding stimulates or inhibits the transcription of specific genes causing changes in protein synthesis. It is this change that brings about the effects of the hormone. This action is shown in Fig. 1.11.

Receptor mediated control

Hormone receptors are an important site of endocrine regulation. The number of active receptors can be increased or decreased to alter the strength of an endocrine signal. This allows the cell to respond to the deficiency or excess of a hormone. This control can be very subtle, for example, GnRH receptors in the pituitary gland are downregulated if GnRH secretion is not pulsatile. This effect is used clinically to suppress the reproductive hormones.

Overview of the Endocrine System

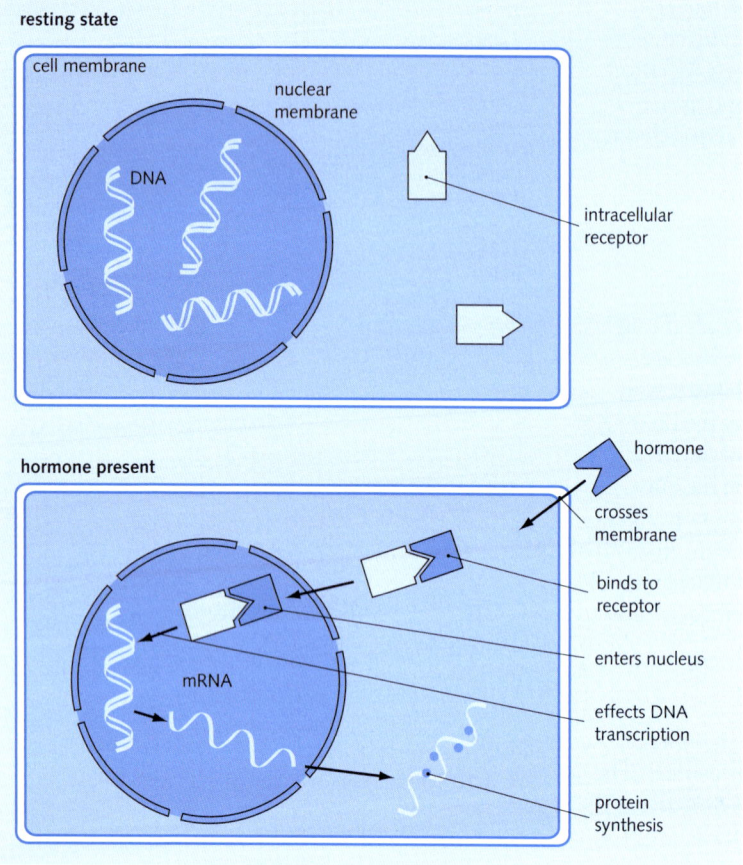

Fig. 1.11 Mechanism of action of an intracellular receptor.

Relationship of the nervous and endocrine systems

Integration
The nervous and endocrine systems have a very close relationship, since they both use chemical signals to communicate between cells, and they may share a common evolutionary origin. The overlap between some hormones and neurotransmitters also supports this idea (e.g. somatostatin is found in both systems). The close relationship allows the two systems to coordinate responses to maintain homeostasis.

Neural control of hormones
The nervous system can control the endocrine system through two routes:
- Hypothalamus.
- Autonomic nervous system (sympathetic and parasympathetic).

The endocrine system often acts as a long-term output from the brain to complement the action of short-term neural responses. This is demonstrated by the three responses to stress listed in the order they take effect:
- Noradrenaline is released from sympathetic nerves.
- Preformed adrenaline is released from the adrenal medulla.
- Cortisol is synthesized by the adrenal cortex.

Hormonal control of neurons
To complete this circuit, the hormones of the endocrine system also affect the nervous system. Negative feedback to the hypothalamus has already been described. However, many hormones affect other areas of the brain, for example:
- Thyroid hormone deficiency causes depression.
- Leptin and insulin regulate feelings of hunger.
- Adrenaline increases mental activity.
- Melatonin regulates the feeling of tiredness.

Relationship of the Nervous and Endocrine Systems

Comparison between the nervous and endocrine systems

While the two systems function closely they have different modes of action. The hypothalamus combines these actions since it is an endocrine tissue composed of nerve cells called neurosecretory cells.

Nervous system

The nervous system uses very localized chemical signals at synapses to transmit membrane depolarization between neurons. The effects of the nervous system are very rapid but of short duration and expensive metabolically (i.e. the neurotransmitters and depolarization require a lot of energy). The specific target cell is determined mostly by the location of chemical release rather than the receptors.

Endocrine system

The endocrine system uses very generalized chemical signals, though a few endocrine tissues can depolarize. These signals require less energy than neural signals. The signals travel throughout the body in the blood stream, and the target cell is determined mainly by the presence and specificity of receptors. The signals of the endocrine system tend to be slower but with a longer duration.

- Define the terms hormone, endocrine, target cell, and paracrine.
- Name five hormones along with the gland that secretes them.
- Outline the role of the hypothalamus in the endocrine system.
- Where do hormones from the hypothalamus act?
- Outline the role of the anterior pituitary gland in the endocrine system.
- Where do hormones from the anterior pituitary gland act?
- Describe the control of thyroid hormones starting in the hypothalamus.
- What is an eicosanoid? Is it a hormone?
- Define negative feedback, and give an example of this process.
- Describe the role of amplification in the endocrine system.
- State the three types of hormone along with an example of each.
- Describe the synthesis and properties of polypeptide hormones.
- Describe the mode of action of a G-protein receptor along with three examples of hormones that act through these receptors.
- Describe the mode of action of a tyrosine kinase receptor along with both hormones that act through these receptors.
- Describe the synthesis and properties of steroid hormones.
- Describe the receptors that steroid hormones act on.
- Describe the properties of modified amino acid hormones.
- Where can neurons affect the endocrine system?
- List examples of hormones that act on neurons.
- Compare the action of hormones and neurons.

13

2. The Hypothalamus and the Pituitary Gland

The hypothalamus and pituitary gland control the endocrine system. To understand how the other endocrine organs function it is essential to know how they relate to these controllers.

The hypothalamus is a part of the brain that controls the secretion of many hormones. It can release small quantities of hormones in response to neuronal signals, and thus it acts as an interface between the nervous and endocrine systems, converting neural signals into chemical signals.

Hypothalamic hormones are carried locally in the blood stream. They are detected by cells in the anterior part of the pituitary gland, which is found just below the hypothalamus. In response, the pituitary gland secretes different hormones in larger quantities. Most of these hormones regulate other endocrine organs (e.g. thyroid stimulating hormone), however, some affect parts of the body directly (e.g. prolactin).

The posterior part of the pituitary gland functions in a slightly different way because it is a direct extension of the hypothalamus. Neurosecretory cells in the hypothalamus synthesize hormones that are transported along their axons. These hormones are released into capillaries within the posterior pituitary gland to affect body parts directly.

The secretory activity of the hypothalamus and pituitary gland can also be affected by hormones released from other endocrine organs (e.g. thyroxine from the thyroid gland). This feedback helps to control hormone levels, and it is a key component of endocrine function.

At first glance the hypothalamus and pituitary gland seem needlessly complicated to perform a simple task. There are two main reasons for this arrangement:
- It allows intricate regulation of hormone levels.
- It amplifies the initial signal so that a few neurons can affect cells throughout the body.

After reading this chapter you should be able to:
- Describe the structure and development of the hypothalamus and pituitary gland.
- List the hormones released by these structures and state their effects.
- Explain the major disorders associated with these structures.

Important words:
Hypothalamus: a part of the brain that controls the endocrine system
Pituitary gland: an endocrine gland beneath the hypothalamus that controls other endocrine glands
Adenohypophysis: the anterior pituitary gland
Neurohypophysis: the posterior pituitary gland
Adenoma: a benign tumour

Anatomy

Hypothalamus

The hypothalamus is located at the base of the forebrain beneath the thalamus and together they form the lateral walls of the third ventricle. The optic chiasma is anterior to the hypothalamus and the mammillary bodies are found posteriorly. The inferior part of the hypothalamus—called the median eminence—gives rise to the pituitary stalk, which is continuous with the posterior pituitary gland. This arrangement is shown in Figs 2.1 and 2.2.

The hypothalamus receives multiple inputs about the homeostatic state of the body. These arrive by two means:
- Circulatory, e.g. temperature, blood glucose, hormone levels.
- Neuronal, e.g. autonomic function, emotional.

It responds to these inputs by the secretion of hormones that either regulate the release of hormones from the anterior pituitary or are released directly from the posterior pituitary (e.g. antidiuretic hormone—ADH). It also responds by neuronal signals to other areas of the central nervous system (CNS).

Pituitary gland

The pituitary gland is divided into two lobes with distinct embryological origins, structure, and function:

The Hypothalamus and the Pituitary Gland

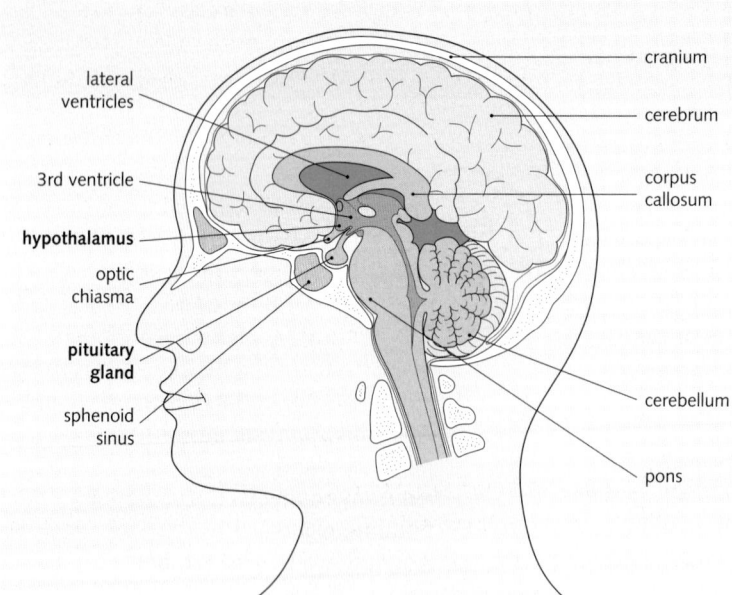

Fig. 2.1 Medial sagittal section of head showing the location of the hypothalamus and pituitary gland.

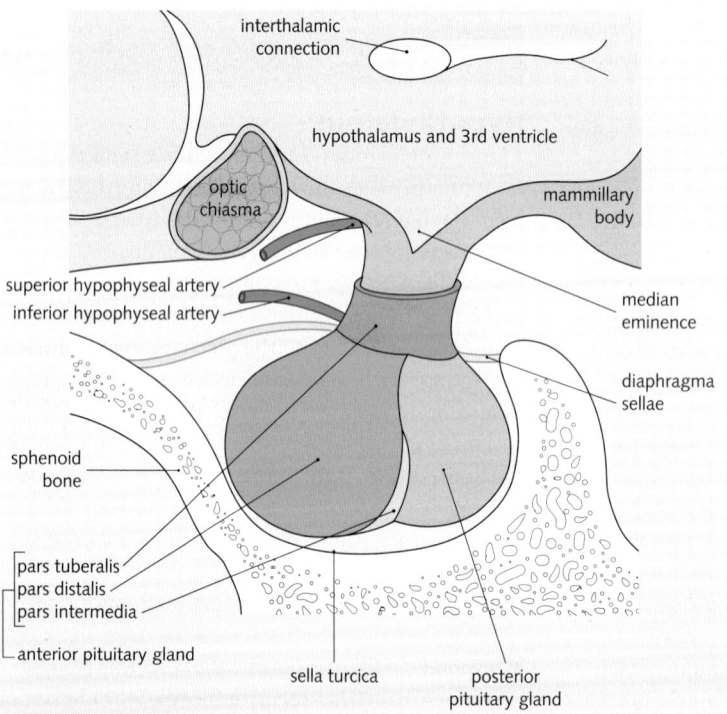

Fig. 2.2 Anatomical relationship of the pituitary gland and the hypothalamus to surrounding structures.

The pituitary gland gets its name from the incorrect belief that it secreted nasal mucus put forward by the Greek physician Galen (129–216 AD). 'Pituitary' means 'mucus'.

The hypothalamus is a 'V' shape because it surrounds the third ventricle.

- Anterior pituitary or adenohypophysis.
- Posterior pituitary or neurohypophysis.

The pituitary gland lies in a deep bony hollow in the sphenoid bone (the sella turcica), and it is covered by the fibrous diaphragma sellae. The optic chiasma lies above this diaphragm directly superior to the anterior lobe. The posterior lobe is connected to the median eminence of the hypothalamus by the pituitary stalk (infundibulum). The cavernous sinuses including the cranial nerves III–VI lie laterally. See Figs 2.1 and 2.2.

 Tumours of the pituitary gland are surrounded by the bone of the sella turcica, so they can only expand upwards into the optic chiasma causing visual field defects. Further expansion compresses cranial nerves III, IV, V, and VI in the wall of the cavernous sinus.

Blood supply

The blood supply to the anterior pituitary gland first passes through the median eminence, creating a vascular communication. The median eminence of the hypothalamus is supplied by the superior hypophyseal artery, which forms a plexus within it. The blood then drains into portal veins, which supply the anterior pituitary gland. Hormones secreted from the median eminence pass directly to the anterior pituitary in the blood stream as shown in Fig. 2.3.

The posterior pituitary gland is supplied by the inferior hypophyseal arteries. These vessels do not communicate with the median eminence.

The veins draining the pituitary gland drain to the cavernous sinuses.

Development

Hypothalamus
The hypothalamus develops from the embryological forebrain; it can be identified at week six of gestation.

Anterior pituitary
The anterior pituitary develops as an outgrowth of the ectoderm of the primitive mouth called Rathke's pouch. It grows upwards until it fuses with the down-growing infundibulum of the hypothalamus. The anterior pituitary is composed of non-neural secretory epithelial tissue, and it is not directly connected to the hypothalamus.

The connection to the roof of the primitive mouth is gradually lost along with its blood supply. The

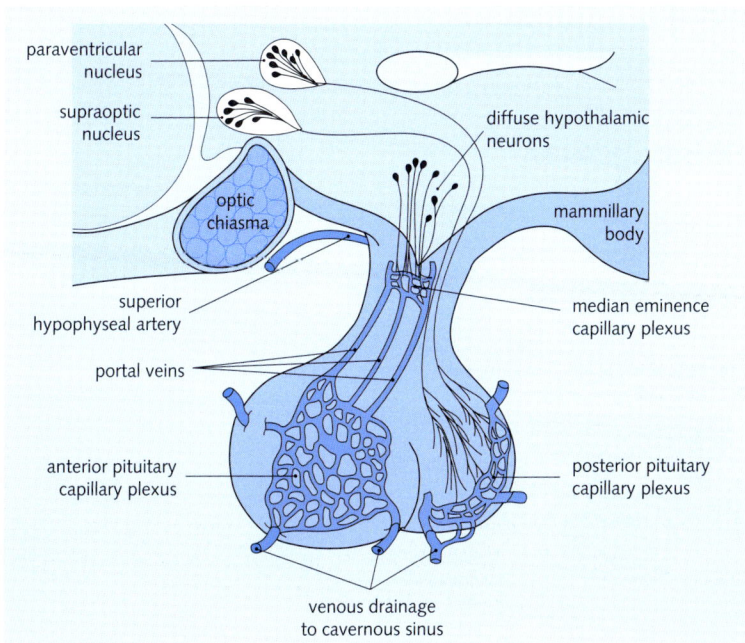

Fig. 2.3 Communication between the hypothalamus and pituitary gland; note the difference between the anterior and posterior pituitary gland.

17

portal veins from the hypothalamus grow down to replace this blood supply, and these are the only communication between the hypothalamus and the anterior pituitary.

As the connection to the primitive mouth is lost, nests of epithelial cells may be left behind. These can give rise to cysts or tumours, which may secrete ectopic hormones (e.g. craniopharyngiomas—see the disorders section).

The embryology of the pituitary gland is shown in Fig. 2.4.

Posterior pituitary

The posterior pituitary is derived from the neuroectoderm of the primitive brain tissue. It develops as an outgrowth from the hypothalamus called the infundibulum. Axons from neurosecretory cells in the hypothalamus pass downwards in the stalk of the pituitary gland and terminate in the posterior pituitary. A direct neuronal connection between the hypothalamus and posterior pituitary is formed, and this is the only means of communication between these structures.

Microstructure

Hypothalamus

There are a number of different secretory neurons in the hypothalamus, each specialized to secrete specific hormones. Neurons that secrete the same chemical may be arranged in clusters called nuclei, or they may be scattered diffusely. Some neurons can secrete more than one hormone.

Anterior pituitary

The anterior pituitary is composed of cords of secretory cells in a rich network of capillaries. Six types of secretory cells can be distinguished using immuno-histochemistry. These are listed with the hormones that they synthesize:

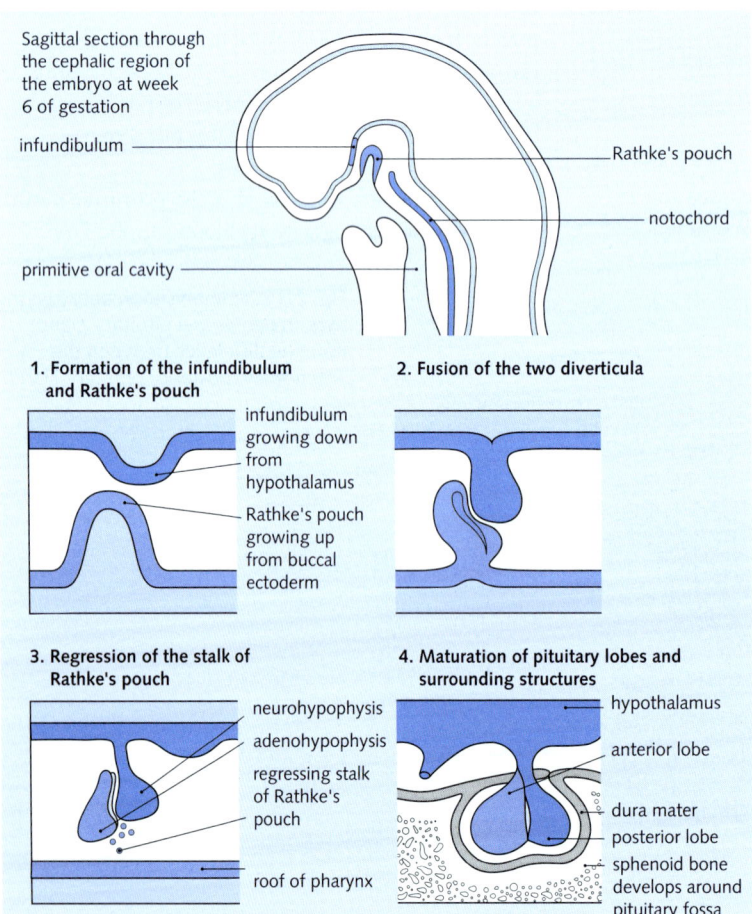

Fig. 2.4 Embryological development of the anterior and posterior lobes of the pituitary gland.

- Somatotrophs—growth hormone (GH).
- Gonadotrophs—luteinizing hormone (LH) and follicle stimulating hormone (FSH).
- Corticotrophs—adrenocorticotrophic hormone (ACTH) and melanocyte stimulating hormone (MSH).
- Thyrotrophs—thyroid stimulating hormone (TSH).
- Lactotrophs—prolactin.
- Chromophobes—inactive secretory cells.

In the past, cells were differentiated by their pH; the terms acidophil and basophil in older textbooks refer to this.

The anterior pituitary is divided into three distinct areas (see Fig. 2.2):
- Pars distalis—the majority of the gland.
- Pars tuberalis—a layer of mostly gonadotroph cells around the pituitary stalk.
- Pars intermedia—a thin layer of corticotroph cells between the anterior and posterior pituitary. It is very small in humans.

Posterior pituitary

The posterior pituitary is composed of two cell types, but it contains no secretory cells:
- Non-myelinated axons, originating from the hypothalamus.
- Pituicytes, which are stellate (star-shaped) glial support cells.

Within the axons there are microtubules and mitochondria that are involved in the transport of neurosecretory granules. These granules travel from the hypothalamus to the axon terminals in the posterior pituitary, where they are stored before release. The axon terminals lie close to blood sinusoids, where the neurosecretory granules are released into the systemic circulation (Fig. 2.5).

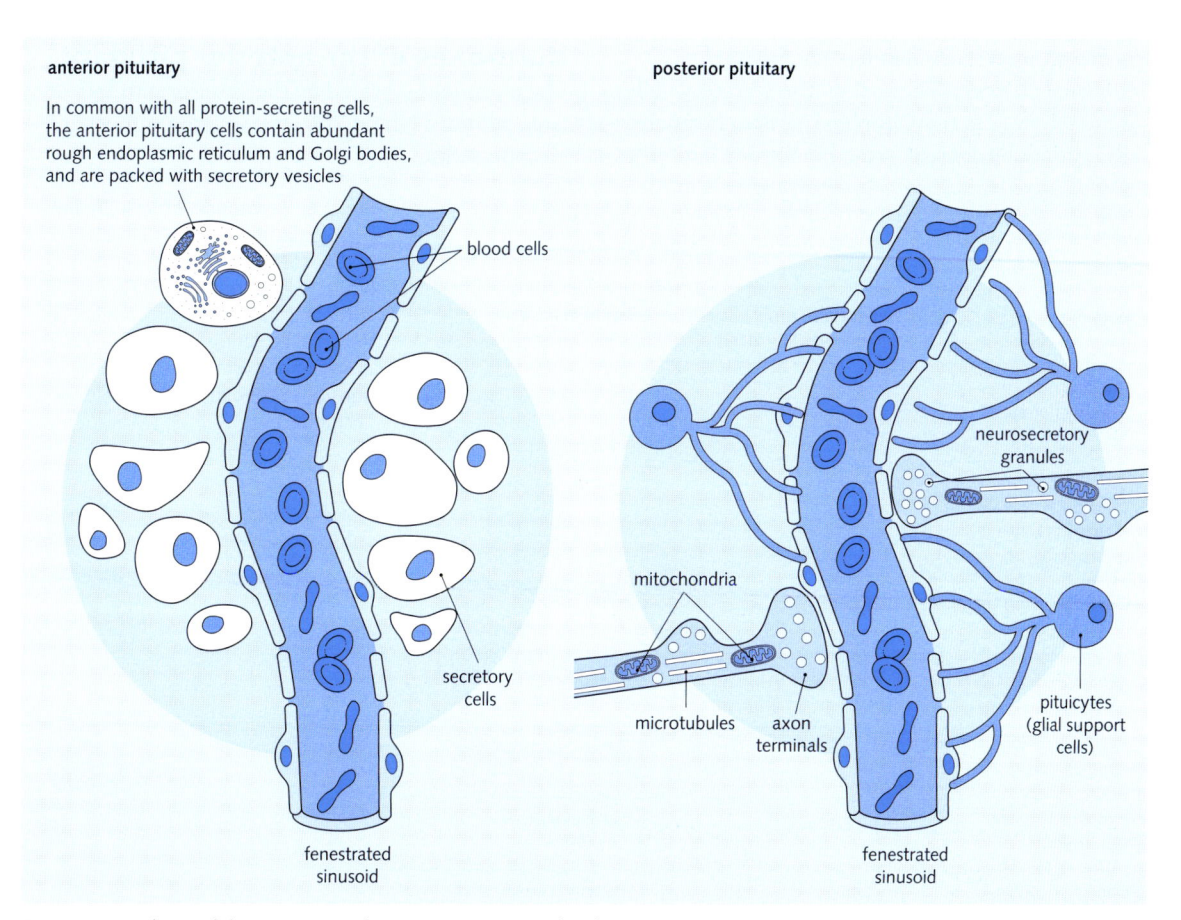

Fig. 2.5 Histology of the anterior and posterior pituitary gland.

The Hypothalamus and the Pituitary Gland

Hormones

Hormones of the hypothalamus

The hypothalamus secretes very small quantities of hormones into the portal veins to exert control over the anterior pituitary. The quantity is so small that the hormones can rarely be detected in systemic blood, but by travelling in the portal veins directly to the anterior pituitary, their concentration is high enough to produce an effect. This system allows for a rapid response and amplification of the signal. The hypothalamic hormones are often released in a pulsatile manner. The pulses vary in amplitude and rate, often with a circadian rhythm (see Ch. 6).

Hormones that regulate anterior pituitary function

The hormones secreted by the hypothalamus are small peptides (between 3 and 44 amino acid residues), except for dopamine, which is derived from the amino acid tyrosine. These hormones are shown in Fig. 2.6 along with their effects.

The factors that regulate the secretion of these hormones are discussed independently in the individual chapters. They act on the secretory cells in an excitatory (e.g. thyrotrophin releasing hormone—TRH) or inhibitory (e.g. growth hormone inhibiting hormone—GHIH) manner. There is some overlap in function between these peptides; for example, TRH can stimulate prolactin release. It is likely that other hypothalamic hormones will be discovered in the future.

Hormones released from the posterior pituitary

The small peptides ADH and oxytocin are synthesized in the cell bodies of magnocellular neurons arranged into two nuclei in the hypothalamus:
- Supraoptic nucleus.
- Paraventricular nucleus.

While both nuclei secrete both hormones, the supraoptic tends to secrete more ADH whereas the paraventricular favours oxytocin. The hormones pass along the axons bound to glycoproteins. They pass through the median eminence to the posterior pituitary where they are stored before release. Their actions are described in the section 'Hormones of the posterior pituitary' (p.22).

Hormones of the anterior pituitary

The hormones secreted by the anterior pituitary are large peptides (about 200 amino acid residues) or glycopeptides.

Hormones of the hypothalamus

Hormone	Target cells in the anterior pituitary gland	Effect on the anterior pituitary gland
Growth hormone releasing hormone (GHRH)	Somatotrophs	↑ GH release
Growth hormone inhibiting hormone (GHIH also called somatostatin)	Somatotrophs and thyrotrophs	↓ GH and TSH release
Corticotrophin releasing hormone (CRH)	Corticotrophs	↑ ACTH release
Gonadotrophin releasing hormone (GnRH)	Gonadotrophs	↑ LH and FSH release
Thyrotrophin releasing hormone (TRH)	Thyrotrophs and lactotrophs	↑ TSH and prolactin release
Prolactin releasing factors (PRF)	Lactotrophs	↑ Prolactin release
Dopamine (prolactin inhibiting hormone)	Lactotrophs	↓ Prolactin release

Fig. 2.6 Hormones secreted by the hypothalamus and their effects on the secretion of the anterior pituitary hormones. (ACTH, adrenocorticotrophic hormone; FSH, follicle stimulating hormone; GH, growth hormone; LH, luteinizing hormone; TSH, thyroid stimulating hormone.)

Hormones

The six main hormones are:
- Growth hormone (GH).
- Thyroid stimulating hormone (TSH also called thyrotrophin).
- Adrenocorticotrophic hormone (ACTH).
- Luteinizing hormone (LH).
- Follicle stimulating hormone (FSH).
- Prolactin (PRL).

 The hormones of the anterior pituitary can be remembered using the mnemonic 'Fresh Pituitary Tastes Almost Like Guinness'.

The hormones synthesized by the anterior pituitary are released into the systemic circulation. They act in two ways:
- Regulation of other endocrine organs—TSH, ACTH, GH, LH, and FSH.
- Direct effects on distant organs—prolactin.

GH and prolactin are large peptides whereas the others are glycopeptides. The individual hormones are discussed in more detail in later chapters (see also Figs 2.7 and 2.8). The secretion and release of these hormones often follows the pulsatile pattern of the releasing hormones from the hypothalamus.

The pars intermedia of the anterior pituitary gland also secretes a number of less important hormones including:
- Melanocyte-stimulating hormone (MSH), which stimulates melanocytes in the skin.
- Beta-endorphin, an *endo*genous mo*rphin*e, which may have a role in the control of pain.

Hormonal feedback

In response to the small quantities of releasing hormones secreted by the hypothalamus, the anterior pituitary is stimulated to secrete hormones in quantities large enough to act on endocrine organs throughout the body. The release of pituitary hormones is also regulated by hormones from other endocrine glands mainly through negative feedback mechanisms, e.g. thyroxine from the thyroid inhibits the release of TSH from the anterior pituitary.

Hypothalamic regulation of prolactin release is unique because the main control is inhibitory: dopamine secreted from the hypothalamus inhibits release of prolactin from the pituitary. This is

Hormones of the anterior pituitary gland

Hormone	Synthesized by	Stimulated by	Inhibited by	Target organ	Effect	Chapter
GH	Somatotrophs	GHRH	GHIH and IGF-1	Liver	Stimulates IGF-1 production and opposes insulin	9
TSH	Thyrotrophs	TRH	T_3	Thyroid gland	Stimulates thyroxine release	3
ACTH	Corticotrophs	CRH	Glucocorticoids	Adrenal cortex	Stimulates glucocorticoid and androgen release	4
LH+FSH	Gonadotrophs	GnRH, sex steroids	Prolactin, sex steroids	Reproductive organs	Release of sex steroids	11
Prolactin	Lactotrophs	PRF and TRH	Dopamine	Mammary glands and reproductive organs	Promotes growth of these organs and initiates lactation	11
MSH	Corticotrophs	–	–	Melanocytes in skin	Stimulates melanin synthesis	–
Beta-endorphin	Corticotrophs	–	–	Unknown	May be involved in pain control	–

Fig. 2.7 Hormones synthesized and secreted by the anterior pituitary and their effects. (ACTH, adrenocorticotrophic hormone; CRH, corticotrophin releasing hormone; FSH, follicle stimulating hormone; GH, growth hormone; GHRH, growth hormone releasing hormone; GnRH, gonadotrophin releasing hormone; GHIH, growth hormone inhibiting hormone; LH, luteinizing hormone; MSH, melanocyte stimulating hormone; TRH, thyrotrophin releasing hormone; TSH, thyroid stimulating hormone.)

The Hypothalamus and the Pituitary Gland

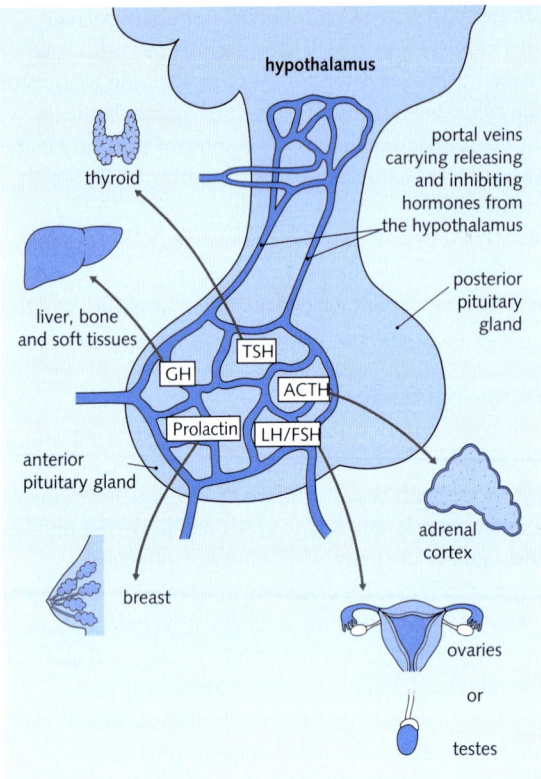

Fig. 2.8 Hormones of the anterior pituitary gland and their respective target organs.

Hormones of the posterior pituitary

Two major hormones are synthesized in the hypothalamus and released into the systemic circulation from the posterior pituitary:
- Antidiuretic hormone (ADH) also called arginine vasopressin (AVP).
- Oxytocin.

Both hormones are peptides consisting of nine amino acid residues that vary by a single residue. The main actions of these hormones are shown in Figs 2.9 and 2.10.

Antidiuretic hormone

ADH acts mainly on the collecting duct of the kidney to prevent water excretion. It also has a vasoconstricting action at high doses, hence the name vasopressin. Low blood volume detected by peripheral baroreceptors stimulates very high ADH release to increase blood pressure.

A small proportion of ADH is released into the portal veins where it stimulates corticotrophs in the anterior pituitary gland to secrete ACTH.

Oxytocin

When a baby suckles the mother's breast, stretch receptors in the nipple send signals to the brain via sensory nerves. These signals reach the paraventricular neurons causing depolarization and oxytocin release from the posterior pituitary. The oxytocin reaches the myoepithelial cells of the breast, which contract pushing milk out of the breast. This reflex is illustrated in Fig. 14.19 (p.191).

important if a tumour stops the hypothalamic releasing hormones from reaching the anterior pituitary. The levels of most pituitary hormones will fall, while the levels of prolactin will increase.

Hormones secreted by the posterior pituitary gland

Hormone	Synthesized by	Stimulated by	Inhibited by	Target organ	Effect	Chapter
Adreno-cortico-trophic hormone (ADH)	Supraoptic vasopressinergic neurons	Raised osmolarity; low blood volume	Lowered osmolarity	Kidney	Increases the permeability of the collecting duct to reabsorb water	7
Oxytocin	Paraventricular oxytocinergic neurons	Stretch receptors in the nipple and cervix; oestrogen	Stress	Uterus and mammary glands	Smooth muscle contraction leading to birth or milk ejection	14

Fig. 2.9 Hormones secreted by the posterior pituitary and their effects.

Disorders of the Anterior Pituitary

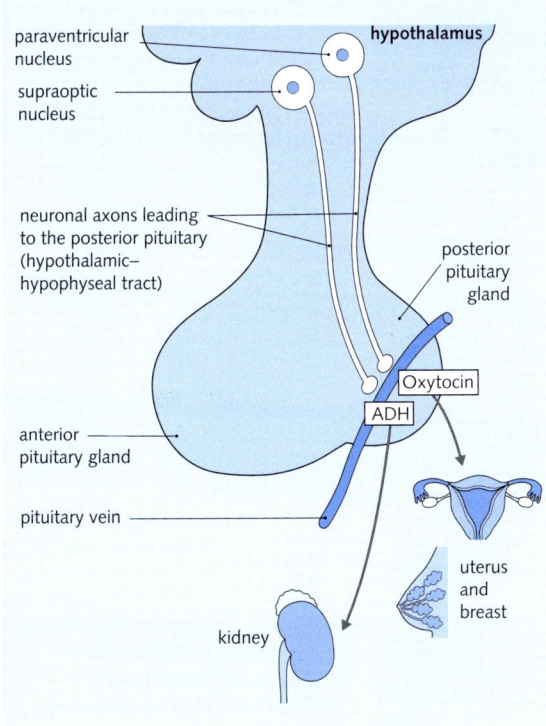

Fig. 2.10 Hormones of the posterior pituitary gland and their respective target organs.

Disorders of the anterior pituitary

Aetiology
The nine I's of pituitary pathology are:
- Iatrogenic e.g. surgery or radiotherapy.
- Invasion i.e. tumours.
- Infarction e.g. Sheehan's syndrome.
- Idiopathic i.e. no underlying cause known.
- Injury e.g. severe head trauma.
- Infection e.g. tuberculosis (very rare).
- Infiltration e.g. sarcoidosis (very rare).
- Immunological e.g. lymphocytic hypophysitis (very rare).
- Inherited e.g. congenital hormone deficiency (very rare).

Tumours
The majority of pituitary gland disorders are caused by benign tumours of the secretory cells called adenomas. Disease occurs as a result of three processes:
- Hyperpituitarism—excess pituitary hormone secretion.
- Hypopituitarism—insufficient pituitary hormone secretion.
- Compression of surrounding structures—caused by space-occupying lesions.

These tumours are classified into two groups: functioning and non-functioning adenomas. Functioning adenomas present early whilst still very small. These microadenomas cause disease by excess hormone release, which can be fatal if untreated. They also cause compression, so other pituitary hormones may be deficient (Fig. 2.11).

Non-functioning adenomas usually present at a later stage as larger macroadenomas. These cause disease indirectly by compressing surrounding structures, often causing insufficient pituitary hormone release when they compress the portal vessels or secretory cells.

Hyperpituitarism
Prolactinomas
All the anterior pituitary secretory cells can form tumours, however, the vast majority are prolactinomas i.e. tumours of the prolactin secreting cells (Fig. 2.12). They are more common in women in whom they tend to present earlier, before visual disturbance occurs. Excess prolactin secretion—hyperprolactinaemia—causes galactorrhoea and

Disorders of the hypothalamus

Primary diseases of the hypothalamus are very rare, but they tend to cause deficiency of hypothalamic hormones and the corresponding pituitary hormones. Dopamine deficiency has the opposite effect, resulting in excessive prolactin secretion from the anterior pituitary. The main causes of hypothalamic hormone deficiency are:
- Trauma/surgery.
- Radiotherapy.
- Congenital gonadotrophin releasing hormone (GnRH) deficiency (Kallmann's syndrome) causing infertility.
- Congenital GHRH deficiency causing dwarfism.
- Primary glial cell tumours of the hypothalamus.

Lesions in the hypothalamus can cause many other abnormalities, including disorders of consciousness, behaviour, thirst, satiety, and temperature regulation. These disorders usually occur together with hypopituitarism and diabetes insipidus.

23

The Hypothalamus and the Pituitary Gland

Anterior pituitary hormones and the disorders caused by their deficiency and excess

Hormone	Deficiency	Excess
GH	Dwarfism in children or adult GH deficiency syndrome	Gigantism in children, acromegaly in adults
LH and FSH	Gonadal insufficiency (decreased sex steroids)	Extremely rare but causes infertility
ACTH	Adrenocortical insufficiency (decreased cortisol and adrenal androgens)	Cushing's disease (increased cortisol and adrenal androgens)
TSH	Hypothyroidism (decreased thyroid hormones)	Extremely rare but causes hyperthyroidism (increased thyroid hormones)
Prolactin	Hypoprolactinaemia (failure in postpartum lactation)	Hyperprolactinaemia (impotence in males, amenorrhoea in females and decreased libido)

Fig. 2.11 Disorders caused by the deficiency or excess of anterior pituitary hormones. (ACTH, adrenocorticotrophic hormone; FSH, follicle stimulating hormone; GH, growth hormone; LH, luteinizing hormone; TSH, thyroid stimulating hormone.)

Adenomas of the anterior pituitary gland and their effects

Tumour	Hormone excess	Percentage of all pituitary tumours	Disease	Chapter
Prolactinoma	Prolactin	50%	Hyperprolactinaemia	2
Non-secretory prolactinoma	None	20%	Hypopituitarism	2
Somatotrophic cell adenoma	GH	20%	In children: gigantism In adults: acromegaly	9
Corticotrophic cell adenoma	ACTH	5%	Cushing's disease	4
Gonadotrophic cell adenoma	LH and FSH	Very rare	Infertility	–
Thyrotrophic cell adenoma	TSH	Very rare	Hyperthyroidism	3

Fig. 2.12 Adenomas of the anterior pituitary gland and their effects. (ACTH, adrenocorticotrophic hormone; FSH, follicle stimulating hormone; GH, growth hormone; LH, luteinizing hormone; TSH, thyroid stimulating hormone.)

hypogonadism, the symptoms of which are shown in Fig. 2.13.

Investigations

A number of symptoms and investigations are assessed to achieve a diagnosis:
- Are there symptoms of a specific endocrine abnormality?
- Visual field assessment is carried out to detect compression of the optic chiasma.
- Are there abnormal hormone levels in the blood? If an excess is suspected, all pituitary hormones should be tested as an adenoma can cause related deficiencies.
- Suppression tests are carried out—these assess pituitary response to hormone analogues or inhibiting factors to locate the lesion on the

MRI is the best method for the detection of pituitary adenomas. It has replaced plain X-rays of the skull.

Disorders of the Anterior Pituitary

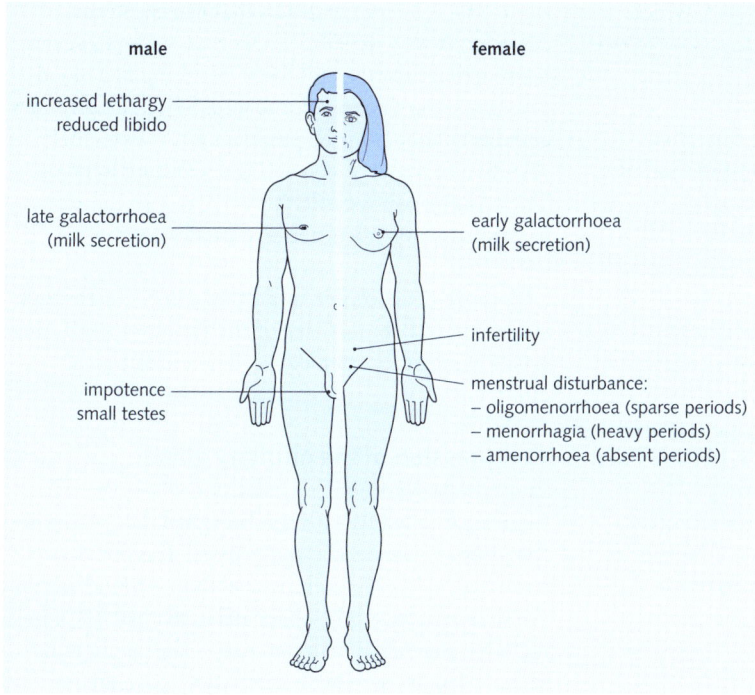

Fig. 2.13 Symptoms and signs of hyperprolactinaemia.

endocrine axis. Generally adenomas display reduced negative feedback.
- Magnetic resonance imaging (MRI) or computed tomography (CT) scans are used to detect abnormal anatomy.

The main suppression tests for anterior pituitary levels are to measure:
- GH in response to an oral glucose tolerance test, which normally suppresses GH levels.
- ACTH in response to dexamethasone, a steroid that normally suppresses CRH and ACTH release.

Hypogonadism caused by hyperprolactinaemia is a very common feature of pituitary adenomas. Prolactin secreting adenomas are the most common type of functioning adenoma and secrete excess prolactin by definition. Non-functioning adenomas prevent hypothalamic dopamine inhibition of prolactin release by compression so that excess prolactin is released.

Treatment
There are three methods of treating excess hormone production, but they all carry the risk of causing hypopituitarism:
- Bromocriptine (dopamine agonist) to reduce prolactin secretion.
- Octreotide (synthetic somatostatin) to reduce GH secretion.
- Surgical removal of pituitary adenoma.
- Irradiation to prevent adenoma recurrence.

Hypopituitarism
Pituitary insufficiency (hypopituitarism) often presents with insidious onset depression, tiredness, and hypogonadism as most hormone levels fall and prolactin levels rise. When there is a deficiency of more than one pituitary hormone it is called panhypopituitarism. The causes of pituitary insufficiency are more varied than those of hyperpituitarism. However, the most common cause is the treatment of hyperpituitarism.

Non-functioning adenomas
About 20% of pituitary tumours do not produce hormones so they are called non-functioning adenomas (sometimes called chromophobe adenomas). They are almost always non-active prolactinomas. Since there is no excess hormone production they present late with symptoms caused

The Hypothalamus and the Pituitary Gland

by compression of surrounding structures, which become progressively severe:
- Headaches.
- Pituitary hormone deficiencies.
- Hypogonadism due to hyperprolactinaemia.
- Loss of peripheral vision due to compression of the optic nerve (bitemporal hemianopia).
- Cranial nerve palsies (starting with nerve IV).
- Raised intracranial pressure.

The tumour can cause hormone deficiencies by direct compression of the secretory cells or by compressing the portal veins that bring the hypothalamic releasing factors. Secretion of anterior pituitary hormones is inhibited in a characteristic order: GH, LH and FSH, ACTH, TSH (see Fig. 2.11 for symptoms). Unless the compression is very severe prolactin secretion often increases initially since dopamine inhibition is lost. This excess prolactin secretion is not from the cells of the adenoma but it causes the symptoms shown in Fig. 2.13.

The Bible according to endocrinology: Goliath was a giant due to excessive GH secretion from a somatotroph tumour. This grew to compress his optic nerve causing a bilateral hemianopia so he failed to see David's stone coming from the side.

Other tumours
Tumours in surrounding tissues can also compress the pituitary gland causing panhypopituitarism, the most common being:
- Craniopharyngiomas.
- Gliomas (especially in the optic chiasma).

Craniopharyngiomas are rare tumours formed in the remnants of Rathke's pouch left behind when the connection with the pharynx regressed. They can form above, below, or within the sella turcica.
  Gliomas are primary tumours of the glial support cells found throughout the brain.

Infarction of the pituitary gland
Infarction of the pituitary gland causes necrosis of all secretory cells causing complete panhypopituitarism, including loss of prolactin secretion. Sheehan's syndrome is a rare condition that can cause this disorder. It may develop if a woman suffers a severe haemorrhage during childbirth resulting in a blood pressure drop due to volume loss. Since the pituitary gland is enlarged during pregnancy, it is extremely sensitive to hypotension and the resulting hypoxia. The ensuing panhypopituitarism causes a failure to lactate, amenorrhoea, and eventually death if untreated.

A more catastrophic infarction can occur through spontaneous haemorrhage of the pituitary itself, this is called pituitary apoplexy; it is a neurosurgical emergency.

Compression of the pituitary gland
Empty sella syndrome is a condition where the sella turcica partially fills with cerebrospinal fluid, causing the pituitary gland to be compressed. It is not always pathological (some cerebrospinal fluid is found within the sella in at least 50% of normal individuals). Causes include congenital incompetence of the diaphragma sellae, pituitary surgery or irradiation, postpartum pituitary infarction (Sheehan's syndrome), and coexisting pituitary tumour.

The pituitary has a great secretory reserve—more than 75% must be affected before clinical manifestations are evident.

Diagnosis of hypopituitarism
Diagnosis of hypopituitarism involves the same steps as for hyperpituitarism, however, stimulation is used instead of suppression:
- Symptoms.
- Visual field assessment.
- Basal hormone levels in the blood.
- Stimulation tests.
- MRI or CT scan.

The main stimulation tests for anterior pituitary levels are to measure:
- GH in response to an insulin tolerance test, which normally increases GH levels.
- Cortisol in response to hypoglycaemia or the ACTH analogue Synacthen®.
- LH and FSH in response to GnRH or the anti-oestrogen clomiphene.

Treatment of hypopituitarism
The main treatment of hypopituitarism is hormone replacement, which requires frequent monitoring. All the major anterior pituitary hormones can be replaced, though prolactin is not readily available since it is rarely needed:
- Subcutaneous GH replacement using human recombinant GH.
- Oral cortisol replacement.
- Oral thyroxine once cortisol replacement has begun.
- Oral or intramuscular testosterone in males.
- Oral oestrogen and progesterone cyclically in females.
- Intramuscular human chorionic gonadotrophin, LH, and FSH are given if male or female fertility is required.

Surgery may be required to remove adenomas, gliomas, or craniopharyngiomas.

Disorders of the posterior pituitary

Diabetes insipidus
A deficiency of ADH (vasopressin) secretion prevents osmotic control of the kidney so that very dilute polyuria occurs. Up to 20 litres of urine can be passed in a day, causing a potentially fatal dehydration and constant thirst. It is rare and usually idiopathic, but it can be caused by trauma, surgery or tumours (Fig. 2.14).

An ADH stimulation test is used to distinguish between deficient ADH and unresponsive kidneys. The condition is treated with desmopressin, a long acting vasopressin analogue, to control fluid loss.

Excess antidiuretic hormone secretion
The syndrome of inappropriate secretion of ADH (SIADH) can be caused by neurological, endocrine, malignant, or infective diseases, but it can also be idiopathic, postoperative, or caused by medications. Excess ADH causes water retention resulting in hypo-osmotic hyponatraemia (low sodium). The symptoms progress from malaise and weakness to confusion and coma. If untreated it can be fatal. Oedema does not occur (see Fig. 2.14).

Posterior pituitary hormones and the disorders caused by their deficiency and excess		
Hormone	Deficiency	Excess
ADH	Diabetes insipidus (polyuria, hypotension)	Syndrome of inappropriate ADH secretion (SIADH)
Oxytocin	Failure to progress in labour and difficulty with breastfeeding	No effect

Fig. 2.14 Disorders caused by the deficiency or excess of posterior pituitary hormones. (ADH, adrenocorticotrophic hormone.)

The Hypothalamus and the Pituitary Gland

- Describe the anatomical location of the hypothalamus.
- Explain how the hypothalamus regulates the anterior pituitary gland including relevant vasculature.
- List the hormones secreted by the hypothalamus along with their effects.
- Describe the role of the hypothalamus in the release of posterior pituitary hormones.
- Describe the anatomical relationships of the pituitary gland.
- Explain the embryological development of the pituitary gland and how this affects hypothalamic control.
- List the hormones of the anterior pituitary gland along with their effects.
- List the hormones of the posterior pituitary gland along with their effects.
- Name the most common types of pituitary adenoma.
- List the effects of compression by a pituitary tumour. In what order do they occur?
- List the anterior pituitary hormones in the order that they are affected by compression.
- List the symptoms of hyperprolactinaemia in males and females. Which sex presents to the doctor earliest?
- List three treatment options for adenomas.
- Describe the investigations of pituitary hormone excess and deficiency.
- Describe the difference between a suppression and stimulation test along with an example of each.
- The deficiency of which pituitary hormone is life threatening?
- Perform a musical interpretation of a pituitary gland undergoing infarction.
- Describe the treatment of panhypopituitarism.
- Name and describe the disorders of excess and deficient ADH secretion.
- Draw a diagram to illustrate the reflex that initiates lactation.

3. The Thyroid Gland

Diseases of the thyroid gland are the second most common endocrine disorder, after diabetes. The main diseases are caused by excess or deficiency of thyroid hormones, and they tend to affect middle-aged or elderly women.

The thyroid gland regulates the body's metabolism (Fig. 3.1). It is the largest endocrine organ in the body, and it is found anterior to the trachea in the lower neck. The gland is composed of two lobes that are joined by a narrow 'isthmus' in the centre.

Within the gland the cells are arranged in spherical follicles that surround a thyroid hormone store. The thyroid is the only endocrine gland to store large quantities of preformed hormones. These thyroid hormones are released in response to thyroid stimulating hormone (TSH) from the anterior pituitary gland. The follicles release two hormones:

- T_4—a prohormone that acts as a plasma reservoir.
- T_3—the active hormone.

These hormones increase the rate of metabolism in almost every cell in the body.

Excessive release is called hyperthyroidism and this causes abnormally fast metabolism. The patients feel hot and sweaty; they often lose weight.

Deficient release is called hypothyroidism. Metabolism is slow, making the patients feel lethargic; they frequently gain weight.

After reading this chapter you should be able to:

- Explain the structure and development of thyroid gland.
- Describe the synthesis of thyroid hormones.
- Understand the regulation and physiological effects of thyroid hormones.
- Discuss the major disorders associated with thyroid function.

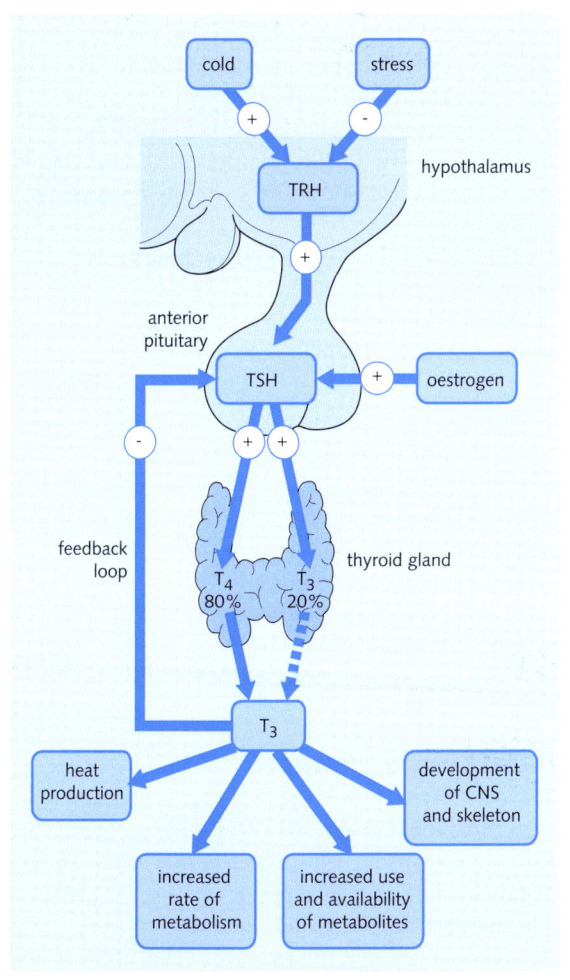

Fig. 3.1 Hormonal regulation of the thyroid hormones. (T_3, tri-iodothyronine; T_4, thyroxine; TRH, thyrotrophin releasing hormone; TSH, thyroid stimulating hormone.)

Important words:
Thyrotrophin: the same as thyroid stimulating hormone (TSH)
Thyroglobulin: the store of thyroid hormones found in the thyroid follicles
Parathyroid glands: endocrine organs found behind the thyroid gland that regulate calcium metabolism and are essential to life
Deiodination: removal of an iodine molecule
Goitre: an enlarged thyroid gland

The Thyroid Gland

Anatomy

Thyroid gland

The thyroid gland is palpable in about 50% of women and 25% of men. It is located in the neck, inferior to the larynx and cricoid cartilage. It has two lobes, each about 5 cm long and joined by a narrow isthmus. The lobes lie either side of the trachea and oesophagus, and the isthmus crosses the trachea anteriorly, usually over the second and third tracheal cartilages (Fig. 3.2).

The thyroid gland is bound to the trachea by the pretracheal fascia so that it moves with the trachea and larynx on swallowing, but not when the tongue is protruded. The gland is surrounded by a fibrous capsule within the pretracheal fascia (Fig. 3.3).

> The thyroid gland derives its name from the Greek word 'thyreos' meaning 'shield' and 'oid' meaning 'like'. It is unclear whether this is because the Greeks had shields with slits in the middle or that the anatomists knew nothing about shields.

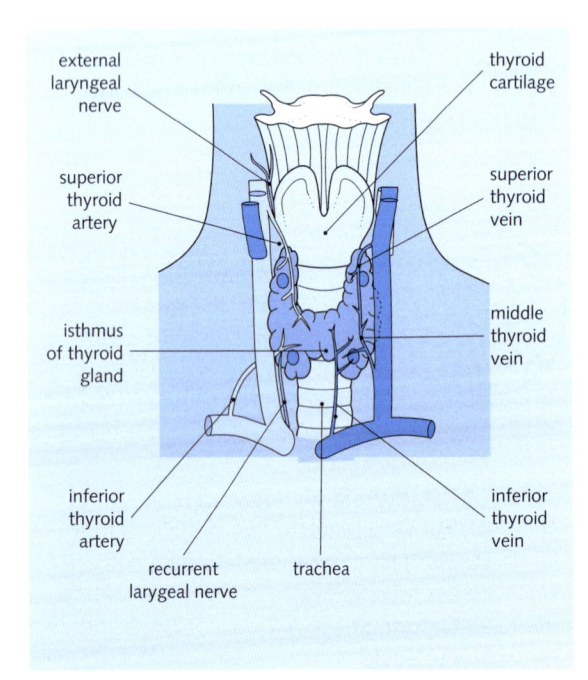

Fig. 3.2 Anterior view of the neck, showing the location and blood supply of the thyroid gland.

Blood supply, nerves, and lymphatics

The thyroid gland is highly vascular, in fact the blood flow per gram is greater than that to the kidneys. It is supplied by two arteries that anastomose (join) within the gland: the inferior and superior thyroid arteries.

The inferior thyroid artery is a branch of the thyrocervical trunk that arises from the subclavian arteries. It ascends behind the carotid sheath to enter the thyroid posteriorly. The right recurrent laryngeal nerve is intimately related to this artery near the inferior pole of the thyroid gland. Surgery to the thyroid gland can bruise this nerve causing temporary difficulty with speaking. To minimize the risk to this nerve, the artery is ligated far away from the thyroid gland during thyroidectomy.

The superior thyroid artery is the first branch of the external carotid artery. The external laryngeal nerve is related to this artery, but it is at less risk than the recurrent laryngeal during thyroid surgery. The superior thyroid artery is ligated close to the thyroid gland to reduce this risk (Fig. 3.2).

A third artery, called the thyroid ima artery, is present in 10% of people. It supplies the isthmus and it arises near the aortic arch though the exact origin varies.

The thyroid gland is drained by three veins:
- Superior thyroid vein.
- Middle thyroid vein.
- Inferior thyroid vein.

The first two veins drain into the internal jugular, whereas the inferior vein drains into the brachiocephalic veins.

Thyroid lymphatics drain into four groups of nodes:
- Prelaryngeal lymph nodes.
- Pretracheal lymph nodes.
- Paratracheal lymph nodes.
- Deep cervical lymph nodes.

Parathyroid glands

The parathyroid glands are four oval shaped structures about 5 mm across; they are embedded in the thyroid capsule behind both lateral lobes. These glands are described as superior and inferior pairs. The number of parathyroid glands often varies between 2 and 6 and the location of the inferior pair differs widely. They are supplied by branches of the inferior thyroid artery. Their function is described in Chapter 8.

Development

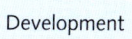

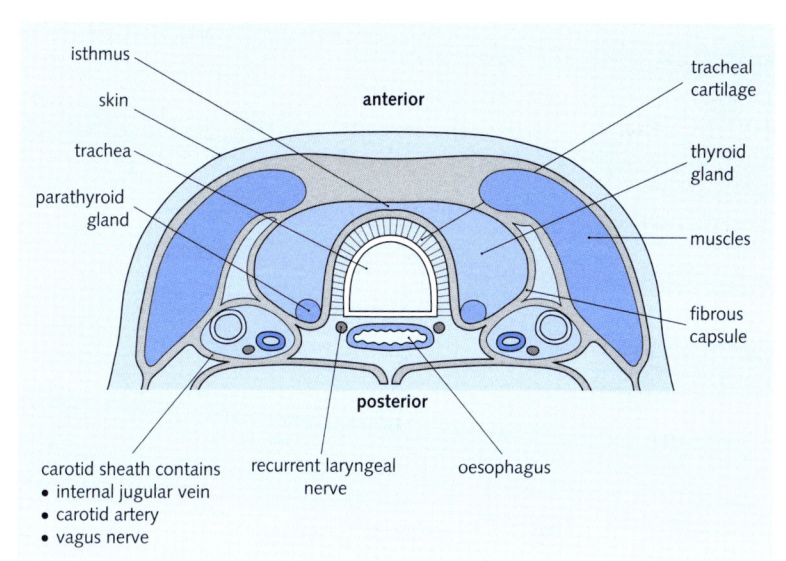

Fig. 3.3 Horizontal section of the anterior neck at the level of the sixth cervical vertebra, showing the location of the thyroid and parathyroid glands and their surrounding structures.

Microstructure

Thyroid gland

The thyroid is composed of about one million spherical follicles or acini. Each follicle is lined by a single layer of secretory epithelial cells (follicular cells) around a colloid-filled space. These cells secrete thyroglobin—a storage form of thyroid hormone—into the colloid. When the thyroid gland is not actively secreting hormones the size of the colloid store and the follicle itself increase in diameter.

When the follicular cells enter an active secretory phase microvilli form on their inner surface and thyroglobin is absorbed. The colloid store shrinks as a result. The absorbed thyroglobin is broken down to release thyroid hormone. The histology of the thyroid gland is shown in Fig. 3.4.

Another type of secretory cell is found between the follicles. These parafollicular cells (C cells) synthesize and secrete calcitonin.

> The thyroid gland is the only endocrine gland to store its hormone in an extracellular compartment. 2–3 months of thyroid hormone are stored within the follicles, and this delays the onset of symptoms in deficiency diseases.

Parathyroid glands

There are three cell types in the parathyroid glands:
- Chief cells—synthesize parathyroid hormone (PTH).
- Oxyphil cells—exhausted chief cells, their numbers increase with age.
- Adipocytes—contain fat, their numbers also increase with age.

Development

Thyroid gland

The thyroid gland is an endodermal structure that develops as an out-pouching in the floor of the pharynx behind the tongue. It descends in the neck, but it remains attached to the tongue by the thyroglossal duct until it reaches its final position.

In 50% of people a remnant of this duct forms a small pyramidal lobe extending superiorly from the isthmus. Thyroglossal cysts can form anywhere along the course taken by the thyroid. They present as neck lumps that rise when the tongue is protruded.

Parathyroid glands

The parathyroid glands are endodermal structures that develop from the pharyngeal pouches. The inferior parathyroid glands are formed by the dorsal portion of the 3rd pouch while the dorsal portion of the 4th pouch forms the superior parathyroid glands.

31

The Thyroid Gland

Do not confuse the thymus gland with the thyroid gland. The thymus gland is an organ of the immune system that lies in the thorax. The two glands are entirely separate in function and structure.

Hormones

The thyroid gland synthesizes and secretes three hormones:
- Thyroxine (T_4).
- Tri-iodothyronine (T_3).
- Calcitonin.

Calcitonin is involved with calcium homeostasis (discussed in Chapter 8).

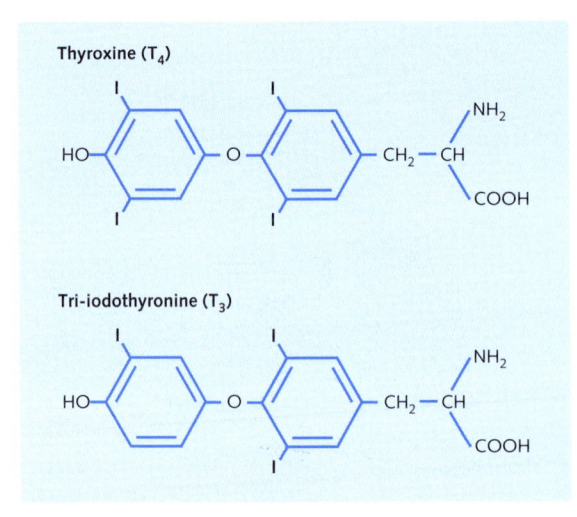

Fig. 3.5 Structures of T_3 and T_4.

synthesis involves the processing of tyrosine and iodine followed by a reaction to bind them together. These steps are also shown in Fig. 3.6.

Tyrosine processing is relatively simple since tyrosine molecules are already within the cell:

Thyroglobulin synthesis   Tyrosine is converted into the glycoprotein thyroglobulin, which contains approximately 110 tyrosine residues.

The processing of iodine involves two stages as plasma iodine concentrations are very low:

Iodine trapping   Plasma iodide ions (I^-) are actively transported from the plasma into the follicular cells against a steep concentration gradient. This is a rate limiting step.

Iodide oxidation I^- is rapidly oxidized into iodine (I_2) by a peroxidase enzyme near the luminal membrane. This increases reactivity.

The two components are then combined in the extracellular follicle lumen:

Iodination of thyroglobulin   Reactive iodine rapidly attaches to the tyrosine molecules within the extracellular thyroglobulin. The reaction requires thyroperoxidase, an enzyme on the luminal membrane of the follicle cells. Monoiodotyrosine (MIT or T_1) and diiodotyrosine (DIT or T_2) are formed.

Coupling   Tyrosine molecules within thyroglobulin are then coupled together. Combinations of T_1 and T_2 can form thyroid hormones:
- T_3 is made from $T_1 + T_2$.
- T_4 is made from $T_2 + T_2$.

Only a small proportion of coupling reactions form T_3 and T_4.

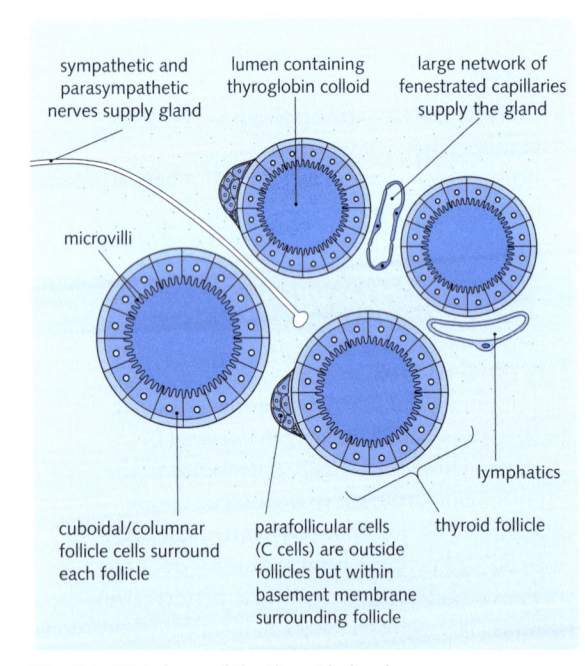

Fig. 3.4 Histology of the thyroid gland.

Synthesis

T_3 and T_4 are derived from two molecules of the amino acid tyrosine and iodine. T_3 contains three iodine atoms and T_4 contains four. Their structures are shown in Fig. 3.5.

Thyroid hormones are formed in the follicle lumen, not in the cells. The process of T_3 and T_4

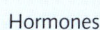

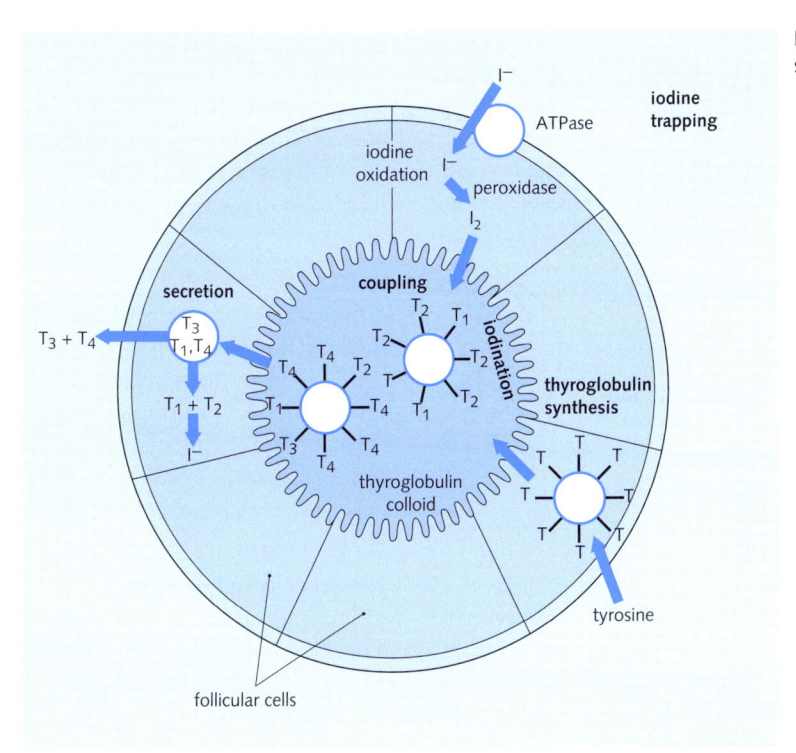

Fig. 3.6 Steps in the synthesis and secretion of T_3 and T_4.

The thyroid hormones can now be released on demand:

Secretion The iodinated thyroglobulin is taken into the follicular cells by pinocytosis and broken down by lysosomal enzymes. Coupled tyrosine molecules are released including some T_3 and T_4, which diffuse into the blood. T_1 and T_2 are also released, but they are broken down to rescue iodine molecules.

Iodine metabolism

Iodine is acquired from the diet mainly from iodized salt, meat and vegetables. About 150 mg of iodine is needed per day though only a fraction of this is absorbed. The thyroid gland cells are the only cells that can actively absorb and utilize plasma iodine; a considerable quantity of iodine is stored in the thyroid as preformed thyroid hormones. Iodine is returned to the plasma by the breakdown of these thyroid hormones. Iodine is excreted mainly via the kidneys.

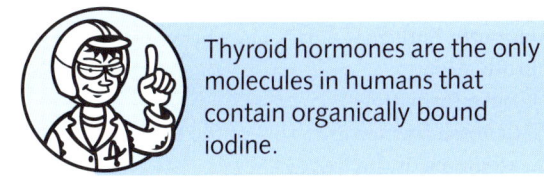

Thyroid hormones are the only molecules in humans that contain organically bound iodine.

Regulation

Hypothalamic thyrotrophin releasing hormone (TRH) stimulates the release of thyroid stimulating hormone (TSH or thyrotrophin) from the anterior pituitary gland. TSH acts on extracellular receptors on the surface of thyroid follicle cells. Cyclic AMP (cAMP) is formed and this stimulates five of the stages of synthesis and secretion:
- Iodine uptake.
- Thyroglobin synthesis.
- Iodination.
- Coupling.
- Pinocytosis for secretion.

As a result T_3 and T_4 are synthesized and secreted more rapidly (see Fig. 3.1). TSH also has long-term actions on the thyroid gland by increasing its size and vascularity to improve hormone synthesis.

A number of factors affect thyroid hormone release. Three main factors stimulate secretion:
- Long-term exposure to cold temperatures acting on the anterior pituitary.
- Oestrogens acting on the anterior pituitary.
- Adrenaline acting directly on the thyroid gland.

Thyroid hormone release is inhibited by excess thyroid hormones and glucocorticoids (e.g. cortisol) by acting on the anterior pituitary to suppress TSH release.

The Thyroid Gland

Transport of thyroid hormones

The thyroid hormones circulate bound to plasma proteins produced in the liver, which protect the hormones from enzymatic attack:
- 70% are bound to thyroid-binding globulin (TBG).
- 30% are bound to albumin.

Only 0.1% of T_4 and 1% of T_3 are carried unbound—it is this free (unbound) fraction that is responsible for their hormonal activities.

Both T_3 and T_4 can cross cell membranes, though a carrier transport may be involved.

The concentration of circulating T_4 is much higher than that of T_3 (50:1). There are two reasons for this:
- The thyroid secretes more T_4 than T_3.
- T_4 has a longer half-life (7 days vs 1 day).

Actions

Fig. 3.7 describes some of the differences between T_4 and T_3. T_4 is a relatively inactive, stable molecule that can be thought of as a prohormone. T_3 is the active hormone, since it is readily available and it has more effect on receptors. The benefit of producing both hormones is that T_4 can maintain a background level of activity whilst T_3 levels can adapt rapidly to changing environments.

Peripheral tissues can regulate local T_3 levels by increasing or decreasing T_3 synthesis. T_4 is converted to T_3 by deiodination, i.e. removal of one iodine atom catalysed by deiodinase enzymes. Two main forms of this enzyme have been found:
- Type 1—found on the cell surface in most tissues. It raises local T_3.
- Type 2—intracellular enzymes that raise cellular T_3 in the central nervous system (CNS) and pituitary gland.

A further deiodinase enzyme can remove a different iodine molecule from T_4 to form reverse T_3 (rT_3). This is an inactive molecule that is rapidly cleared from the circulation by the kidney and liver. Production of rT_3 is favoured by low energy stores and illness, so that energy is conserved.

> The majority of plasma T_3 is formed by the deiodination of T_4 NOT from the thyroid gland. This is important in the treatment of hypothyroidism since only T_4 is given.

Free plasma T_3 enters cells and binds to intracellular T_3 receptors located in the membrane, mitochondria, and nucleus of the target cells. The intracellular actions are described in Fig. 3.8 and related to physiological effects.

In general, T_3 promotes energy production in every cell in the body. This causes heat production and maintains metabolism.

Feedback

T_3 receptors are also found in the pituitary gland, where they inhibit the release of TSH to create a negative feedback loop. Excess T_3 concentrations inhibit TSH release while deficient T_3 concentrations stimulate TSH release. This mechanism helps to maintain T_3 levels and, therefore, stabilizes metabolic rate.

Disorders of the thyroid gland

The thyroid gland is prone to a number of diseases that can alter its function and structure. These diseases frequently have wide ranging systemic effects because thyroid hormones regulate the metabolism of almost every cell in the body. The main categories of disease are:
- Hyperthyroidism—excess of thyroid hormone production.
- Hypothyroidism—deficiency of thyroid hormone production.
- Goitre formation.
- Adenoma (benign growths) of the thyroid.
- Carcinoma of the thyroid.

Comparison of T_3 and T_4

	T_3	T_4
Proportion of secreted thyroid hormone	10%	90%
Percentage free in plasma	1%	0.1%
Relative activity	10	1
Half-life (days)	1	7

Fig. 3.7 Comparison of T_3 and T_4.

Actions of T_3		
Site of action	Intracellular effects	Physiological results
Cell membrane	Stimulates the Na^+/K^+ATPase pump	Increased demand for metabolites, e.g. glucose
Mitochondria	Stimulates growth, replication and activity; basal metabolic rate is raised	Increased heat production, oxygen demand, heart rate, and stroke volume
Nucleus	Increases expression of enzymes necessary for energy production	Lipolysis, glycolysis, and gluconeogenesis increased to raise blood metabolite levels and cellular metabolite use
Neonatal cells	Essential for cell division and maturation	Essential for normal development of CNS and skeleton

Fig. 3.8 Intracellular and physiological actions of T_3.

Hyperthyroidism

Hyperthyroidism is defined as an overactive thyroid gland leading to excess thyroid hormones (T_4 and T_3). When this becomes symptomatic it is called thyrotoxicosis. It is a common disorder affecting 1/50 females and 1/250 males. The symptoms and signs of thyrotoxicosis are illustrated in Fig. 3.9.

The major symptoms of hyperthyroidism can be remembered as: '**D**on't **E**vade **F**eeling **H**ot **A**nd **S**weaty **P**atients' i.e. **D**iarrhoea, **E**motional lability, **F**atigued, **H**eat intolerance, increased **A**ppetite, **S**weating, and **P**alpitations.

Presentation is usually slow with a history lasting over 6 months.

An acute exacerbation of symptoms is called a thyrotoxic crisis; it is usually brought on by infection in previously undiagnosed patients. Surgery or radioactive ablation of the thyroid gland can also be responsible as the damaged thyroid follicles release their contents.

The main causes of hyperthyroidism are:
- Graves' disease—an autoimmune disease.
- Multinodular goitre—nodular enlargement of the thyroid in the elderly.
- Toxic adenoma—a benign thyroid hormone producing tumour.

Diagnosis

Thyroid function tests are the main component of diagnosis. Serum TSH, free T_3, and free T_4 are measured by radioimmunoassay (RIA). Raised T_3 and T_4 indicate that hyperthyroidism is present. Raised TSH suggests the fault lies in or above the pituitary gland, whereas low TSH points to a thyroid organ lesion.

Other tests include:
- Autoantibody detection e.g. Graves' disease.
- Radioisotope scanning to show the size of the thyroid gland and any abnormal 'hot' areas such as a toxic adenoma.
- ECG for sinus tachycardia or atrial fibrillation.

Treatment

There are three methods of treatment:
- Carbimazole—this drug inhibits the peroxidase reactions of T_3 and T_4 synthesis. It takes 3–4 weeks to have an effect.
- Radioactive iodine therapy—^{131}I is only taken up by thyroid tissue; it kills the cells leading to reduced T_3 and T_4 synthesis. The response is slow and carbimazole may be required.
- Partial thyroidectomy—the thyroid gland is surgically removed leaving some tissue and the parathyroid glands.

Both radioactive iodine and partial thyroidectomy carry a high risk of long-term hypothyroidism. The remaining thyroid tissue may be insufficient to meet the body's needs, especially as the patient ages. Their treatment is described under hypothyroidism.

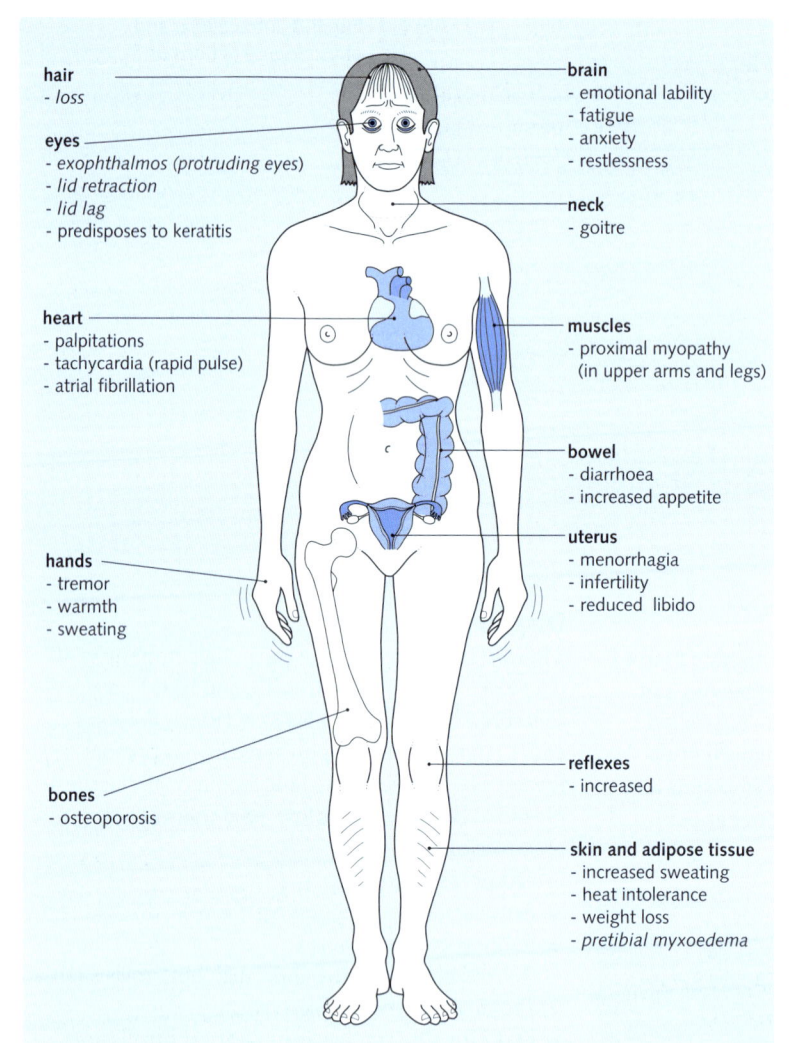

Fig. 3.9 Symptoms and signs of thyrotoxicosis (hyperthyroidism). The features in italic are only found in Graves' disease.

hair
- loss

eyes
- *exophthalmos (protruding eyes)*
- *lid retraction*
- *lid lag*
- predisposes to keratitis

heart
- palpitations
- tachycardia (rapid pulse)
- atrial fibrillation

hands
- tremor
- warmth
- sweating

bones
- osteoporosis

brain
- emotional lability
- fatigue
- anxiety
- restlessness

neck
- goitre

muscles
- proximal myopathy (in upper arms and legs)

bowel
- diarrhoea
- increased appetite

uterus
- menorrhagia
- infertility
- reduced libido

reflexes
- increased

skin and adipose tissue
- increased sweating
- heat intolerance
- weight loss
- *pretibial myxoedema*

Graves' disease

Graves' disease is an autoimmune disease in which autoantibodies against the TSH receptors stimulate the receptors so that thyroid hormones are produced in excess. Graves' disease is the most common cause of hyperthyroidism; it is especially common in middle-aged women and it has a genetic component.

The disease follows either a relapsing-remitting course or one with fluctuating severity. Rarely Graves' disease can progress to hypothyroidism with time.

Graves' disease can cause the classical picture of hyperthyroidism with bulging eyes (exophthalmos), goitre (with bruit) and swollen legs (pretibial myxoedema). It is diagnosed by detection of autoantibodies along with low TSH and raised T_3. The treatment is consistent with other causes of hyperthyroidism, but radioactive iodine and surgery are especially likely to cause hypothyroidism.

Eye disease is an important symptom of Graves' disease. Inflammation of the orbit causes the eye to protrude, which can lead to discomfort and double vision. This symptom may occur before thyroid hormone levels rise. It does not always respond to treatment, and it may develop in patients with well controlled disease.

In Graves' disease, the vasculature of the thyroid gland can increase to such an extent that a bruit can be heard over the goitre using a stethoscope.

Hypothyroidism

Hypothyroidism is defined as an underactive thyroid gland leading to deficient thyroid hormones (T_4 and T_3). When this becomes symptomatic it is called myxoedema. It is slightly less common than hyperthyroidism affecting 1/100 females and 1/500 males. The symptoms and signs of myxoedema are illustrated in Fig. 3.10. Presentation is even more gradual than in hyperthyroidism with many symptoms frequently being ignored.

Thyroid hormones are essential between birth and puberty for the normal development of the CNS. Deficiency can cause irreversible mental retardation called cretinism. TSH levels are checked in all newborns for this relatively common abnormality; the levels will be raised if the thyroid gland is not functioning correctly.

Under-treated hypothyroidism can progress to a life threatening myxoedema coma. This rare condition is characterized by bradycardia and hypotension. Plasma levels of glucose and sodium can also drop, and type II respiratory failure may develop.

Diagnosis

Hypothyroidism is not investigated as thoroughly as hyperthyroidism, since treatment does not vary. Free T_3 and T_4 levels are low, whereas TSH levels are usually raised. If TSH is low then a lesion of the hypothalamus or pituitary is likely. Autoantibodies can be detected in Hashimoto's thyroiditis.

Treatment

All hypothyroidism is treated with thyroxine (T_4) administered as an oral tablet in varying doses. The dose is increased over several months with regular monitoring of TSH levels until TSH levels are within the normal boundaries. This process is slow since it takes 4 weeks for TSH levels to reflect an increased dose due to the long half-life of thyroxine. Thyroxine therapy is usually maintained for life.

Over-treatment of hyperthyroidism

Radioactive ablation and surgical removal of the thyroid gland initially cure hyperthyroidism but, with time, the remaining thyroid tissue is often insufficient. Hypothyroidism can develop and life long thyroxine treatment is required.

Many drugs can also cause reversible hypothyroidism including lithium, amiodarone, and excess iodine.

Hashimoto's thyroiditis

When the thyroid gland is inflamed the disease is called thyroiditis. This can be caused by autoimmune or viral processes. Hashimoto's thyroiditis is a destructive autoimmune disease that is especially common in middle-aged women. It is mediated by autoantibodies against rough endoplasmic reticulum (microsomal antibodies) or thyroglobulin. The presence of these antibodies can be tested to confirm the diagnosis. The thyroid gland is infiltrated by lymphocytes that cause the gland to enlarge, forming a goitre.

The initial destruction of the thyroid gland can release the thyroglobulin colloid causing temporary hyperthyroidism. The patients usually progress to a euthyroid (normal) state and finally develop progressive hypothyroidism.

De Quervain's thyroiditis

De Quervain's thyroiditis is inflammation of the thyroid gland caused by a virus. It is common in young or middle-aged women, in whom it causes a tender swollen gland along with a febrile illness. The inflammation destroys the follicles, which causes hypothyroidism and leakage of the thyroglobulin colloid. An immune reaction against this colloid causes the formation of granulomas so this disease is also called granulomatous thyroiditis.

Primary atrophic hypothyroidism

Spontaneous or primary atrophic hypothyroidism is a disease resulting in hypothyroidism in the elderly. The thyroid gland becomes shrunken and fibrosed so that there is no goitre. It is suspected that this disease is the end-stage of many thyroid diseases including Hashimoto's and de Quervain's thyroiditis.

Goitres

A goitre is a swelling in the neck caused by an enlarged thyroid gland. It is a common finding, and it is usually asymptomatic, however large goitres can compress the oesophagus and trachea. If a goitre is associated with hyperthyroidism it is described as 'toxic'. Non-toxic goitres secrete normal or reduced levels of thyroid hormones. Goitres are treated by correcting the underlying pathology or by surgical removal for cosmetic reasons or to prevent compression of surrounding structures.

The Thyroid Gland

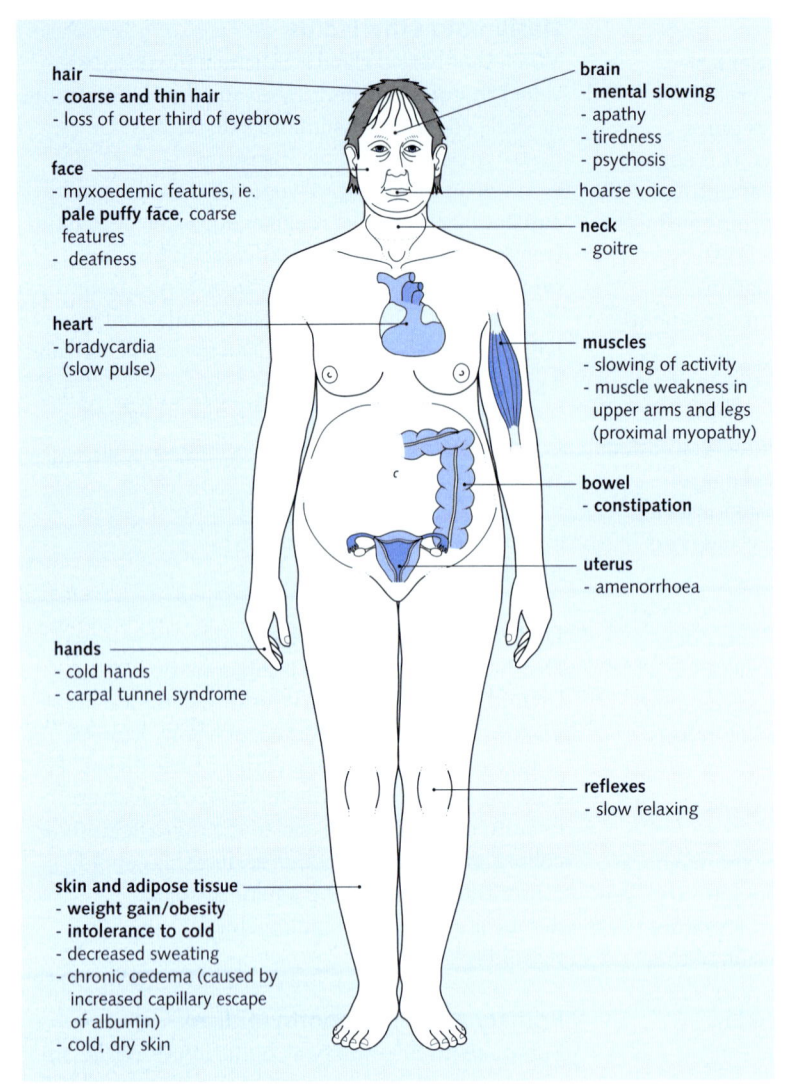

Fig. 3.10 Symptoms and signs of myxoedema (hypothyroidism). The main features are shown in bold.

hair
- **coarse and thin hair**
- loss of outer third of eyebrows

face
- myxoedemic features, ie. **pale puffy face**, coarse features
- deafness

brain
- **mental slowing**
- apathy
- tiredness
- psychosis

hoarse voice

neck
- goitre

heart
- bradycardia (slow pulse)

muscles
- slowing of activity
- muscle weakness in upper arms and legs (proximal myopathy)

bowel
- constipation

uterus
- amenorrhoea

hands
- cold hands
- carpal tunnel syndrome

reflexes
- slow relaxing

skin and adipose tissue
- weight gain/obesity
- intolerance to cold
- decreased sweating
- chronic oedema (caused by increased capillary escape of albumin)
- cold, dry skin

> An enlarged thyroid gland can be distinguished from a thyroglossal cyst by asking the patient to swallow and stick out their tongue. The thyroid gland will move with the trachea on swallowing, whereas a thyroglossal cyst rises when the tongue is protruded.

Iodine deficiency

Iodine deficiency was once a common cause of goitre in regions where the soil lacked iodine (e.g. Derby), but nowadays iodine is added to salt to prevent this.

Deficient iodine means that thyroid hormones cannot be synthesized so TSH levels rise due to a lack of feedback. TSH stimulates follicle and blood vessel growth and development of new blood vessels, so the thyroid gland enlarges. This does not cure the iodine deficiency so the goitre continues to grow.

The goitre formed by this process is diffusely enlarged and smooth. It is sometimes called an endemic goitre because it occurred in certain regions.

Graves' disease

The constant stimulation of TSH receptors in Graves' disease causes a goitre in a similar manner to iodine deficiency with similar characteristics. The gland becomes very vascular, to the extent that a bruit can be heard using a stethoscope.

Puberty and pregnancy

Higher levels of thyroid hormones are required in puberty and pregnancy so the thyroid gland often enlarges to meet the increased demand. This enlargement is a physiological response, not a pathological process. The goitre regresses once the demand lessens.

> During pregnancy the secretion of T_4 increases, however blood levels remain constant due to the increase in thyroid binding globulin controlled by oestrogen.

Multinodular goitre

Many elderly people have an enlarged thyroid that contains many nodules of varying sizes. These nodules are formed from hyperplasia (increased number) of thyroid cells. The excess cells sometimes cause excess thyroid hormone production i.e. hyperthyroidism. The disease is then called *toxic* multinodular goitre.

Thyroiditis

Inflammation of the thyroid gland can cause swelling, and infiltration by lymphocytes can also cause enlargement. The goitre formed is usually slightly nodular, but it may be tender if the inflammation is acute.

Thyroid gland neoplasia

Thyroid lumps are common and usually benign, however, they must be investigated. Solitary thyroid lumps are found in 5% of women and it is very difficult to distinguish between benign (80%) and malignant (20%) on clinical grounds. A fine-needle aspiration should be performed along with thyroid function tests. Aspiration alone will not distinguish a follicular adenoma from a follicular carcinoma but low TSH suggests the former. Ultrasound can be performed to detect cysts.

Causes of solitary thyroid lumps include:
- Thyroid cysts.
- Nodule of multinodular goitre.
- Follicular adenoma.
- Malignancy.

Five separate forms of cancer can arise in the thyroid gland, but three of these are derived from the follicle cells. These tumours are summarized in Fig. 3.11.

Medullary carcinomas of the parafollicular cells often secrete ectopic hormones including:
- Calcitonin—usually asymptomatic.
- Adrenocorticotrophic hormone (ACTH)—Cushing's syndrome.
- 5-hydroxytryptamine (5-HT; serotonin)—carcinoid syndrome.

Characteristics of primary thyroid gland malignancies

Type	Cell type	Age group	Route of metastasis	Prognosis
Papillary	Follicle cells	All	Cervical lymphatics	Excellent
Follicular	Follicle cells	Middle-aged	Blood to bone, lung, and brain	Good
Medullary	Parafollicular cells	Middle-aged and elderly	Cervical lymphatics	Variable but usually good
Malignant lymphoma	Lymphatics	Elderly	Local invasion	Poor
Anaplastic	Follicle cells	Elderly	Local invasion	Very poor

Fig. 3.11 The characteristics of the five primary thyroid gland malignancies.

The Thyroid Gland

- Describe the anatomical shape and location of the thyroid and parathyroid glands. Why does the thyroid gland move during swallowing?
- Describe the blood supply to the thyroid gland, and describe the nerves that are related to these vessels.
- Describe how the cells of the thyroid gland are arranged.
- How do the thyroid and parathyroid glands develop in the embryo?
- List the hormones secreted by the thyroid gland, and describe their actions.
- Describe the six steps in thyroid hormone synthesis.
- Explain the endocrine control of the thyroid gland.
- State the main symptoms and signs of thyrotoxicosis. What is the most common cause?
- State the main symptoms and signs of myxoedema. What is the main cause?
- Describe the treatment of hyperthyroidism and hypothyroidism.

4. The Adrenal Glands

The two adrenal glands allow the body to deal with both the emotional and physical stresses of life. They secrete four groups of hormones that have a wide range of effects.

The adrenal glands are located above each kidney. They are composed of two different types of endocrine tissue, which have different embryological origins.

The adrenal cortex is derived from embryonic mesoderm; it is regulated by adrenocorticotrophic hormone (ACTH) from the pituitary gland, and it responds by secreting three types of steroid hormone:
- Glucocorticoids to deal with stress.
- Mineralocorticoids to regulate blood volume.
- Androgens for sexual development.

The adrenal medulla is derived from ectodermal tissue, and it consists of sympathetic nerve cells; it is under the direct control of the sympathetic nervous system, and it responds by secreting two modified amino acid hormones:
- Adrenaline.
- Noradrenaline.

The most important hormones produced by the adrenal glands are glucocorticoids, such as cortisol. They are regulated by hypothalamic and anterior pituitary hormones to form the hypothalamic–pituitary–adrenal (HPA) axis (Fig. 4.1). Synthetic versions of these hormones are commonly used to treat inflammatory illness and they are referred to as 'steroids' or 'corticosteroids'.

An excess of glucocorticoids causes Cushing's syndrome and a deficiency causes Addison's disease.

Fig. 4.1 Hormonal regulation of cortisol. (ACTH, adrenocorticotrophic hormone; CRH, corticotrophin releasing hormone.)

Important words:
Catecholamine: adrenaline or noradrenaline
Cushing's syndrome: an excess of glucocorticoids
Cushing's disease: Cushing's syndrome caused by an ACTH secreting tumour of the anterior pituitary gland
Conn's syndrome: an excess of aldosterone caused by a tumour of the adrenal cortex
Addison's disease: deficiency of cortisol and aldosterone caused by progressive destruction of the adrenal gland

The Adrenal Glands

After reading this chapter you should be able to:
- Picture the structure and development of adrenal glands.
- Explain the regulation and physiological effects of steroid hormones secreted by the adrenal cortex.
- Describe the major disorders of the adrenal cortex.
- Discuss the physiological effects of hormones secreted by the adrenal medulla.
- Discuss the major disorders of the adrenal medulla.

Anatomy

There are two adrenal glands, one on top of each kidney, which differ significantly in their relations and shape. Both glands are retroperitoneal and are embedded in adipose tissue. Each gland is composed of an internal medulla and external cortex. They are also called the suprarenal glands.

Right adrenal gland

The right adrenal gland is pyramidal in shape, lying between the inferior vena cava and the right crus (a large tendon) of the diaphragm. The liver is located superiorly.

Left adrenal gland

The left adrenal gland is crescent shaped; it lies medially to the left crus of the diaphragm. Anteriorly, the body of the pancreas and the splenic artery are adjacent. The stomach is situated superiorly, separated by the peritoneum. The location of both glands is shown in Fig. 4.2.

Blood supply, nerves, and lymphatics

The outer cortex receives no significant innervation; instead it is regulated by ACTH from the pituitary gland and other blood-borne factors.

> The adrenal gland's name comes from the Latin meaning 'towards the kidney'.

The medulla is innervated directly by the splanchnic nerves, which arise from the thoracic spinal cord and do not synapse before reaching the adrenal medulla. The nerves are, therefore, preganglionic sympathetic nerves that release acetylcholine, while all other tissues receive only postganglionic sympathetic innervation. The cause of this relationship is apparent from their development (see below).

The glands receive a rich blood supply from the adrenal arteries, which are branches of the inferior phrenic arteries and renal arteries. They are drained by the adrenal veins into the inferior vena cava and renal vein. Lymphatic drainage passes to the para-aortic nodes.

Development

Adrenal cortex

The adrenal cortex is formed from an area of mesoderm that surrounds the adrenal medulla; this is close to the origin of the gonads. The cortical layer thickens to form the fetal cortex, which stimulates the differentiation of the adrenal medulla.

Later, more mesodermal cells surround the fetal cortex to form the permanent cortex found in adults. At birth, the permanent cortex has two layers while a third (the zona reticularis) develops by the third year. During this time the fetal cortex regresses until only the developed medulla and permanent cortex are left.

Adrenal medulla

The adrenal medulla is derived from ectodermal neural crest cells of the embryo. These cells contribute to many diverse structures including all the noradrenaline secreting postganglionic neurons in the sympathetic nervous system. The secretory cells in the adrenal medulla secrete either adrenaline or noradrenaline, and they are essentially highly specialized neurons (Fig. 4.3).

Microstructure

Adrenal cortex

The adult adrenal cortex makes up about 90% of the adrenal gland by weight. It is divided into three layers, which secrete the following groups of steroid hormone (Figs 4.4 and 4.5):
- Outer zona glomerulosa secretes mineralocorticoids.
- Middle zona fasciculata secretes glucocorticoids.
- Inner zona reticularis secretes androgens and glucocorticoids.

Microstructure

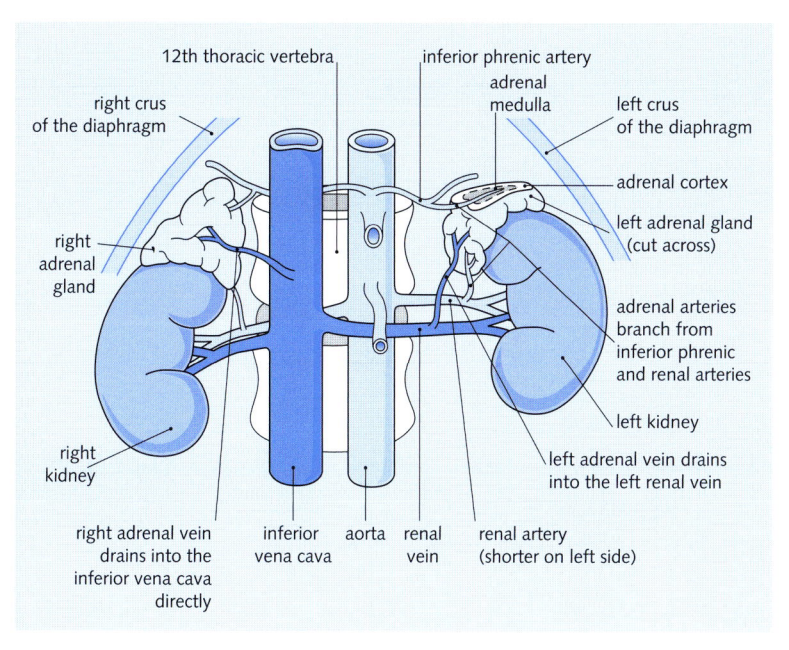

Fig. 4.2 Location and blood supply of the adrenal glands.

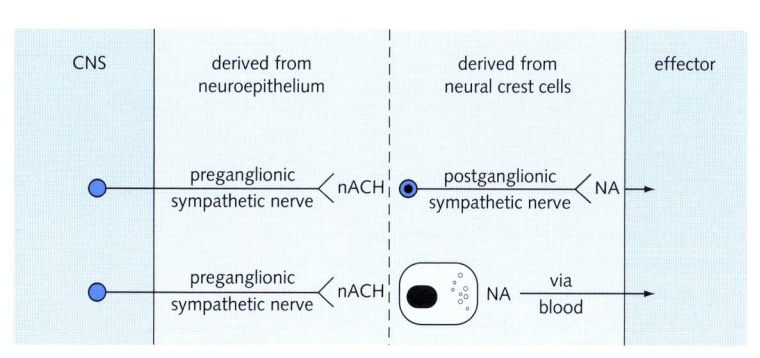

Fig. 4.3 Comparison between the adrenal medulla and the sympathetic nervous system. (ACh, acetylcholine; CNS, central nervous system; NA, noradrenaline.)

Fig. 4.4 Microstructure of the adrenal gland and the major hormones secreted in each region.

Microstructure of the adrenal gland and the major hormones secreted in each region

Region	Name	Cell structure	Hormones synthesized
Outer cortex	Zona glomerulosa	Cells arranged in clumps (Latin, glomerulus: little ball)	Mineralocorticoids (mainly aldosterone)
Middle cortex	Zona fasciculata	Cells arranged in cords alongside blood sinusoids; (Latin, fasciculus: bundle)	Glucocorticoids (mainly cortisol)
Inner cortex	Zona reticularis	Network of smaller cells (Latin, reticularis: network)	Glucocorticoids and androgens (DHEA)
Centre of gland	Adrenal medulla	Loose network of neurosecretory cells surrounded by blood sinusoids	Catecholamines (adrenaline and noradrenaline)

43

The Adrenal Glands

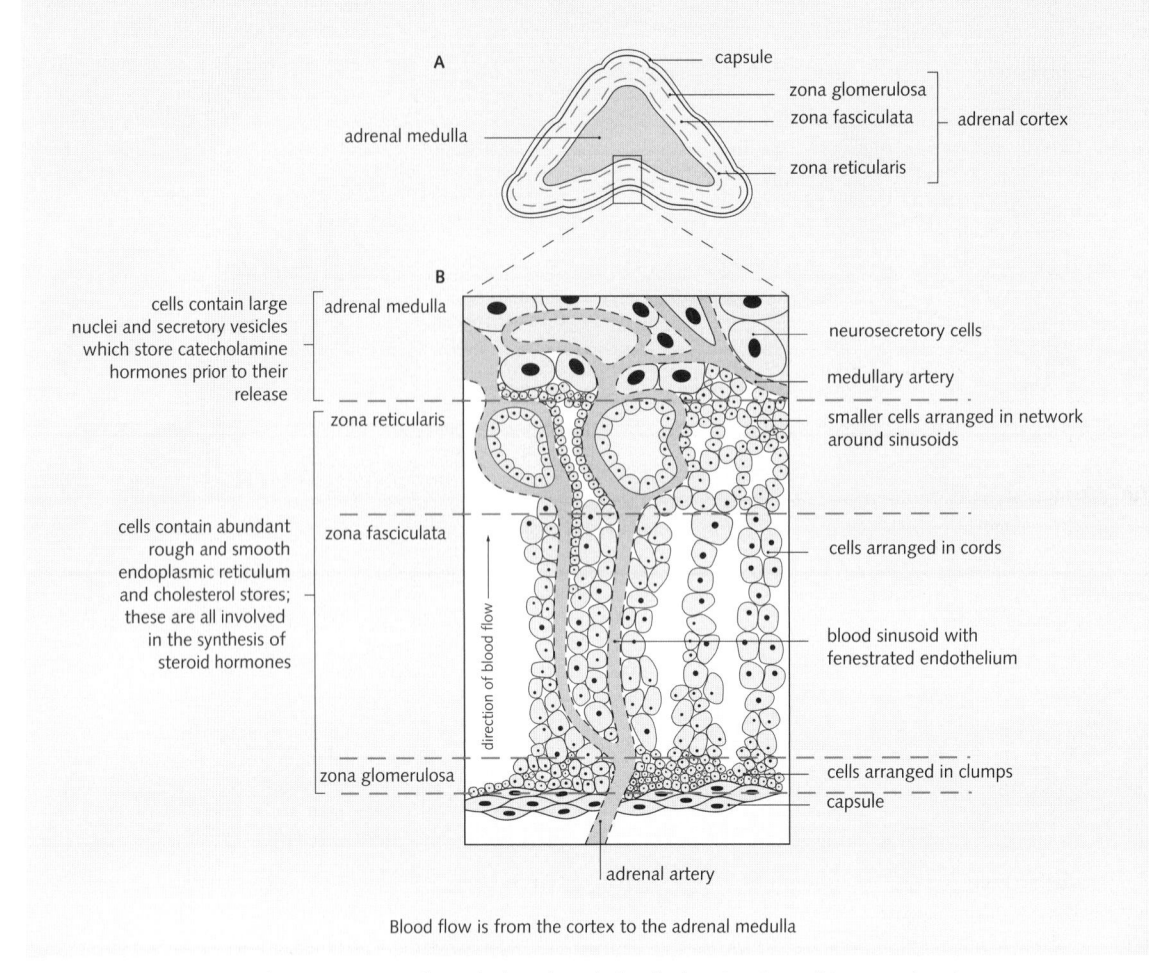

Fig. 4.5 Microstructure of a cross-section through the adrenal glands showing the cell types and regions.

These hormone groups will be explained later in the chapter.

The cells of the zona fasciculata and zona reticularis are arranged in columns around blood sinusoids. The blood in these sinusoids passes directly into the adrenal medulla.

To remember the order of the layers in the adrenal cortex think of GFR that stands for 'glomerular filtration rate' in the nearby kidney.

Adrenal medulla

The adrenal medulla is composed of two types of neuroendocrine cell:
- Noradrenaline secreting cells (20%).
- Adrenaline secreting cells (80%).

Both types contain neuroendocrine granules that store the hormone. In older textbooks these cells are called chromaffin cells because they turn a dark brown colour if exposed to oxygen after fixation in chrome salts. The cells are arranged around blood sinusoids.

Medullary cells require the steroid cortisol to convert noradrenaline to adrenaline. Cortisol is produced in the cortex, and it travels in the cortical capillaries to the medulla. Separate medullary arteries supply oxygenated blood directly.

Hormones of the adrenal cortex

The adrenal cortex secretes three groups of steroid hormones:
- Mineralocorticoids, e.g. aldosterone.
- Glucocorticoids, e.g. cortisol.
- Androgens, e.g. DHEA.

Steroid hormones are synthesized from cholesterol. They are small lipid-soluble molecules that cross membranes readily. Inside cells they act on intracellular receptors to regulate gene expression. The synthesis and mechanism of action of steroid hormone are discussed in more detail in Chapter 1.

Mineralocorticoids and aldosterone
Regulation of aldosterone
Mineralocorticoids help to regulate the electrolyte balance of plasma; their name is derived from this action on the body's minerals. Aldosterone is the main mineralocorticoid secreted by the outer layer of the cortex called the zona glomerulosa. Its release is stimulated by three main factors:
- Angiotensin II.
- High plasma potassium.
- ACTH.

Angiotensin II is released in response to low blood volume as part of the renin-angiotensin system (see Ch. 7 and Fig. 4.6). ACTH from the anterior pituitary gland is less important as a regulator, so pituitary failure does not severely impair aldosterone secretion. An excess of aldosterone due to an adrenal adenoma is called Conn's disease.

Actions of aldosterone
Aldosterone acts mainly on the distal convoluted tubule (DCT) and the collecting duct of the kidney. It causes reabsorption of sodium ions in exchange for potassium and hydrogen ions. Water is also reabsorbed and blood volume is increased. Other hormones are involved in this mechanism and they are discussed in more detail in Chapter 7 on fluid balance.

Intracellular actions of aldosterone
To cause these physiological effects aldosterone acts on the nucleus via an intracellular receptor. Only cells that express this receptor can respond to aldosterone. The expression of four genes is increased in the cells of the DCT and collecting duct.

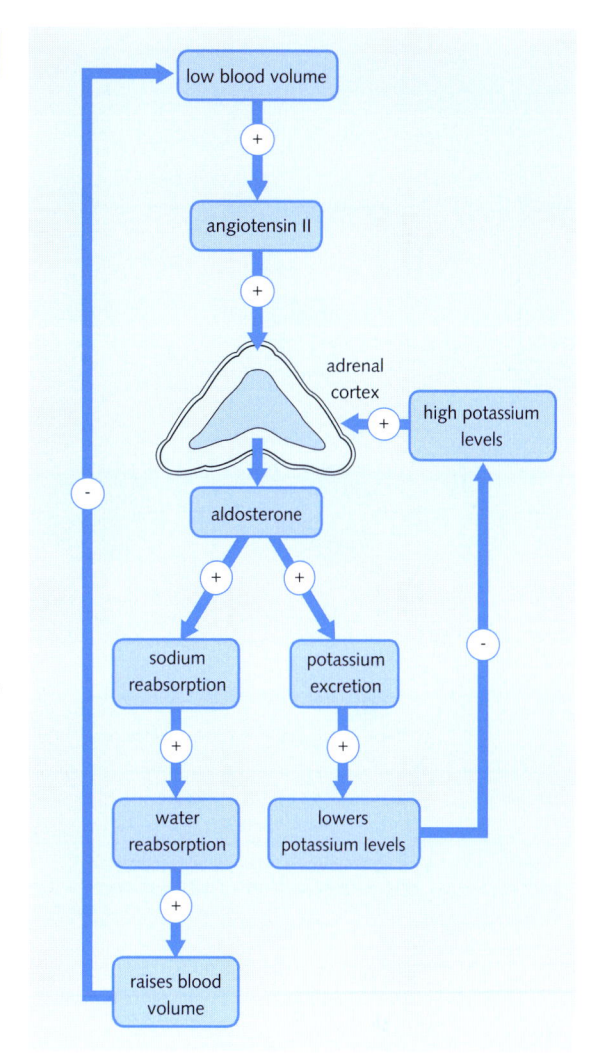

Fig. 4.6 Control of aldosterone secretion.

The actions of the gene products (proteins) are described in Fig. 4.7.

Aldosterone circulates in the plasma with 60% bound to albumin and 40% free and, therefore, active. The high proportion of free hormone causes aldosterone to be rapidly degraded by the liver, giving a short half-life of about 15 minutes.

Glucocorticoids and cortisol
Glucocorticoids act on the metabolism of carbohydrate, protein, and to a lesser extent fat. It is the action on glucose (a carbohydrate) that is responsible for their name. Glucocorticoids also depress the immune system, and it is for this effect that 'steroids' are most often used as medication. Cortisol is the main glucocorticoid. An excess of glucocorticoids due to treatment or pathology is called Cushing's syndrome.

The Adrenal Glands

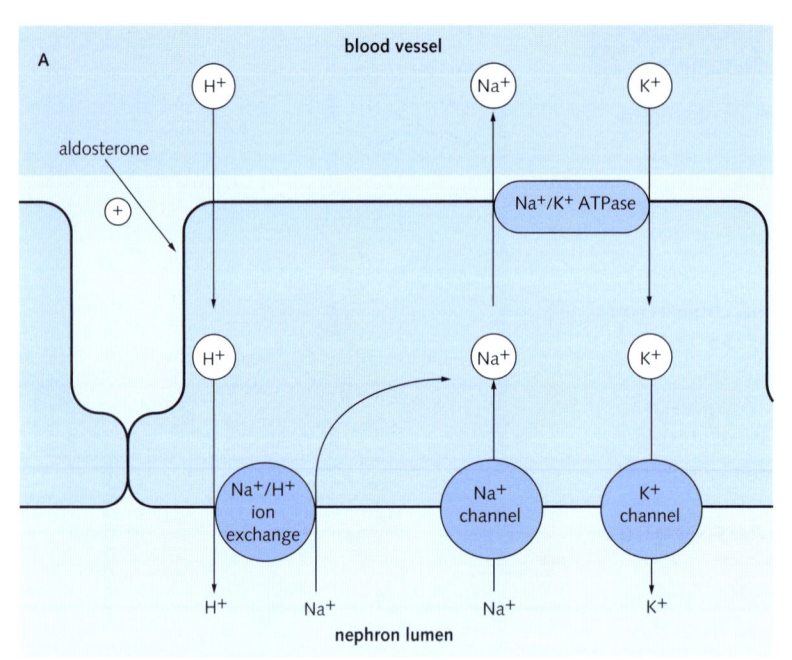

Fig. 4.7 Intracellular and physiological actions of aldosterone in the nephron. Aldosterone acts to increase the levels of the four proteins shown (A) causing the physiological responses (B).

B Effects of proteins induced by aldosterone in the nephron

Protein	Location	Action	Physiological response
Na^+/K^+ ATPase	Cell membrane on the side of the blood supply	Active pump that increases cell potassium and lowers cell sodium levels	Creates an ion gradient that drives the other proteins
Na^+ channel	Cell membrane on the side of the nephron	Reabsorbs sodium from the nephron lumen	Increases plasma sodium and water to increase blood volume
K^+ channel	Cell membrane on the side of the nephron	Excretes potassium into the nephron lumen	Decreases plasma potassium
Na^+/H^+ ion exchanger	Cell membrane on the side of the nephron	Reabsorbs sodium in exchange for hydrogen ions	Makes the plasma more alkaline

Regulation of cortisol

Corticotrophin releasing hormone (CRH) is secreted by the hypothalamus, and it stimulates the anterior pituitary to release ACTH. This acts on the zona fasciculata and zona reticularis of the adrenal cortex, which secrete cortisol. Cortisol has a negative feedback effect on the hypothalamus and anterior pituitary gland to inhibit CRH and ACTH release.

Cortisol release displays a circadian rhythm, i.e. the rate of secretion changes through a 24-hour period (Fig. 4.8). The highest levels of cortisol release are in the early morning, peaking at about 6 am, then falling through the day. This circadian variation is initiated in the hypothalamus by changes in sensitivity to cortisol levels. Cortisol exerts a

> The inhibitory action of cortisol on ACTH release is important clinically. Patients treated with long-term 'steroids' cannot simply stop because ACTH release and, therefore, cortisol production would also stop. Instead the dose must be lowered over a number of months.

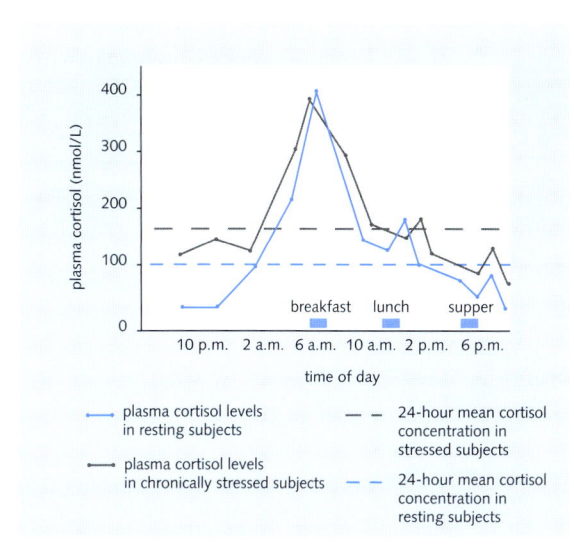

Fig. 4.8 Circadian variation in plasma cortisol in resting and chronically stressed subjects.

weaker negative feedback effect in the morning so CRH release rises.

Actions of cortisol
Like adrenaline, cortisol allows the body to deal with 'stress' such as trauma, haemorrhage, and fever. This effect is very important and cortisol deficiency can rapidly become life threatening. The response to stress is called the general adaptation syndrome (GAS), and it is divided into three phases:

Alarm reaction   A stressful stimulus causes:
- Noradrenaline release from sympathetic nerves.
- Adrenaline and noradrenaline release from adrenal medulla.
- Cortisol release from adrenal cortex.

Resistance   Cortisol has a slower and longer lasting action than adrenaline and noradrenaline; it allows the resistance to stress to be maintained. It also counteracts the effects of other hormones (e.g. insulin) to maintain substrates required to combat stress.

Exhaustion   Prolonged stress causes continued cortisol secretion, and it results in muscle wastage, immune system suppression, and hyperglycaemia.

Cortisol affects almost every cell in the body. The physiological effects are described in Fig. 4.9. The main actions of cortisol are:
- Increase of energy metabolite levels in the blood.
- Suppression of the immune system and inhibition of allergic and inflammatory processes.

There is some overlap between the actions of mineralocorticoids and glucocorticoids. Cortisol can have mineralocorticoid actions, whereas aldosterone can act as a glucocorticoid.

Intracellular actions of cortisol
Like all steroid hormones, cortisol acts via intracellular receptors to regulate gene expression. The receptor and genes vary between cells, and this accounts for the wide range of actions. The anti-inflammatory actions are produced by inhibiting phospholipase A_2, an enzyme that is essential for the production of prostaglandins from arachidonic acid.

Most cortisol (95%) is transported round the body bound to plasma proteins:
- 80% bound to cortisol-binding protein.
- 15% bound to albumin.
- 5% free and active.

It is inactivated in the liver by conjugation and then excreted from the kidney. About 1% of cortisol is excreted into the urine without metabolism. This can be detected by 24-hour urine collection to estimate blood cortisol levels.

Androgens
Androgens are male sex steroids, i.e. hormones involved in the growth and function of the male genital tract. They also stimulate muscle growth (anabolism), hence their use as an illicit drug in sport. They are secreted in both males and females, however in males they account for only a small proportion of total androgen production.

Actions of adrenal androgens
Adrenal androgens are synthesized in the zona reticularis of the adrenal gland; the main adrenal androgens are:
- Dehydroepiandrosterone (DHEA).
- Androstenedione.

Androgens secreted by the adrenal glands have weak biological activity, but they are converted to more active androgens, such as testosterone, by enzymes in peripheral tissues.

Adrenarche
The initiation of androgen secretion from the adrenal glands is called adrenarche. It occurs a few years before puberty (about 7–9 years of age), and it is marked by maturation of the zona reticularis.

The Adrenal Glands

Fig. 4.9 Physiological effects of cortisol related to the symptoms of Cushing's syndrome. (ACTH, adrenocorticotrophic hormone; FSH, follicle stimulating hormone; GH, growth hormone; LH, luteinizing hormone; TSH, thyroid stimulating hormone.)

Physiological effects of cortisol and the symptoms of Cushing's syndrome

Process/system affected	Effect of cortisol	Related pathology in Cushing's syndrome
Carbohydrate metabolism	Raises blood glucose by stimulating gluconeogenesis and preventing glucose uptake	Hyperglycaemia and diabetes
Protein metabolism	Increases breakdown of proteins in skeletal muscle, skin and bone to release amino acids	Muscle weakness and wasting; thin easily bruising skin
Fat metabolism	Stimulates lipolysis and increases fatty acid levels in the blood	Fat redistributed to the face and trunk causing a moon face, buffalo hump, and abdominal stretch marks
Immune system	Suppresses the action and production of immune cells; inhibits the production of cytokines and antibodies	Infections, poor healing, peptic ulceration
Endocrine system	Suppresses the secretion of anterior pituitary hormones: ACTH, LH, FSH, TSH and GH	Suppression of growth in children
Nervous system	Influences fetal and neonatal neuron development; influences behaviour and cognitive function; augments the actions of the sympathetic system	Depression, insomnia, psychosis, and confusion
Water metabolism	Has weak mineralocorticoid actions: raises sodium and water retention	Hypertension and heart failure
Calcium metabolism	Decreases calcium absorption from the gut; increases calcium excretion in the kidneys; increases calcium resorption from bones	Osteoporosis

In males, the early development of the male sex organs may result from adrenal androgens released after adrenarche. In male adult life adrenal androgens account for only 5% of total activity, so they are physiologically negligible. Androgens are discussed in greater detail in Chapter 13.

In the female adrenal androgens are responsible for about 50% of total androgen activity from adrenarche to the end of life. These hormones help to promote the growth of female pubic and axillary hair.

Disorders of the adrenal cortex

The main diseases of the adrenal cortex are caused by an excess or deficiency of mineralocorticoids or glucocorticoids. There are four named diseases affecting the adrenal cortex hormones that are a common cause of pre-exam tremor, however they are rare diseases:

- Cushing's syndrome—excessive cortisol production.
- Cushing's disease—ACTH secreting tumour.
- Conn's syndrome—aldosterone secreting tumour.
- Addison's disease—deficiency of cortisol and aldosterone.

To remember the named diseases of the adrenal hormones think: 'Cushy Cort cons miners into adding both' i.e.:

- Cushing's—Cortisol (excess)
- Conn's—Mineralocorticoids (excess)
- Addison's—Both (deficiency)

Disorders of the Adrenal Cortex

Action of aldosterone	Hyperaldosteronism	Hypoaldosteronism
Increases plasma Na+	Hypernatraemia rarely occurs because of other mechanisms regulating fluid volume	Loss of Na+ is accompanied by loss of water, so plasma Na+ concentration does not change
Decreases plasma K+	Hypokalaemia	Hyperkalaemia
Decreases plasma H+	Metabolic alkalosis	Mild metabolic acidosis
Maintains extracellular fluid volume	Hypertension	Volume depletion and postural hypotension

Fig. 4.10 Clinical symptoms caused by hyperaldosteronism (e.g. Conn's syndrome) and hypoaldosteronism (e.g. Addison's disease).

Hyperaldosteronism

Excess aldosterone production causes sodium ion and water retention with increased excretion of potassium and hydrogen ions. The main symptoms and signs (Fig. 4.10) are:
- Hypertension (high blood pressure).
- Hypokalaemia (low potassium).
- Alkalosis (raised blood pH).
- Polyuria and polydipsia (thirst).
- Muscle weakness and spasm.

A number of blood tests are used for diagnosis:
- Urea and electrolytes (U + Es) for hypokalaemia.
- Aldosterone levels (raised).
- Renin levels (variable).

Aldosterone increases blood volume, which inhibits renin secretion. If renin levels are low then the disorder is primary hyperaldosteronism, i.e. the disease originates in the adrenal glands. The adrenal glands can then be imaged by CT/MRI scanning.

Renin stimulates aldosterone release via angiotensin II, so high renin levels suggest secondary hyperaldosteronism. This disorder is external to the adrenal glands; it is a common response to heart failure and renal disease.

Primary hyperaldosteronism and Conn's syndrome

Primary hyperaldosteronism is a rare disease that is responsible for about 1% of patients with hypertension. The vast majority of primary hyperaldosteronism is caused by Conn's syndrome, in which the patients have an adenoma of the zona glomerulosa. This is discussed later in the chapter.

Secondary hyperaldosteronism

Secondary hyperaldosteronism is a very common problem caused by activation of the renin–angiotensin system. The most common cause is excessive diuretic therapy, but it is also a feature of:
- Congestive heart failure.
- Renal artery stenosis.
- Nephritic syndrome.
- Cirrhosis with ascites.

All these conditions result in decreased renal perfusion, which stimulates renin release.

Excess cortisol
Cushing's syndrome

Cushing's syndrome is a rare condition caused by a chronic excess of glucocorticoids. The disorder can be in the anterior pituitary gland or the adrenal cortex, or it may result from excess medication. It has a five-year mortality of about 50% if it is not treated. The symptoms and signs of Cushing's syndrome are shown in Figs 4.9 and 4.11; it is most common in adult women.

Diagnosing excess cortisol is complicated by the circadian variation in cortisol secretion. Two main tests are employed to overcome this problem:
- 24-hour urinary free cortisol: 1% of free cortisol is excreted unmetabolized, and this can be measured to give an accurate reflection of plasma cortisol.
- Dexamethasone suppression test: plasma cortisol is measured before an oral dexamethasone (a synthetic glucocorticoid) dose and then at 8 am the next morning. In a normal person plasma cortisol would be suppressed.

49

The Adrenal Glands

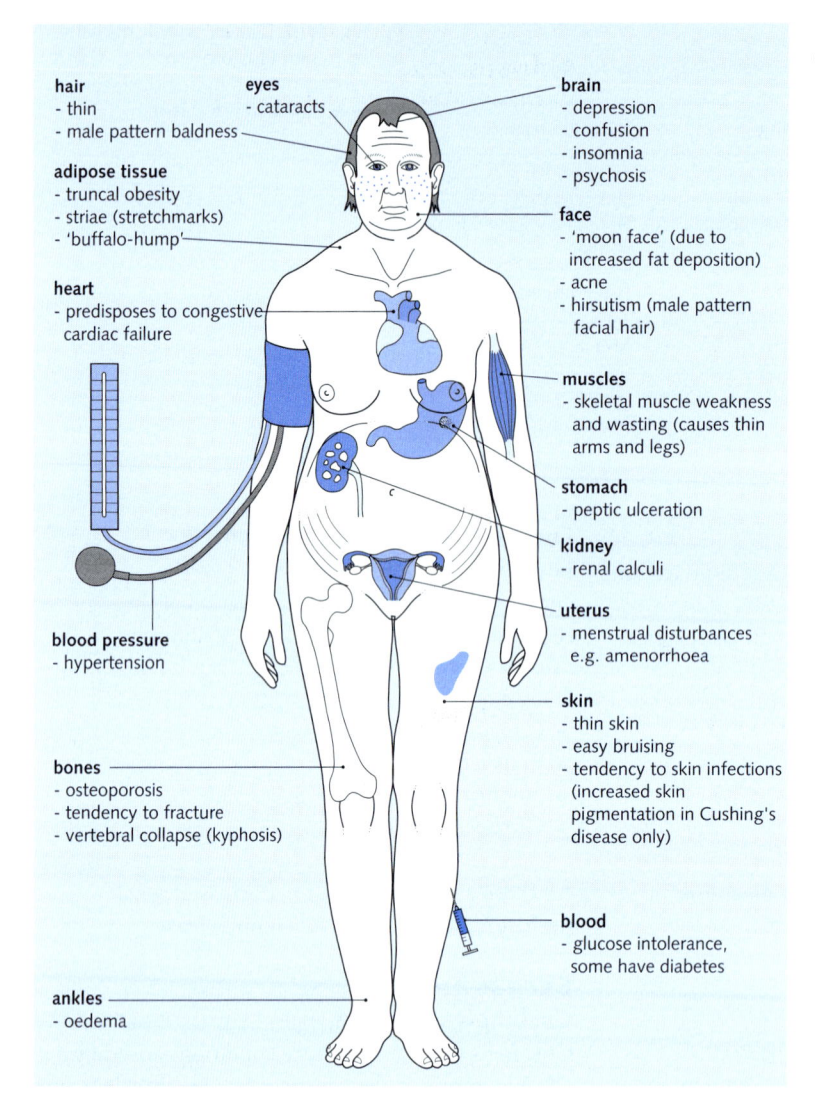

Fig. 4.11 Symptoms and signs of Cushing's syndrome.

hair
- thin
- male pattern baldness

eyes
- cataracts

adipose tissue
- truncal obesity
- striae (stretchmarks)
- 'buffalo-hump'

heart
- predisposes to congestive cardiac failure

blood pressure
- hypertension

bones
- osteoporosis
- tendency to fracture
- vertebral collapse (kyphosis)

ankles
- oedema

brain
- depression
- confusion
- insomnia
- psychosis

face
- 'moon face' (due to increased fat deposition)
- acne
- hirsutism (male pattern facial hair)

muscles
- skeletal muscle weakness and wasting (causes thin arms and legs)

stomach
- peptic ulceration

kidney
- renal calculi

uterus
- menstrual disturbances e.g. amenorrhoea

skin
- thin skin
- easy bruising
- tendency to skin infections (increased skin pigmentation in Cushing's disease only)

blood
- glucose intolerance, some have diabetes

Once cortisol excess has been confirmed, further tests using higher doses of dexamethasone and measuring ACTH levels can locate the source. CT scans are used once a source has been identified.

Treatment with glucocorticoids
Glucocorticoids (often simply called 'steroids') are used to treat a wide range of medical conditions, usually to reduce immune reactions. These conditions include asthma, inflammatory bowel disease, rheumatoid arthritis, and post-transplantation. Patients are treated with the lowest dose that will control their condition because prolonged use can cause the features of Cushing's syndrome. Inhaled steroids are used in asthma to reduce the systemic dose, especially in children where growth retardation may occur.

Cushing's disease
Adenomas of the corticotroph cells in the anterior pituitary can release excess ACTH. This stimulates the adrenal cortex to secrete excess cortisol, leading to bilateral enlargement of the cortex. The negative feedback that normally prevents excess ACTH release is absent in the tumour.

This type of tumour causes Cushing's syndrome with the additional sign of pigmented skin. This is due to an excess of melanocyte-stimulating hormone (α-MSH)—formed by the same gene that makes ACTH; it stimulates the activity of melanocytes in the skin. Cushing's disease occurs most frequently in young adult women.

Cushing's disease is treated by surgical removal of the pituitary adenoma. This may result in panhypopituitarism (see Chapter 2 for more details).

Ectopic adrenocorticotrophic hormone production

Ectopic ACTH can be secreted by the rare, but highly malignant, small-cell anaplastic carcinoma of the lung (also called oat-cell carcinoma). This carcinoma displays the characteristics of a neuroendocrine cell despite developing from bronchial epithelium. Even more rarely, tumours of the thymus, ovary, pancreas and carcinoid tumours can secrete ACTH or CRH. The excess production is so dramatic that patients rarely exhibit features of Cushing's syndrome before death. Ectopic hormones are discussed in Chapter 10.

> Cushing's disease and syndrome are named after the American neurosurgeon Harvey Cushing (1869–1939), who clarified the function of the anterior pituitary gland and described the effects of excess cortisol. In his spare time he also developed the first surgical technique for removing pituitary tumours.

Neoplasia of the adrenal cortex

Benign adenoma of the adrenal cortex is relatively common, but only a small proportion secrete hormones. If cortisol is secreted then Cushing's syndrome develops; aldosterone-secreting adenomas cause Conn's syndrome.

Adrenal adenomas are the most common cause of Cushing's syndrome in children, but they account for only 10% of adult disease. In Conn's syndrome, adenomas of the adrenal cortex are the most common cause of primary hyperaldosteronism in all age groups. Adenomas associated with either syndrome are removed surgically, but cortisol replacement is necessary due to long-term ACTH inhibition.

Carcinoma of the adrenal cortex is a very rare condition. They secrete vast excesses of glucocorticoids and androgens. The patient usually dies before the physical features of Cushing's syndrome develop.

Deficiency of cortisol and aldosterone
Congenital adrenal hyperplasia

This is an autosomal recessive condition causing deficiency of 21-hydroxylase, an enzyme normally found in the adrenal cortex. Since this enzyme is required for the synthesis of aldosterone and cortisol both hormones are deficient. Low cortisol triggers ACTH release resulting in hyperplasia of the adrenal cortex. Low aldosterone results in salt loss and neonatal shock in some babies. Full blown congenital adrenal hyperplasia is rare (1 in 10 000 births), however, 1 in 100 births partially express this condition.

The enlarged adrenal cortex secretes excess androgens causing adrenogenital syndrome. This presents differently in each sex. In males, it causes early (precocious) pseudopuberty; signs of secondary sexual development can be found by 6 months of age, but the child is not fertile. Early bone maturation causes short adult height.

In females androgen excess causes masculinization (also called virilization). The symptoms are similar to those found in polycystic ovarian syndrome as described in Chapter 12. They include:
- Masculine body shape.
- Balding of temporal skull.
- Increased muscle bulk.
- Deepening of the voice.
- Enlargement of the clitoris.

Adrenal cortex insufficiency

Adrenal cortex insufficiency tends to affect the whole adrenal cortex rather than specific layers. Accordingly, deficiency of glucocorticoids, mineralocorticoids, and androgens occur together though clinical effects are due to cortisol and aldosterone deficiency. These effects are shown in Fig. 4.12. Hydrocortisone (cortisol) and fludrocortisone (a mineralocorticoid) therapy must be initiated before the underlying disease process is treated.

Suspected adrenal cortex insufficiency is investigated using the ACTH stimulation test. A synthetic ACTH analogue is injected and plasma cortisol levels are measured every 30 minutes. If the cortisol levels do not rise sufficiently then the disease is of the adrenal cortex (i.e. Addison's disease).

Addison's disease

Primary insufficiency of the adrenal cortex is called Addison's disease; it is characterized by deficient secretion of glucocorticoids and mineralocorticoids.

The Adrenal Glands

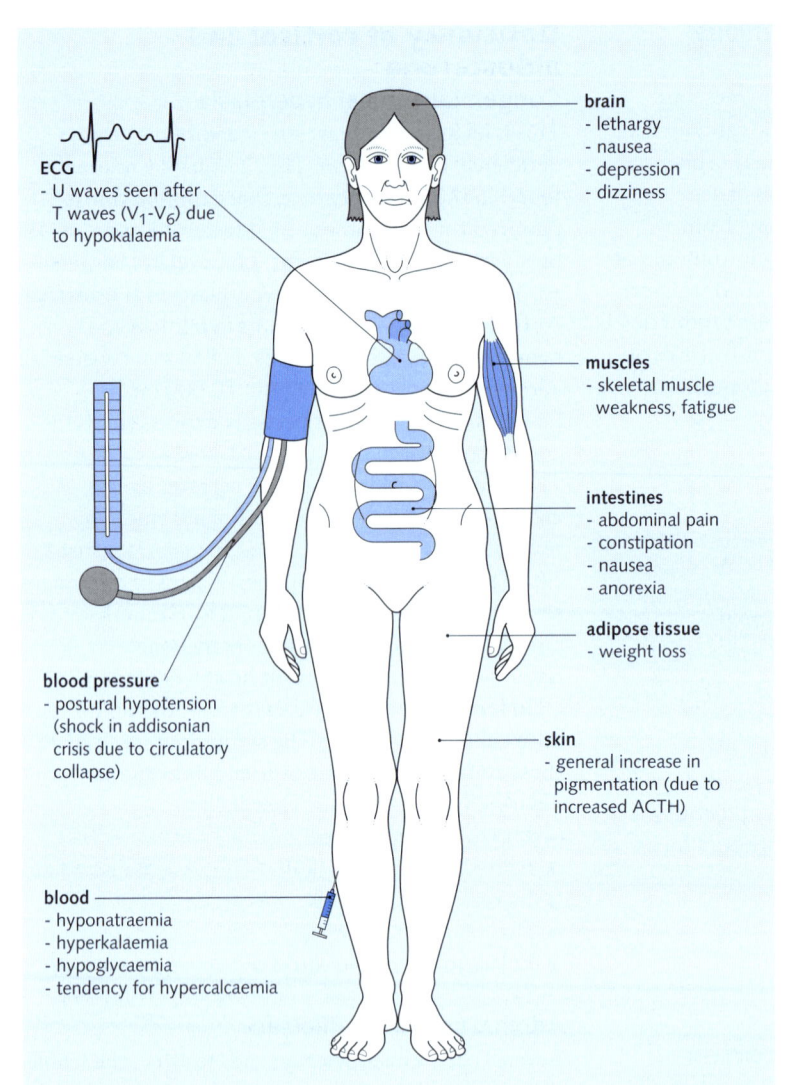

Fig. 4.12 Symptoms and signs of Addison's disease.

ECG
- U waves seen after T waves (V_1-V_6) due to hypokalaemia

brain
- lethargy
- nausea
- depression
- dizziness

muscles
- skeletal muscle weakness, fatigue

intestines
- abdominal pain
- constipation
- nausea
- anorexia

adipose tissue
- weight loss

skin
- general increase in pigmentation (due to increased ACTH)

blood pressure
- postural hypotension (shock in addisonian crisis due to circulatory collapse)

blood
- hyponatraemia
- hyperkalaemia
- hypoglycaemia
- tendency for hypercalcaemia

It is a rare chronic condition caused by progressive destruction of the adrenal cortex. This destruction can result from autoimmune adrenalitis, infection (e.g. tuberculosis, fungi), or tumour. Addison's disease presents with adrenal cortex insufficiency, but the high levels of circulating ACTH can cause skin pigmentation too.

An acute exacerbation of Addison's disease is called an adrenal crisis. It is a life threatening emergency caused by stressful events such as infection. Its presentation is the same as acute adrenal cortical failure.

Acute adrenal cortical failure

Acute adrenal cortex failure is a life threatening condition characterized by:
- Hypotensive shock.

The English authoress Jane Austen is believed to have suffered from Addison's disease.

- Hypovolaemic shock.
- Hypoglycaemia.

The adrenal cortex can be destroyed acutely by bilateral haemorrhagic necrosis following disseminated intravascular coagulation. Essentially blood clots block the venous drainage of the adrenal cortex causing cell death. These clots can form following severe septicaemia. Meningococcal septicaemia is the most common cause and this is called Waterhouse–Friderichsen syndrome.

A similar situation can occur if long-term high-dose steroid treatment is stopped abruptly. The prolonged treatment chronically suppresses ACTH release from the anterior pituitary gland so that no cortisol is secreted from the adrenal cortex for a number of weeks.

Secondary adrenocortical insufficiency
Disorders of the hypothalamus and anterior pituitary gland can also cause deficiency of adrenal cortex steroid hormones. Any condition that causes a reduction in CRH or ACTH release will prevent the synthesis of glucocorticoids especially. These conditions are described in more detail in Chapter 2.

Hormones of the adrenal medulla

The adrenal medulla secretes two hormones: noradrenaline and adrenaline. Both of these hormones are described as catecholamines, a term that is derived from their chemical structure (Fig. 4.13). Eighty per cent of catecholamine released from the adrenal glands is adrenaline.

Regulation
Catecholamines are released in response to stress (e.g. exercise, pain, shock, hypoglycaemia and imminent exams). Stress stimulates an area of the hypothalamus that activates the sympathetic system, including the adrenal medulla. It receives no direct regulation from the pituitary gland.

Actions
Catecholamines from the adrenal medulla perform similar functions to direct sympathetic neuronal connections, in that they prepare the body for fight-or-flight. Their effects last longer than the neuronal signals, so they help to minimize the harm caused by repeated stress. Adrenaline and noradrenaline have similar effects to each other.

Their main actions are described in Fig. 4.14. Generally they improve mental and physical abilities to overcome the stressor; for more detail on the sympathetic nervous system see *Crash Course Nervous System*.

Intracellular actions
Adrenal catecholamines bind to receptors in a similar manner to neuronal signals. These extracellular receptors are linked to intracellular G-proteins that initiate a signal cascade. The effect of the signal depends on the receptor present and the cell type.

Synthesis
Noradrenaline is synthesized from the amino acid tyrosine, which is then converted to adrenaline in response to cortisol from the adrenal cortex.

The medullary cells store catecholamines in cytoplasmic granules. They are released into blood sinusoids by exocytosis in response to acetylcholine from preganglionic sympathetic neurons.

Breakdown
Catecholamines circulate bound to albumin. They are degraded by two enzymes in the liver:
- Monoamine oxidase (MAO).
- Catechol-O-methyl transferase (COMT).

Adrenaline and noradrenaline are converted to vanillyl mandelic acid (VMA or HMMA), which is released into the urine. Urinary VMA levels are measured to

> Monoamine oxidase (MAO)—the enzyme that degrades catecholamines in the liver—is also present in the brain. Inhibitors of this enzyme are used as antidepressants.

Fig. 4.13 Structure of adrenaline and noradrenaline.

The Adrenal Glands

Brain

Effect	Receptor
1. Causes alertness, agitation, fear, anxiety 2. Stimulates release of ACTH	Unknown mechanism

Eyes

Effect	Receptor
Dilates pupils so more light reaches the retina	α_1

Liver

Effect	Receptor
Increases glycogenolysis (release of glucose from glycogen stores) → energy source	α and β

Kidney

Effect	Receptor
Increases renin release	β

Adipose tissue

Effect	Receptor
Increases lipolysis (release of fatty acids from triglyceride store) → energy source	β

Skin

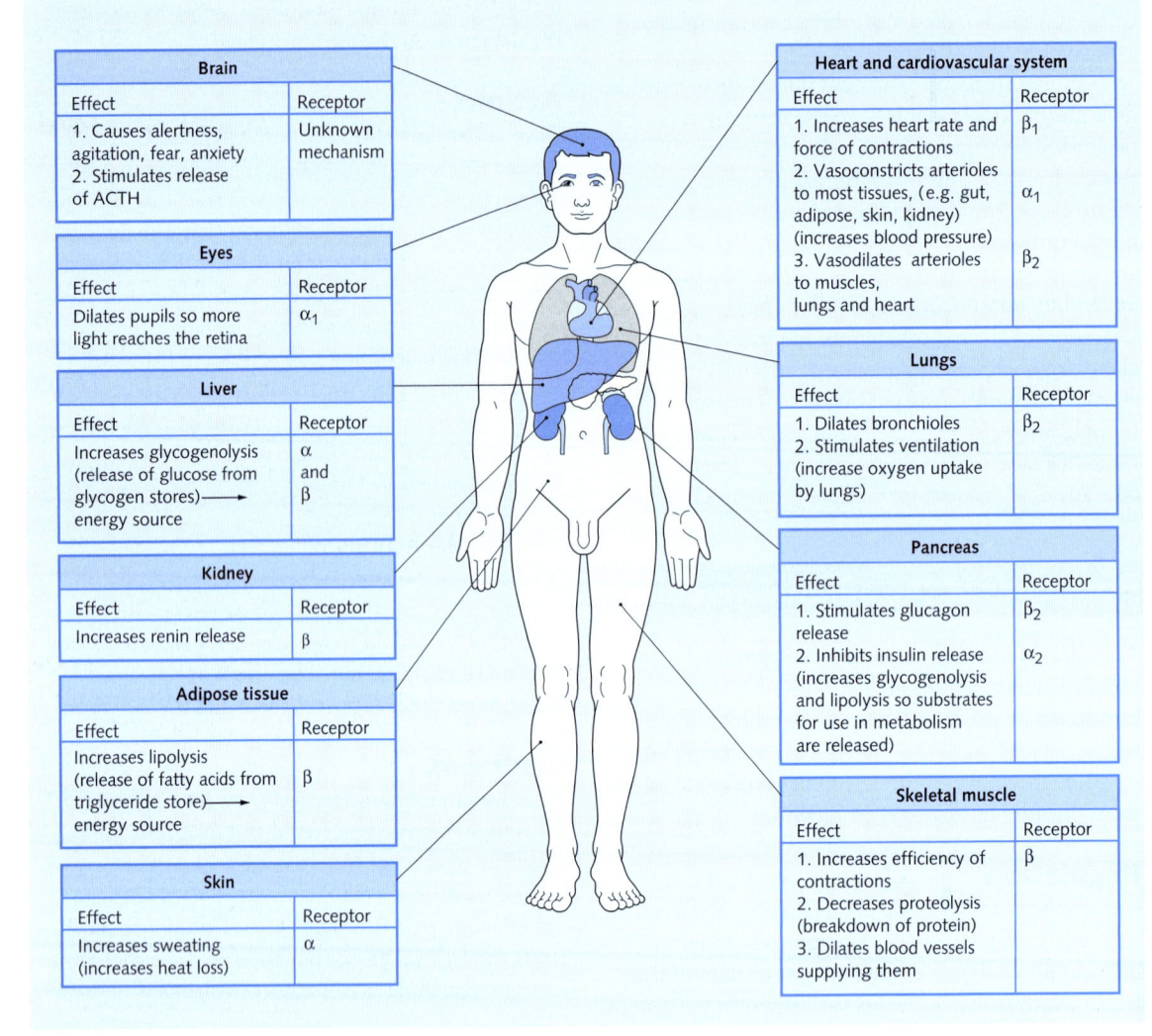

Effect	Receptor
Increases sweating (increases heat loss)	α

Heart and cardiovascular system

Effect	Receptor
1. Increases heart rate and force of contractions	β_1
2. Vasoconstricts arterioles to most tissues, (e.g. gut, adipose, skin, kidney) (increases blood pressure)	α_1
3. Vasodilates arterioles to muscles, lungs and heart	β_2

Lungs

Effect	Receptor
1. Dilates bronchioles 2. Stimulates ventilation (increase oxygen uptake by lungs)	β_2

Pancreas

Effect	Receptor
1. Stimulates glucagon release	β_2
2. Inhibits insulin release (increases glycogenolysis and lipolysis so substrates for use in metabolism are released)	α_2

Skeletal muscle

Effect	Receptor
1. Increases efficiency of contractions 2. Decreases proteolysis (breakdown of protein) 3. Dilates blood vessels supplying them	β

Fig. 4.14 Physiological effects of adrenaline and noradrenaline and the receptors present in each tissue/organ.

detect phaeochromocytomas, a rare tumour of the adrenal medulla that is discussed below.

Disorders of the adrenal medulla

Phaeochromocytomas

Phaeochromocytomas are very rare tumours of the catecholamine-producing cells in the adrenal medulla. They are usually benign and present in only one gland (unilateral). Adrenaline and noradrenaline are secreted in large quantities causing severe, sporadic (paroxysmal) hypertension that can produce headaches. With time the hypertension can become constant leading to heart failure.

The tumour is detected by the high levels of catecholamine breakdown products in the urine, e.g. VMA. It is treated by surgical excision. The surgery has a high peri-operative mortality due to the unstable blood pressure.

Catecholamine-producing tumours can also develop in sympathetic ganglia. These usually occur beside the abdominal aorta near the bifurcation.

> It is an essential component of your medical education to diagnose at least one phaeochromocytoma in yourself or a friend. Thankfully these usually resolve spontaneously with practice at taking blood pressures!

Multiple endocrine neoplasia syndromes

A very rare autosomal dominant mutation causes inheritable phaeochromocytoma. These tumours can also develop in both glands (bilaterally) as a component of multiple endocrine neoplasia syndromes (MEN type II), described in Chapter 10.

- List the five major hormones secreted by the adrenal gland.
- Describe the shape and location of each adrenal gland.
- Describe the innervation of the adrenal medulla and the developmental origins of this relationship.
- State the functions of the adrenal cortex.
- Name the blood vessels that supply and drain the adrenal gland.
- List the three layers of the adrenal cortex and state which group of hormones each layer secretes. Bonus mark: which layer develops last?
- Describe the regulation and actions of mineralocorticoids.
- State the main target tissues and intracellular actions of mineralocorticoids.
- Describe the regulation of glucocorticoid release including variation through the day.
- List the physiological actions of glucocorticoids.
- Describe the actions of adrenal androgens in males and females; how important are these effects in each sex?
- List the symptoms of Conn's syndrome.
- List the main causes of hyperaldosteronism; which are most common?
- Describe the diagnosis of excess glucocorticoids. What makes this more complicated?
- State the difference between Cushing's disease and Cushing's syndrome. List the symptoms of these conditions.
- What is the most common cause of Cushing's syndrome? Ward round bonus point: what is the rarest and most obscure cause that you can think of?
- Describe congenital adrenal hyperplasia including the symptoms in males and females.
- List the symptoms of Addison's disease. Which hormones are deficient?
- Describe the regulation and action of catecholamine hormones from the adrenal medulla.
- What is a phaeochromocytoma? What symptoms does it cause?

5. The Pancreas

The pancreas is an important exocrine gland that secretes digestive enzymes; it also has a significant endocrine function. This chapter focuses on islands of cells within the pancreas that secrete insulin. Insulin is an important hormone essential for the regulation of glucose levels. Its secretion is not controlled by the hypothalamus and pituitary gland in the same manner as many other important hormones.

The pancreas is a retroperitoneal organ found between the duodenum and the spleen. The endocrine cells are arranged within the pancreas in clusters called the islets of Langerhans. These clusters contain four types of cell, the most important and numerous of which are the insulin producing β-cells. After a meal, glucose enters the blood from the gastrointestinal (GI) tract. Insulin is released from the pancreas to promote the uptake and use of glucose by cells and to keep blood glucose within tightly controlled limits. When blood glucose levels drop, insulin secretion is inhibited in favour of another pancreatic hormone called glucagon. This hormone opposes many of the actions of insulin, for example, it causes cells to release glucose into the blood. Fig. 5.1 shows how these hormones regulate blood glucose.

> **Important words:**
> **Metabolite:** a molecule that can be broken down to release energy (ATP)
> **Glucose:** an important metabolite that is a type of simple sugar or monosaccharide (i.e. carbon atoms)
> **Anabolism:** processes that build large molecules
> **Catabolism:** processes that break down large molecules
> **Glycosuria:** glucose in the urine
> **Polyuria:** large volume of urine

Insulin deficiency or insulin resistance causes diabetes mellitus. It is the most common endocrine disorder, affecting at least 2% of people. Blood glucose rises, causing hyperglycaemia, and glucose is excreted in the urine (glycosuria). Water follows the movement of glucose, so the patient produces excess urine and becomes dehydrated.

There are two types of diabetes mellitus. Type 1 (insulin-dependent diabetes mellitus; IDDM) is

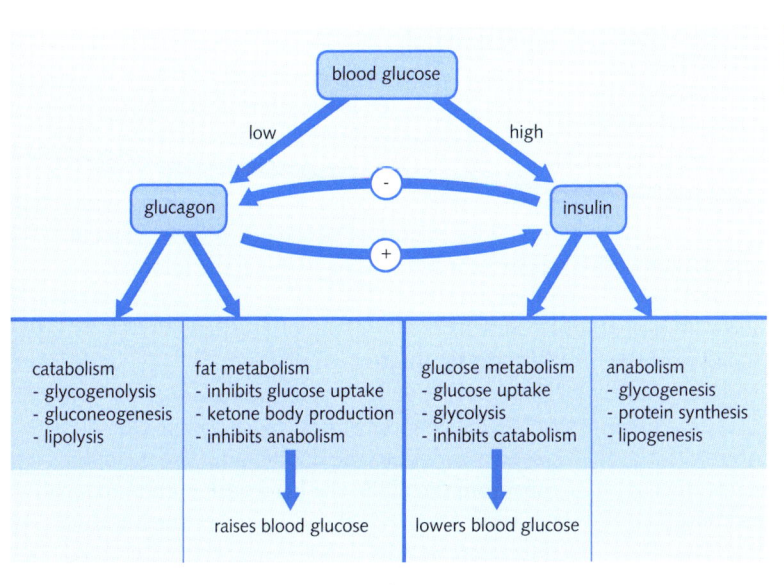

Fig. 5.1 Hormonal regulation of blood glucose and metabolism by insulin and glucagon.

57

The Pancreas

more common in the young and always requires insulin injections. Type 2 (non-insulin dependent diabetes mellitus; NIDDM) is very common in the elderly, and it can sometimes be controlled through diet alone. Poor control of diabetes causes a number of serious and potentially life threatening complications.

After reading this chapter you should be able to:
- Visualize the structure and development of the pancreas and its ducts.
- Understand the regulation and physiological effects of insulin and glucagon.
- Describe the control of glucose homeostasis.
- Explain the aetiology, symptoms, complications and treatment of diabetes mellitus.
- Briefly discuss neoplasia of the endocrine pancreas.

Location and anatomy

The pancreas is a long, flat organ that lies on the posterior of the abdominal wall, anterior to the vertebral bodies, aorta, and inferior vena cava. It is a retroperitoneal structure situated between the duodenum and spleen. For descriptive purposes it is divided into four sections (Fig. 5.2):
- Head.
- Uncinate process.
- Body.
- Tail.

The word 'pancreas' is derived from the Greek words 'pan' meaning 'all' and 'kreas' meaning 'flesh'. Uncinate is the Latin word for 'hooked'.

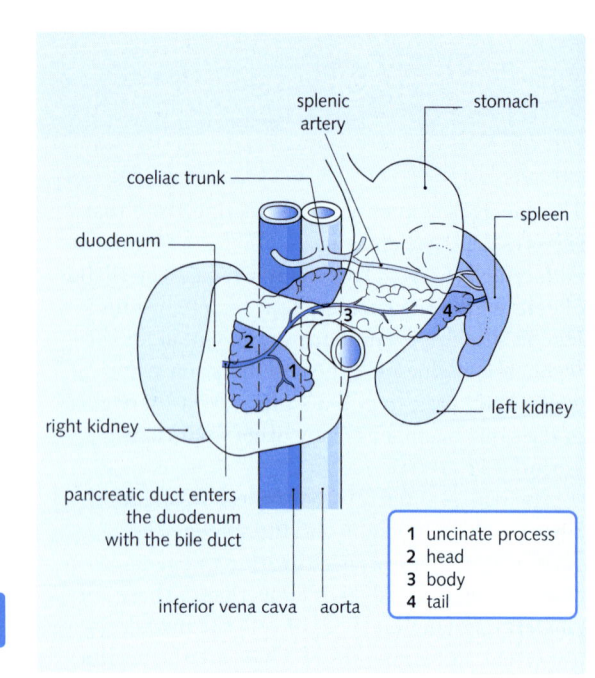

Fig. 5.2 Location of the pancreas in the retroperitoneal abdomen.

The shape of the uncinate process joining the pancreas is similar to the shape of a thumb (uncinate process) and a flat hand (head). The superior mesenteric vessels would lie between the thumb and index finger.

Head and uncinate process
The head lies within the curve of the duodenum with the uncinate process located posteriorly and inferiorly. The uncinate process is a 'hook' of pancreatic tissue that lies left and posteriorly to the head of the pancreas; it is separated from the head by the superior mesenteric vessels. The inferior vena cava and bile duct lie posteriorly; a clinical consequence of this is that the bile duct can be obstructed by masses in the head of the pancreas.

Body and tail
The body of the pancreas passes over the aorta and the left kidney, while the stomach lies in front. The body slopes upwards as it passes from right to left. The coeliac trunk (a large branch of the aorta) is a superior relation, giving rise to the splenic artery that runs along the upper pancreatic border. The tail crosses the left kidney to touch the hilum of the spleen.

Pancreatic duct
The exocrine secretions of the pancreas, which are rich in digestive enzymes, are carried in the pancreatic duct to the duodenum. The main duct runs from the tail to the head with numerous small branches joining on the way. It joins the bile duct and they open into the duodenum at the major duodenal

papilla (ampulla of Vater). In some individuals a smaller accessory duct drains the superior part of the head of the pancreas. It opens into the duodenum separately, at the minor duodenal papilla, about 2 cm proximal to the main papilla.

Blood, lymphatics, and nerves

The pancreas is supplied with blood from branches of the splenic artery. Blood drains into the splenic vein and the portal vein to the liver. Both splenic vessels lie along the upper border of the body and tail. Lymph drains to preaortic lymph nodes via a number of routes.

The main control of the endocrine pancreas is hormonal; however, a few autonomic nerves reach the pancreas via the coeliac plexus and splanchnic nerves.

Microstructure

The pancreas contains exocrine (enzyme secreting) and endocrine (hormone secreting) tissue. The endocrine cells are arranged in spherical clusters called islets of Langerhans within the exocrine tissue (see Fig. 5.3). Each islet has a rich network of fenestrated capillaries, however only 10% of endocrine cells are innervated by the autonomic nervous system.

The islets are made up of endocrine cells containing dense secretory granules. These cells are APUD (amine precursor uptake and decarboxylation) cells (see Chapter 6). There are four types of endocrine cell:
- Glucagon-secreting α-cells (20%).
- Insulin-secreting β-cells (70%).
- Somatostatin-secreting δ-cells (8%).
- Pancreatic polypeptide-secreting F-cells (2%).

Insulin and glucagon help regulate blood glucose levels. Somatostatin inhibits the release of insulin and glucagon. Pancreatic polypeptide inhibits the exocrine (i.e. non-endocrine) functions of the pancreas.

Development

The pancreas is an endodermal structure that develops from two buds derived from the foregut:
- Dorsal bud—the larger bud that forms the majority of the gland.

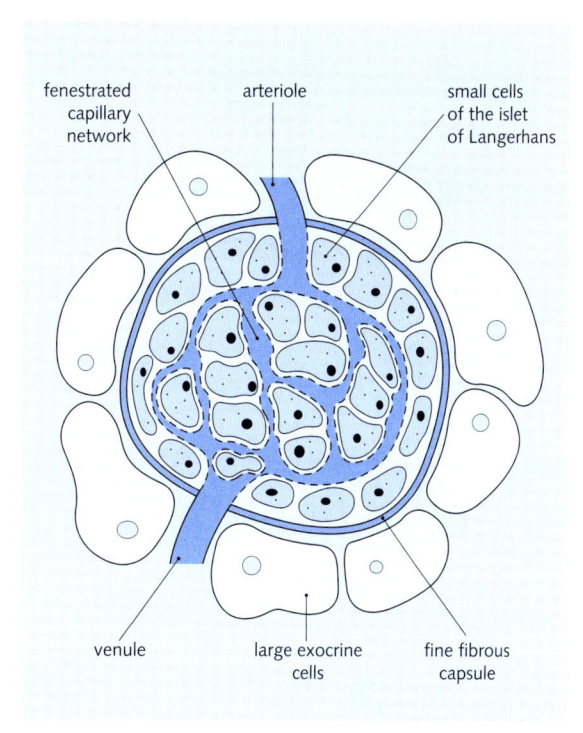

Fig. 5.3 Microstructure of the pancreas showing an islet of Langerhans surrounded by exocrine tissue.

- Ventral bud—the smaller bud from the right side near the bile duct.

The ventral bud rotates behind the duodenum, along with the bile duct, to lie posterior to the dorsal bud. This smaller ventral bud forms the uncinate process as it fuses with the larger dorsal bud. The ducts usually fuse so that the end of the pancreatic duct is formed from the smaller ventral bud. The duct of the dorsal bud may persist as the accessory pancreatic duct. This sequence of events is shown in Fig. 5.4.

Hormones

Insulin

Insulin is a hormone that promotes the uptake, storage and use of glucose. The pancreas secretes insulin when glucose levels are high, for example, after a meal.

As glucose levels fall a few hours after a meal, insulin secretion is reduced. The stored glucose can then be released to maintain blood levels. Insulin secretion never ceases completely; there is always a basal level of insulin in the blood.

The Pancreas

Control of insulin secretion

Insulin secretion is increased by high blood glucose levels. The pancreatic cells detect this stimulus directly because it raises ATP production. Other metabolites (energy molecules such as amino acids and triglycerides—i.e. fat) have a similar but weaker effect. The raised intracellular ATP levels inhibit membrane-bound potassium channels, causing the β-cell to depolarize. The depolarization opens voltage-sensitive calcium channels raising intracellular calcium, which promotes the secretion of preformed insulin secretory granules by exocytosis. This pathway is shown in Fig. 5.5.

Although metabolite concentrations are the main regulators of insulin release, a number of stimuli can also affect this pathway. These stimuli can have an inhibitory or stimulatory effect, but insulin secretion cannot be totally inhibited. The hormone glucagon, which is released when metabolite levels fall, acts as an important inhibitor of insulin's action; however, it stimulates insulin secretion. Fig. 5.6 shows the main factors that control secretion.

Insulin receptors

Insulin causes a cascade of activity by acting through a tyrosine kinase receptor located on the cell membrane. It must act via cell-surface receptors because it is a polypeptide hormone and cannot readily cross the cell membrane. Insulin receptors are present in most cells, and they can be withdrawn into the cell to inactivate them.

When insulin binds to the tyrosine kinase receptors, it causes phosphorylation of tyrosine side chains within the receptor. The phosphorylated receptor attracts and phosphorylates insulin receptor substrate 1 (IRS-1). This activated molecule then initiates a cascade of phosphorylation and aggregation of other proteins to bring about intracellular effects of insulin.

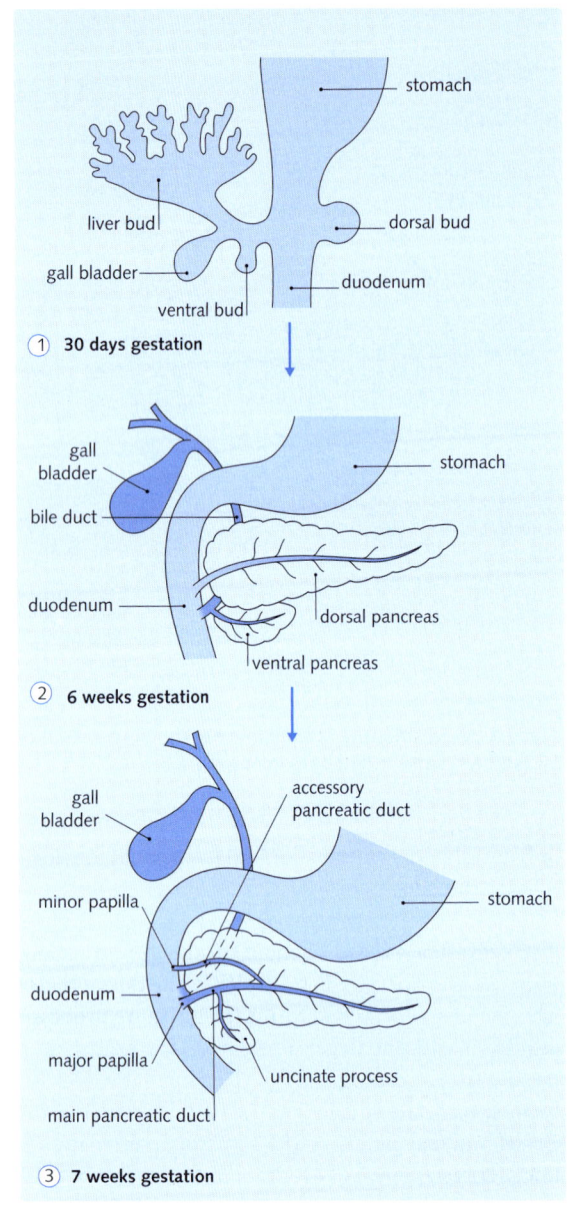

Fig. 5.4 Embryological development of the pancreas.

Synthesis

Insulin is a polypeptide hormone consisting of two short chains (A and B) linked by disulphide bonds. A single gene controls the production of pre-proinsulin, which is broken down to form proinsulin. Further cleavage occurs within the secretory vesicles resulting in two molecules: insulin and C peptide.

Since equal quantities of insulin and C peptide are produced, C peptide acts as a useful marker for β-cell activity in diabetics who receive insulin treatment.

> Insulin receptors are unusual because they phosphorylate the tyrosine side chains of proteins using tyrosine kinase. Most receptors phosphorylate serine and threonine side chains.

Actions of insulin

Insulin has an anabolic effect; it promotes the synthesis of larger molecules. The stimulation of

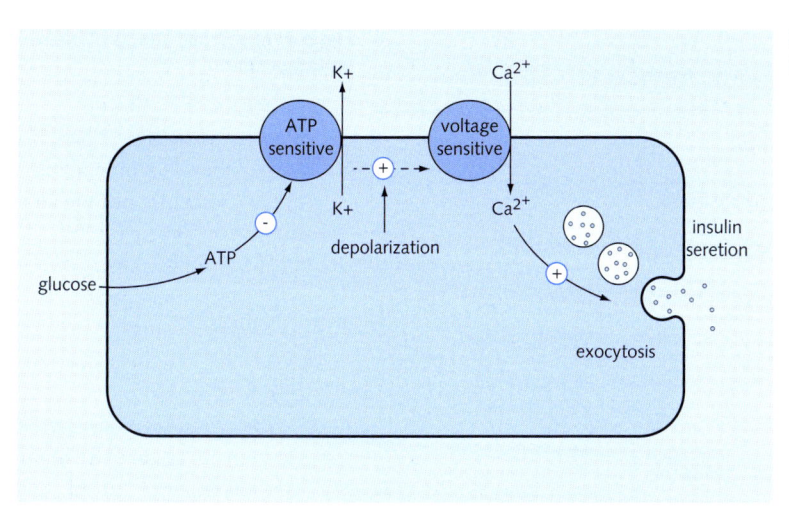

Fig. 5.5 Intracellular stimulation of insulin secretion by glucose.

Fig. 5.6 Factors that control insulin and glucagon secretion.

Factors controlling insulin and glucagon secretion

	Insulin		Glucagon	
	Stimulants	**Inhibitors**	**Stimulants**	**Inhibitors**
Blood glucose	High	Low	Low	High
Metabolites	Amino acids, fatty acids and ketones	–	Amino acids	Fatty acids and ketones
Hormones	Glucagon Some gastrointestinal tract peptides Growth hormone Adrenocorticotrophic hormone (ACTH), thyroid stimulating hormone (TSH)	Adrenaline Somatostatin	Adrenaline Some gastrointestinal tract peptides	Insulin Somatostatin
Innervation	Parasympathetic	Sympathetic	Parasympathetic and sympathetic	–
Other	–	Hypocalcaemia	–	–

insulin receptors regulates many enzymes concerned with metabolites. The specific enzymes vary between cells (Fig. 5.7), but the overall effects are:
- Increased uptake of metabolites.
- Conversion of metabolites to stored forms (this is an anabolic effect).
- Decreased breakdown of stored metabolites.
- Use of glucose for energy over other metabolites.

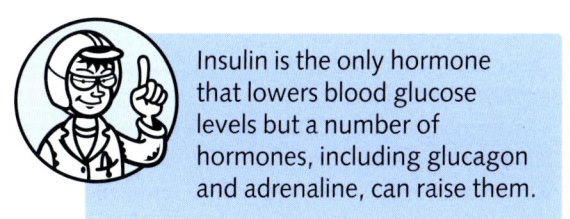

> Insulin is the only hormone that lowers blood glucose levels but a number of hormones, including glucagon and adrenaline, can raise them.

Glucagon

Glucagon is released when blood levels of metabolites are low, causing the release of stored metabolites. In many respects glucagon has the opposite effect of insulin. Its secretion from α-cells is stimulated by a number of factors, of which falling blood glucose is the most important. The main factors are shown in Fig. 5.6.

Synthesis and actions

Glucagon is a single chain polypeptide hormone formed from a larger precursor in a similar manner to insulin. The precursor is pre-proglucagon, which is cleaved in the storage vesicles to yield proglucagon and, finally, glucagon.

Glucagon is a catabolic hormone; it promotes the breakdown of large molecules.

The Pancreas

Fig. 5.7 Metabolic effects of insulin on target cells.

Metabolic effects of insulin on target cells

Target cells	Action of insulin
Muscle cells and many other cells	Stimulates glucose uptake
	Stimulates glycogenesis (glucose → glycogen)
	Stimulates glycolysis (glucose → energy)
	Stimulates amino acid uptake and protein synthesis
	Inhibits glycogenolysis (glycogen → glucose)
	Inhibits proteolysis (protein → amino acids)
Adipose cells	Stimulates glucose uptake
	Stimulates lipogenesis (glucose → fatty acids)
	Inhibits lipolysis (fatty acids → energy)
Liver cells	Stimulates glycogenesis (glucose → glycogen)
	Inhibits glycogenolysis (glycogen → glucose)
	Inhibits gluconeogenesis (amino acids → glucose)
Hypothalamus	May stimulate satiety (fullness)

Glucagon binds to a G-protein-coupled receptor on the cell membrane, and cAMP acts as a second messenger to initiate a cascade effect. Its effects vary between tissues (Fig. 5.8) but broadly its actions are:
- Inhibition of glucose and amino acid uptake.
- Breakdown of stored metabolites into useable metabolites (catabolism).
- Use of fatty acids for energy over other metabolites.

Endocrine control of glucose homeostasis

All cells in the body are capable of using glucose as an energy source by the process of glycolysis. Most cells can also use fatty acids with two important exceptions:
- Neurons (particularly in the CNS) though they can adapt to use ketone bodies.
- Blood cells.

If blood glucose levels drop too low (hypoglycaemia) then the brain is starved of energy. If levels rise too high (hyperglycaemia) then glucose can become toxic. Blood glucose is tightly controlled within narrow limits to prevent either scenario. Fasting glucose levels are normally 3.5–5.5 mmol/L.

For glucose levels to be maintained (glucose homeostasis), the body must be able to increase or decrease these levels in response to changes. There are a number of ways that the body can respond (Fig. 5.9). The liver is especially important in raising blood glucose.

Insulin and glucagon

Glucose homeostasis is maintained by the interplay between insulin and glucagon. These two hormones act as antagonists of each other. Their blood concentrations mirror each other because they are secreted in opposite conditions (Fig. 5.10).
- Insulin lowers blood glucose by stimulating uptake, metabolism and anabolism. It also inhibits the actions of glucagon.
- Glucagon raises blood glucose by simulating gluconeogenesis (synthesis of glucose from amino acids) and glycogenolysis (breakdown of glycogen to release glucose). It also inhibits the actions of insulin but stimulates insulin secretion.

Other hormones

Three non-pancreatic hormones also significantly increase blood glucose:
- Adrenaline—released in response to stress; it inhibits insulin.
- Cortisol—released in response to stress; it reduces sensitivity to insulin.

Disorders

Fig. 5.8 Metabolic effects of glucagon on target cells.

Metabolic effects of glucagon on target cells

Target cells	Action of glucagon
Muscle cells and many other cells	Stimulates glycogenolysis (glycogen → glucose)
	Inhibits glucose uptake
	Inhibits glycolysis (glucose → energy)
	Inhibits amino acid uptake and protein synthesis
Adipose cells	Stimulates lipolysis (fatty acids → energy)
Liver cells	Stimulates glycogenolysis (glycogen → glucose)
	Stimulates gluconeogenesis (amino acids → glucose)
	Stimulates ketogenesis (fatty acids → ketone bodies)

Fig. 5.9 Responses that alter blood glucose levels.

Responses that alter blood glucose levels

Responses that raise blood glucose	Responses that lower blood glucose
Ingestion of glucose in the diet	Increased uptake in cells
Gluconeogenesis—the irreversible conversion of amino acids to glucose (liver)	Metabolism to produce energy
Glycogenolysis—the reversible breakdown of glycogen to release glucose (liver)	Glycogenesis—the reversible conversion of glucose to glycogen
	Lipogenesis, the irreversible conversion of glucose to fatty acids

- Growth hormone—released at night; it reduces sensitivity to insulin.

All three hormones can stimulate glycogenolysis and gluconeogenesis to raise blood glucose levels directly. Neural signals and other hormones can cause less significant rises in blood glucose.

Hyperglycaemia

Hyperglycaemia is an excess of glucose in the blood; it is defined as a fasting concentration >7.8 mmol/L. This can occur in:
- Diabetes mellitus—a common disease caused by insulin deficiency or insulin resistance (reduced sensitivity).
- Glucagonoma—a very rare tumour of the α cells that secrete glucagon.

Hypoglycaemia

Hypoglycaemia is a deficiency of blood glucose; it is defined as a concentration <2.5 mmol/L. It can be caused by:

- Overtreatment of diabetes mellitus, either excess insulin or β-cell stimulating drugs.
- Non-diabetic disorder causing fasting hypoglycaemia.
- Idiopathic excessive insulin secretion causing hypoglycaemia after glucose ingestion.

Disorders

Diabetes mellitus
Types of diabetes mellitus

Diabetes mellitus (DM) is caused by insulin deficiency or insulin resistance (reduced sensitivity). These abnormalities result in chronic hyperglycaemia (excess blood glucose) and metabolic chaos. It is a very common disease affecting about 2% of the population. There are two types:
- Type 1—insulin dependent DM (IDDM) caused by insulin deficiency.
- Type 2—non-insulin dependent DM (NIDDM) caused by insulin deficiency and/or resistance.

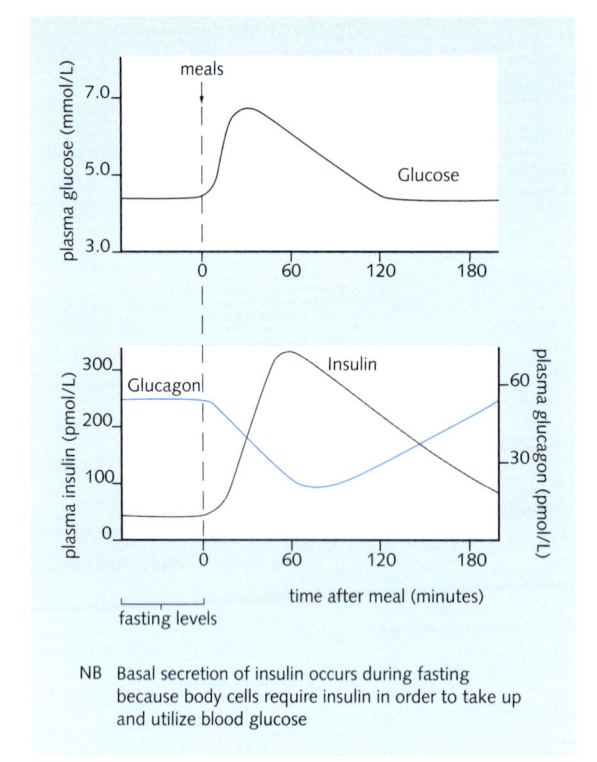

Fig. 5.10 Changes in blood levels of glucose, insulin and glucagon after a carbohydrate-rich meal.

NB Basal secretion of insulin occurs during fasting because body cells require insulin in order to take up and utilize blood glucose

'Diabetes' means 'siphon', a name originating in Roman times when Aretaeus described the disease as 'the liquefaction of the flesh and bone into urine'. The term 'mellitus' referring to the 'honey or sweet-tasting' urine was added in the 17th century by Thomas Willis (of the circle), who also observed that diabetics 'piss a great deal'. Next time you meet a diabetic why not try a sample?

Type 1 diabetes (IDDM)

IDDM is an autoimmune disease resulting in destruction of the islet β-cells, and so causing insulin deficiency. Autoantibodies can be detected in the blood. It is most common in the young, but it can occur at any age. There is a genetic component of about 30% mostly due to HLA genes.

Type 2 diabetes (NIDDM)

NIDDM is usually a disease of the elderly, especially those who are obese; it is about twice as common as IDDM. It can be caused by insulin deficiency, insulin resistance, or a combination of both. Family history is very important since there is almost 100% concordance in identical twins. NIDDM often occurs alongside obesity, hypertension, and hyperlipidaemia (excess fatty acids in the blood); together these features are called 'syndrome X'.

NIDDM can develop in young people in whom it is called maturity-onset diabetes in the young (MODY). This is becoming more common with rising juvenile obesity.

Other causes of diabetes mellitus

Diabetes can also develop secondary to a number of diseases. These include pancreatitis, prolonged corticosteroid use (Cushing's syndrome), acromegaly and thyrotoxicosis.

Do not confuse diabetes insipidus with diabetes mellitus. They both cause polyuria, hence the similar name, but diabetes insipidus is a disorder of antidiuretic hormone (ADH) causing excess, pale (insipid) urine.

Symptoms and presentation

Insulin has many important effects on the regulation of glucose and metabolism. Accordingly the effects of diabetes can appear quite complicated. The effects make more sense if they are thought of in four categories:
- Symptoms of **hyperglycaemia** in both IDDM and NIDDM.
- Symptoms of **starvation** particularly in IDDM.
- Symptoms of **ketoacidosis** in IDDM.
- Symptoms of chronic **complications** in NIDDM and IDDM.

Symptoms of hyperglycaemia

Hyperglycaemia causes dehydration because glucose is an osmotically active substance, i.e. it draws water towards it. In hyperglycaemia, glucose concentration is high in the blood and low in the cells so the cells become dehydrated. The excess glucose is also

Symptoms of diabetes mellitus

Symptoms due to hyperglycaemia	Symptoms due to starvation	Symptoms due to ketoacidosis	Symptoms due to chronic complications
Polyuria (increased urine volume)	Weight loss	Vomiting	Decreased visual acuity
Glycosuria (glucose in the urine)	Wasting	Acetone smell on the breath	Reduced sensation in the limbs
Polydipsia (thirst)	Weakness	Ketonuria, polyuria, and dehydration	Proteinuria
Tiredness		Hyperventilation	Oedema
Tendency to infections		Reduced consciousness	Intermittent claudication
Dehydration (loose skin, hypotension, and tachycardia)		Convulsions	Ischaemic heart disease
Coma		Coma	Hypertension

Fig. 5.11 Symptoms of diabetes mellitus.

excreted in the kidney and again water follows this movement. Excess water is lost from the body along with electrolytes. The resulting symptoms are shown in Fig. 5.11.

Symptoms of starvation
In IDDM the body enters a state of starvation because cells cannot use the excess glucose; this is caused by the high levels of glucagon. The lack of insulin prevents glucose entering cells and being metabolized. Muscle protein and adipose tissue are broken down to release metabolites and this causes the symptoms shown in Fig. 5.11.

Symptoms of ketoacidosis
Ketoacidosis only develops in IDDM. In IDDM, lipolysis (fat breakdown) is a major component resulting in raised blood fatty acid levels. These fatty acids are converted to acetyl coenzyme A (acetyl CoA). The excess acetyl CoA can overload the tricarboxylic acid (TCA) or Krebs cycle, so the liver uses the excess to synthesize ketone bodies. The ketone bodies are synthesized more quickly than they are metabolized by peripheral tissues so they build up in the blood.

The excess ketone bodies are osmotically active, so they compound the dehydration caused by glucose. Since they are acidic, they can cause a metabolic acidosis (ketoacidosis), which causes severe symptoms (see Fig. 5.11).

Presentation of IDDM
IDDM presents with a short history of polyuria, tiredness, and weight loss followed by dehydration and ketoacidosis (Fig. 5.12). The onset is relatively quick (over a number of weeks) so complications have not developed by the time it presents. Diabetes should be suspected in young patients complaining of tiredness or with frequent skin infections (e.g. boils). Failure to treat an IDDM patient with insulin can result in rapid death from cerebral oedema following ketoacidosis.

Presentation of NIDDM
The onset of NIDDM is much slower than that of IDDM, with hyperglycaemia developing over a number of years. By the time of presentation, the patient has often been exposed to excess glucose for so long that complications are already present; in fact, they may be the presenting feature (see Fig. 5.12). The earliest complications are retinopathy and peripheral neuropathy. NIDDM patients can present in a coma due to dehydration instead of ketoacidosis. Severe hypoglycaemia and dehydration can result in hypovolaemic (low blood volume) shock. In the worst cases this can result in a HyperOsmolar Non-Ketotic coma (HONK coma).

Ketoacidosis never occurs in NIDDM because some insulin activity is maintained:
- Anabolic actions (glucose uptake and metabolism) require high levels of insulin.

The Pancreas

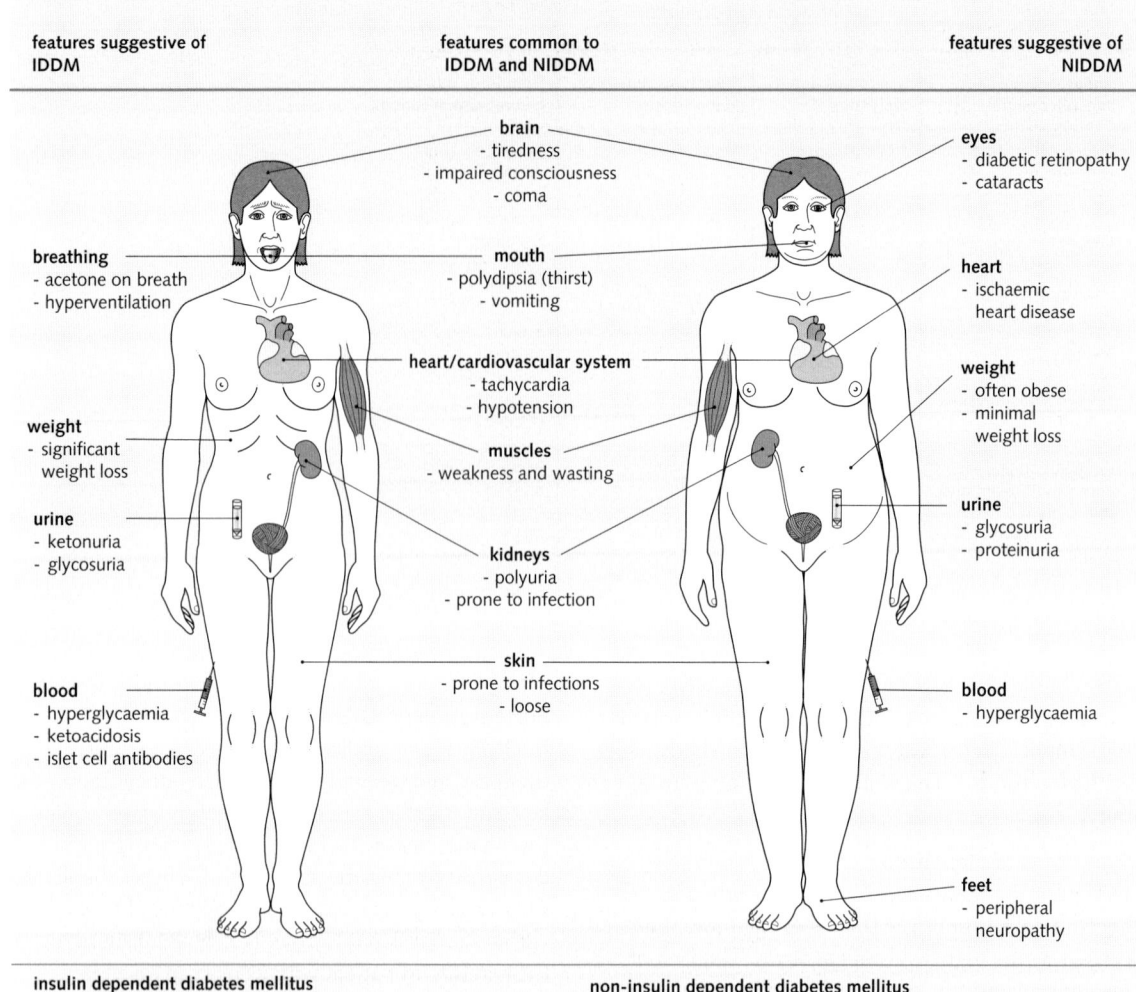

features suggestive of IDDM

- **breathing**
  - acetone on breath
  - hyperventilation
- **weight**
  - significant weight loss
- **urine**
  - ketonuria
  - glycosuria
- **blood**
  - hyperglycaemia
  - ketoacidosis
  - islet cell antibodies

features common to IDDM and NIDDM

- **brain**
  - tiredness
  - impaired consciousness
  - coma
- **mouth**
  - polydipsia (thirst)
  - vomiting
- **heart/cardiovascular system**
  - tachycardia
  - hypotension
- **muscles**
  - weakness and wasting
- **kidneys**
  - polyuria
  - prone to infection
- **skin**
  - prone to infections
  - loose

features suggestive of NIDDM

- **eyes**
  - diabetic retinopathy
  - cataracts
- **heart**
  - ischaemic heart disease
- **weight**
  - often obese
  - minimal weight loss
- **urine**
  - glycosuria
  - proteinuria
- **blood**
  - hyperglycaemia
- **feet**
  - peripheral neuropathy

insulin dependent diabetes mellitus
- patients usually thin
- usually present with a short history of acute symptoms

non-insulin dependent diabetes mellitus
- patients usually overweight (85% obese)
- usually present with a longer history
- may be asymptomatic
- many cases are discovered only by routine testing

Fig. 5.12 Presentation of insulin dependent diabetes mellitus (IDDM) and non-insulin dependent diabetes mellitus (NIDDM).

- Anticatabolic actions (inhibition of lipolysis and protein breakdown) only require low levels of insulin.

In NIDDM, insulin deficiency or resistance does not fall below this lower level so the lipolysis and excess fatty acid release that cause ketoacidosis do not occur.

Complications

Both types of diabetes can produce complications despite treatment; however, good glucose control lowers the risk of most complications. Chronic complications are grouped according to the size of blood vessel they affect:
- Macrovascular—large vessel disease due to accelerated atherosclerosis.
- Microvascular—small vessel disease due to hyaline arteriolosclerosis.

Macrovascular complications

Diabetes causes accelerated atherosclerosis due to chronically raised fatty acid levels (hyperlipidaemia) following low insulin levels. Atheroma develops more rapidly and more severely than in non-diabetics, and it can block arteries causing ischaemia and a high risk

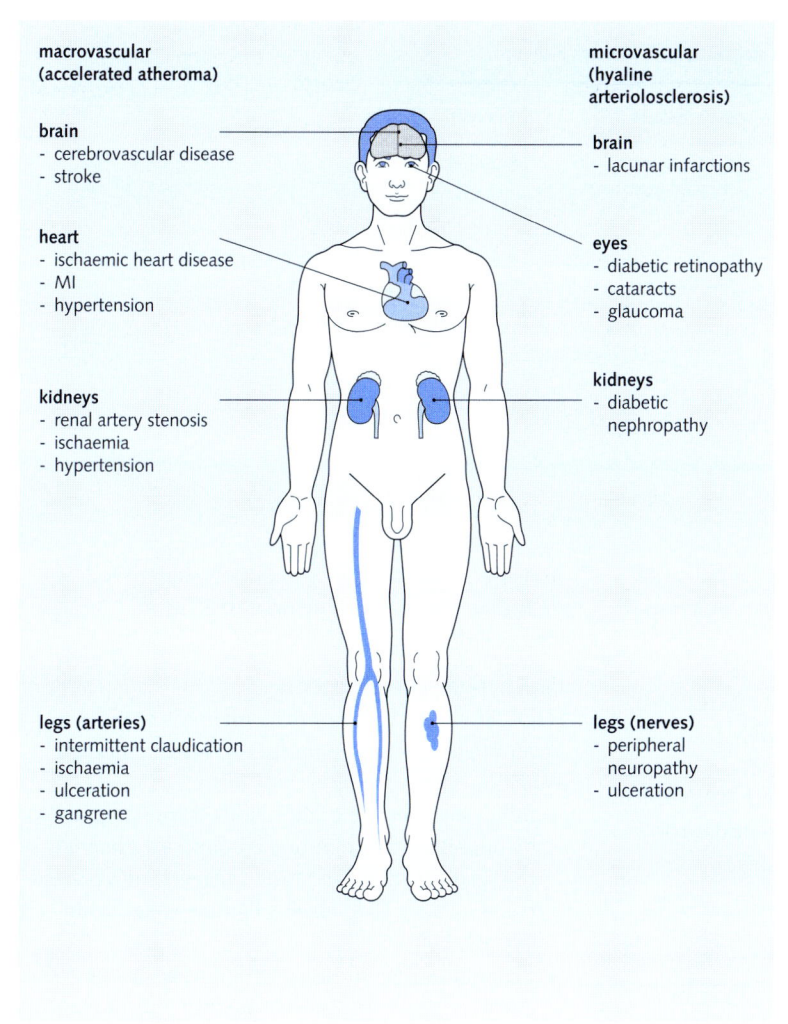

Fig. 5.13 Chronic complications of diabetes mellitus.

macrovascular (accelerated atheroma)

brain
- cerebrovascular disease
- stroke

heart
- ischaemic heart disease
- MI
- hypertension

kidneys
- renal artery stenosis
- ischaemia
- hypertension

legs (arteries)
- intermittent claudication
- ischaemia
- ulceration
- gangrene

microvascular (hyaline arteriolosclerosis)

brain
- lacunar infarctions

eyes
- diabetic retinopathy
- cataracts
- glaucoma

kidneys
- diabetic nephropathy

legs (nerves)
- peripheral neuropathy
- ulceration

of infarction. The major sites of macrovascular disease are shown in Fig. 5.13.

Microvascular complications

While atherosclerosis develops in major arteries, the smaller arterioles and capillaries are at risk of hyaline arteriolosclerosis. This is characterized by thickening of the vessel wall and basement membrane. The vessel lumen is reduced, causing localized ischaemia. The vessel also becomes 'leaky'. This response may be a direct reaction to excess glucose. There are four main patterns of microvascular disease (Figs 5.13 and 5.14). Diabetic retinopathy is the most common cause of blindness in the 30–65 year age group.

Diagnosis

Blood glucose levels are normally 3.5–5.5 mmol/L after an overnight fast. Diabetes is diagnosed if this fasting blood glucose is above 7.8 mmol/L on two occasions. Since NIDDM can develop gradually, a spectrum of disease is seen between normal blood glucose and diabetic blood glucose. A number of markers suggest the need for further observation:
- Glycosuria (glucose in urine).
- Fasting blood glucose of 6–7.8 mmol/L.
- Random blood glucose of >11.1 mmol/L.

To clarify the diagnosis, a glucose tolerance test is sometimes used. The fasting blood glucose is measured as normal, and the patient is then given a drink containing 75 g glucose. Blood glucose is measured 2 hours later, and diabetes is diagnosed if this second measurement is above 11.1 mmol/L.

Treatment

The treatment of diabetes aims to lower blood glucose and normalize metabolism with the least possible interference. Patient education is essential;

The Pancreas

Fig. 5.14 Progressive changes caused by the complications of diabetes mellitus.

Condition	Early changes	Late changes	End result
Retinopathy	Microaneurysms, haemorrhages, hard exudates	Soft exudates, neovascularization	Blindness
Nephropathy	Proteinuria, oedema	Decline of glomerular filtration rate	Renal failure and death
Peripheral neuropathy	Reduced reflexes, reduced sensation (glove and stocking pattern)	Burning or aching sensation, joint deformity	Ulceration and amputation
Lacunar infarcts	Microinfarcts in the brain, asymptomatic	Progressive neurological deficits	Dementia and Parkinsonism

Progressive changes caused by complications of diabetes mellitus

ideally patients will regulate their own medication according to their lifestyle. There are three types of treatment:
- Diet alone (NIDDM).
- Diet and oral hypoglycaemic agents (NIDDM).
- Diet and insulin (IDDM and NIDDM).

All patients with IDDM and many with NIDDM are treated with subcutaneous insulin injections to reduce acute and chronic complications.

> Patients with NIDDM are often treated with insulin in a similar manner to IDDM. This does not mean that they have developed IDDM. Insulin treatment helps to control blood glucose, which has been proven to reduce complications.

Diet
Regulation of diet is essential in all diabetics to help maintain blood glucose levels. Four principles govern this treatment:
- Avoid carbohydrates that can be rapidly absorbed (e.g. glucose) to prevent hyperglycaemia.
- Eat regular, small meals to prevent hypoglycaemia.
- Control calorie intake to lose/stabilize weight (especially NIDDM).
- Eat a low fat, healthy diet to reduce atherosclerosis.

In reality, more than 50% of patients fail to follow their diet and very few manage to achieve long-standing weight loss.

Oral hypoglycaemic agents
NIDDM can be treated with oral medication to lower blood glucose. A number of medications are available:
- Sulphonylureas (e.g. gliclazide) stimulate β-cells by inhibiting the membrane bound K^+ channel; the resulting depolarization causes insulin release (see Fig. 5.5). Side effects include weight gain and hypoglycaemia.
- Biguanides (e.g. metformin) increase peripheral glucose uptake and reduce glucose output from the liver. Their mechanism of action is not understood. Side effects include nausea and diarrhoea.
- Acarbose inhibits intestinal enzymes preventing the digestion of starch; blood glucose rises more slowly after a meal as a result. Side effects include flatulence and diarrhoea.
- Thiazolidinediones (e.g. pioglitazone) are new drugs that reduce insulin resistance, and they are usually used in combination with other drugs. They may have a risk of heart failure and liver toxicity.

Subcutaneous insulin injections
Insulin must be injected because it is digested, and thus inactivated, if taken orally. Subcutaneous injections of insulin are used to treat all IDDM patients and some with NIDDM. It is usually injected into the thigh, upper arm or abdomen. Intensive monitored therapy to maintain low blood

glucose has been proven to reduce long-term complications.

Many different types of insulin are available. In the past, porcine (from pigs) insulin was used, but recombinant human insulin is now favoured since it reduces immunological reactions. Preparations of insulin vary in their duration of action. Clear insulin is short acting and administered just before meals. Cloudy insulin has an intermediate or long effect; it replaces basal insulin secretion.

Patients regulate the type, dose and frequency of injections to meet their requirements and lifestyle. This can vary on a day-to-day basis or a regular dosing schedule can be followed.

Monitoring glucose control

Diabetics use monitoring to assess current blood glucose levels and long-term control. The following tests are available:

- Urine testing for glucose: this can be performed at home but it is unreliable.
- Capillary blood spot testing: this simple test gives an instant digital reading of blood glucose, it can be performed at home and is used to determine insulin or sugar doses.
- HbA_{1C}: glucose binds directly and irreversibly to haemoglobin to form HbA_{1C}. The proportion of HbA_{1C} in the blood gives a measure of glucose control over the previous 2 months. It is used very commonly in clinical practice.
- Fructosamine: this is formed when glucose binds to serum albumin. It gives a measure of glucose control in the last two weeks but is rarely used in clinical practice.

Hypoglycaemia

Treatment for diabetes mellitus may cause hypoglycaemia in a number of situations:
- Low carbohydrate intake (e.g. missed meal).
- Unexpected exercise.
- Insulin overdose.
- Malabsorption.

Symptoms

Diabetics must know the symptoms of their hypoglycaemic attacks, and they should always carry a sugary snack in case they feel a 'hypo' coming on. If action is not taken to prevent it, then recognizing the symptoms and signs of hypoglycaemia can be life saving; they are shown in Fig. 5.15. These symptoms are caused by raised adrenaline secretion and low cerebral glucose. Severe hypoglycaemia can result in a coma. This is treated with intravenous infusion of 50 mL 50% dextrose, and sugary drinks once the patient regains consciousness. Glucagon pens are now available for intramuscular injection in the event of a 'hypo'.

Screening to prevent complications

Along with preventing short-term hypo/hyperglycaemia, diabetic clinics aim to minimize the long-term complications. Many of these complications are treatable if they are detected in their early stages, i.e. before symptoms develop.

Eyes

Retinopathy is a very common cause of blindness, and the early stages are completely treatable by laser photocoagulation. The eyes of diabetics should be tested and inspected regularly for signs of deterioration or retinopathy.

Kidneys

Urine samples should be checked regularly for microalbuminuria caused by early nephropathy (kidney damage). If present, then blood glucose should be regulated more aggressively. An angiotensin-converting enzyme (ACE) inhibitor can be used to prevent the resulting hypertension and to slow the progression to terminal renal failure.

Feet

The feet of elderly or immobile diabetics should be examined regularly by chiropodists. The combination of peripheral neuropathy and peripheral vascular disease can lead to injury, ulceration, infection and, eventually, gangrene. Poorly perfused gangrenous feet may need to be amputated. This can be prevented by early treatment and dressing or vascular surgery in severe cases.

Hypoglycaemia in non-diabetic patients

Most people should be able to tolerate fasting for several days without developing hypoglycaemia. If a patient is unable to do so then the mnemonic 'EXPLAIN' lists the possible causes:

- **EX**ogenous drugs (e.g. alcohol and insulin).
- **P**ituitary insufficiency, growth hormone deficiency.
- **L**iver failure or defective liver enzymes.
- **A**ddison's disease—deficiency of cortisol that raises blood glucose levels (also Autoimmune causes).

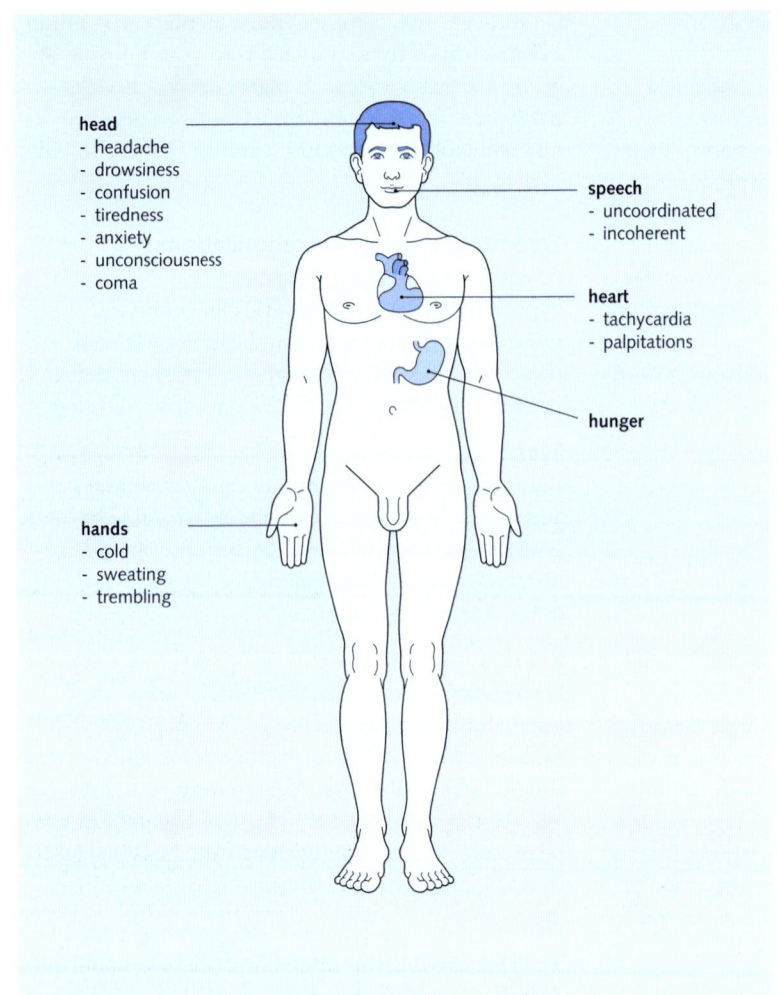

Fig. 5.15 Symptoms of hypoglycaemia.

head
- headache
- drowsiness
- confusion
- tiredness
- anxiety
- unconsciousness
- coma

speech
- uncoordinated
- incoherent

heart
- tachycardia
- palpitations

hunger

hands
- cold
- sweating
- trembling

- **I**nsulinomas—a type of islet-cell tumour (see below).
- Non-pancreatic tumour—by ectopic insulin secretion or simply consuming glucose.

Endocrine pancreatic neoplasia

Tumours of the endocrine cells in the pancreas are called islet-cell tumours; they are usually benign and solitary, but they often secrete a specific hormone. Tumours are named according to the hormone they secrete:

- Insulinomas are the most common type of islet-cell tumour. The excess insulin that they secrete causes severe hypoglycaemic attacks, which can lead to coma.
- Glucagonomas are very rare tumours that secrete glucagon. They are often asymptomatic, but they may cause diabetes mellitus.

Extremely infrequently islet-cell tumours can secrete other hormones such as gastrin in Zollinger–Ellison syndrome. This syndrome is characterized by recurrent, severe and multiple peptic ulcers following excessive stomach acid secretion. Other rare tumours can produce vasoactive intestinal polypeptide (VIP) or adrenocorticotrophic hormone (ACTH).

Disorders

- Describe the anatomical location of the pancreas.
- Describe how the endocrine cells are arranged within the pancreas.
- List the hormones secreted by the pancreas along with the cell type responsible.
- Describe the development of the pancreas.
- How is the hormone insulin synthesized?
- Describe insulin receptors and how they function. What are the main differences from other types of receptors?
- Why is a basal level of insulin always secreted?
- State the dietary principles that should be followed by all diabetics.
- Describe the intracellular mechanism that allows high glucose levels to trigger insulin secretion.
- Describe the action of insulin on glucose uptake and metabolism.
- Describe the action of insulin on amino acid and fatty acid metabolism, including the processes it inhibits.
- List the actions of glucagon and the factors that stimulate its secretion.
- List three hormones that can raise blood glucose levels.
- Compare the two types of diabetes mellitus.
- Explain how ketoacidosis develops in IDDM and why this does not occur in NIDDM.
- List the symptoms caused by hyperglycaemia and starvation; how does dehydration develop?
- Describe the complications of diabetes mellitus along with preventative screening.
- How is diabetes mellitus diagnosed and monitored?
- List the symptoms of hypoglycaemia; how can this be prevented?
- Briefly describe the tumours that can develop in the endocrine cells of the pancreas.

6. Up and Coming Hormones

This chapter describes several tissues whose endocrine functions are still being realized:
- Gastrointestinal (GI) tract—secretes many hormones that regulate digestive function. Some act on distant organs by travelling through the blood, while others act locally as mediators.
- Pineal gland—secretes melatonin which regulates circadian rhythms.
- Adipose tissue—secretes leptin which regulates food intake and fertility.

The GI tract and adipose tissue are examples of diffuse endocrine tissue.

> **Important words:**
> **APUD cells:** peptide-secreting endocrine cells found throughout the body
> **Circadian rhythm:** variations in physiological processes that are repeated over a 24-hour cycle
> **Pineal gland:** melatonin-secreting gland found in the brain, important for regulating circadian rhythms
> **Adipose tissue:** cells that store fat

Endocrine role of the gastrointestinal tract

Endocrine cells that are distributed along the length of the GI tract are called enteroendocrine cells and are a type of APUD cell (see below). They secrete at least 20 different peptide hormones that coordinate the digestion and absorption of food.

Some GI tract peptides (e.g. vasoactive intestinal peptide (VIP), cholecystokinin (CCK), gastrin) also act as neurotransmitters in the CNS and the neurons that innervate the GI tract (called the enteric nervous system). This overlap demonstrates the close relationship and common origin of the endocrine and nervous systems. The following bodily functions are regulated in the CNS by these peptide neurotransmitters:
- Biological rhythms—VIP.
- Satiety (fullness after food)—CCK.
- Thermoregulation—bombesin.
- Growth—somatostatin.

The APUD concept

APUD cells are a group of endocrine cells that secrete small peptide hormones in many tissues throughout the body. They are linked by three features:
- Similar appearance under an electron microscope (e.g. neurosecretory granules).
- Similar biochemical pathway for amine or peptide hormone synthesis.
- Embryological origin from the endodermal GI tract.

They are also called neuroendocrine cells due to their secretion of both neurotransmitters and 'hormones'.

The name 'APUD' is an abbreviation of the method by which the peptide hormones are synthesized: it stands for Amine Precursor Uptake and Decarboxylation. This means that APUD cells convert actively absorbed amine precursors into amino acids, which are used to make the peptide hormones.

The following cells are examples of APUD cells:
- Islets of Langerhans cells that secrete insulin and glucagon.
- Enteroendocrine cells (see below).
- Parafollicular cells that secrete calcitonin; they are found in the thyroid gland.
- Juxtaglomerular complex that secretes renin; they are found in the kidneys.
- Neuroendocrine cells of the respiratory tract that secrete 5-hydroxytryptamine (5-HT; serotonin) and calcitonin.

Gastrointestinal tract peptides

The major peptides secreted by the enteroendocrine cells of the GI tract are described below. Figs 6.1 and 6.2 show the location of GI tract peptide release and the major actions of four GI tract hormones. The actions of other peptides are shown in Fig. 6.3.

Delivery

GI tract peptides reach their target cells by two means:

Up and Coming Hormones

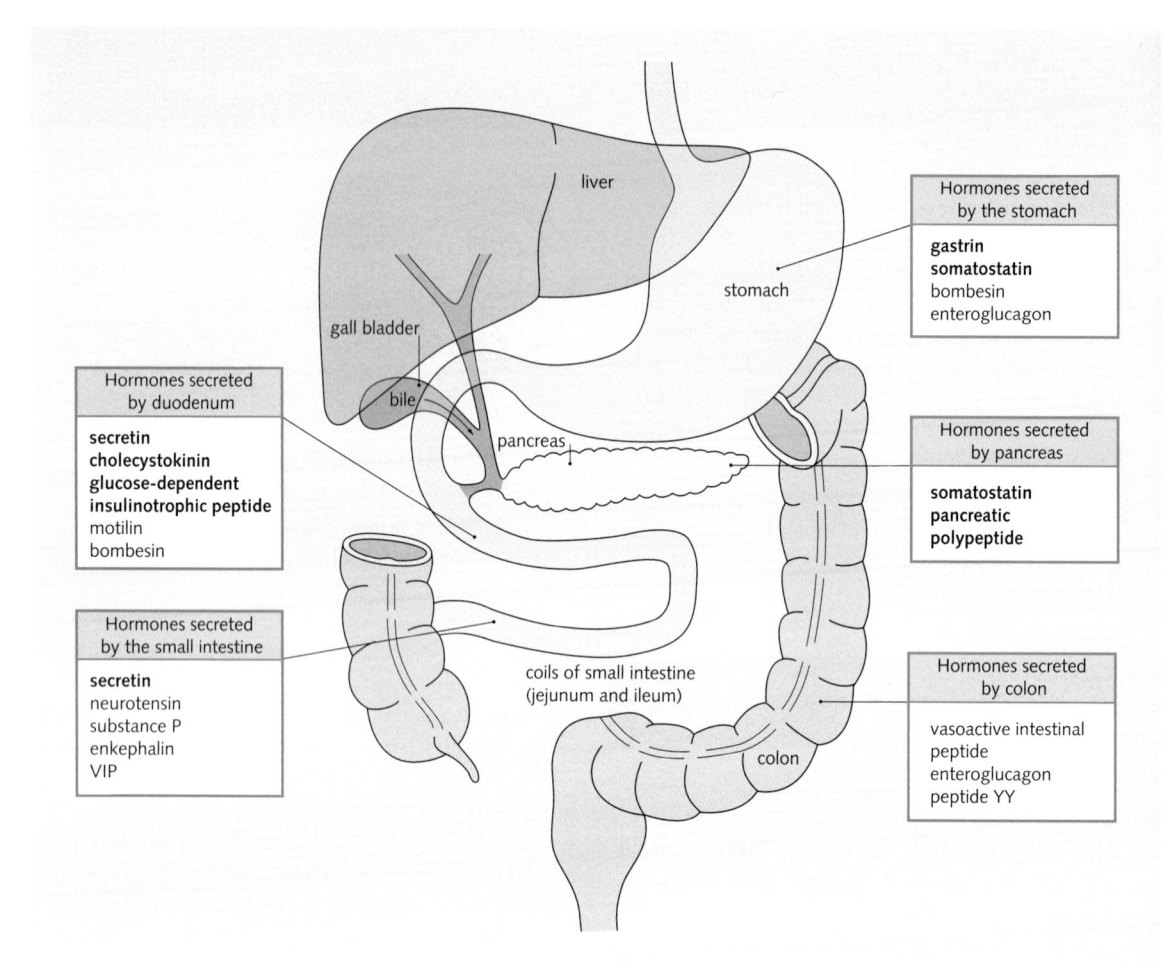

Fig. 6.1 Sites at which the gastrointestinal tract peptides are secreted.

Hormones secreted by the stomach
- **gastrin**
- **somatostatin**
- bombesin
- enteroglucagon

Hormones secreted by duodenum
- **secretin**
- **cholecystokinin**
- **glucose-dependent insulinotrophic peptide**
- motilin
- bombesin

Hormones secreted by pancreas
- somatostatin
- **pancreatic polypeptide**

Hormones secreted by the small intestine
- **secretin**
- neurotensin
- substance P
- enkephalin
- VIP

Hormones secreted by colon
- vasoactive intestinal peptide
- enteroglucagon
- peptide YY

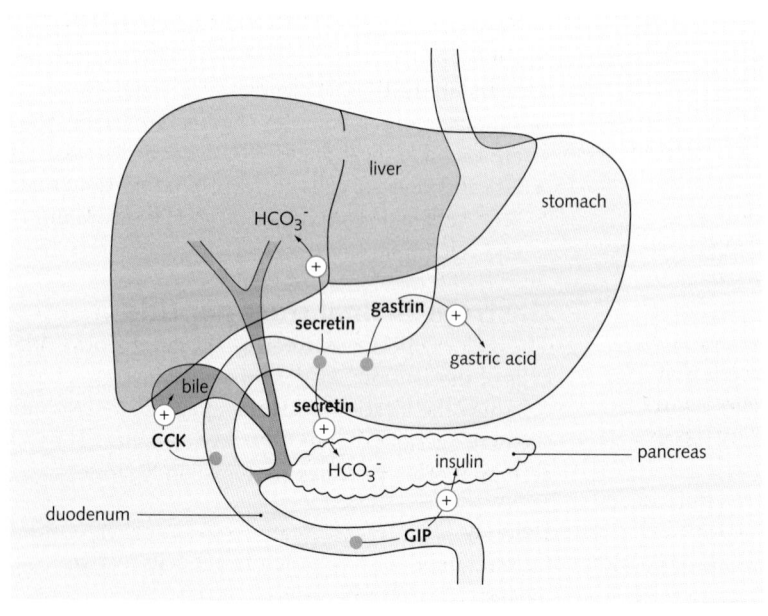

Fig. 6.2 Major actions of four gastrointestinal tract hormones. (CCK, cholecystokinin; GIP, glucose-dependent insulinotrophic peptide.)

Endocrine Role of the Gastrointestinal Tract

The sites of secretion, stimuli for secretion, and actions of the minor gut peptides

Gut peptide	Site of secretion	Stimulus for secretion	Action of peptide
Enteroglucagon	A cells in the stomach and L cells in the colon	Presence of glucose and fat in the stomach	Reduces gastric-acid secretion and gut motility
Bombesin	P cells in the stomach and duodenum	Fasting	Stimulates gastrin release
Motilin	EC cells in the duodenum	Absence of food in the duodenum	Speeds gastric emptying and stimulates colonic motility
Vasoactive intestinal polypeptide (VIP)	D1 cells and neurons in the small intestine and colon	Gut distension	Stimulates local gut secretion, motility, and blood flow
Peptide YY (related to pancreatic polypeptide)	PYY cells of the colon	Presence of intestinal fat	Inhibits gastric motility and acid secretion (peptide YY is elevated in coeliac disease and cystic fibrosis)
Substance P	Enteric neurons in the small intestine	Cholecystokinin (CCK), 5-hydroxy-tryptamine (5-HT)	Stimulates gut motility, secretion, and immune response; may have a role in inflammatory bowel disease
Enkephalin	Enteric neurons in the small intestine	Unknown	Inhibits gut motility and secretion
Neurotensin	N cells of the small intestine	Presence of intestinal fat	Stimulates local gut motility, secretion, and immune response

Fig. 6.3 Actions of the minor gastrointestinal tract peptides. The other peptides are described in the text.

- Endocrine, via the blood.
- Paracrine, act locally to affect nearby cells.

Some peptides act in both manners e.g. somatostatin.

Gastrin

Gastrin is secreted by enteroendocrine cells (G-cells) in the pylorus of the stomach after a meal. It is secreted in response to:
- Peptides or amino acids in the stomach.
- Vagal stimulation (i.e. parasympathetic).
- Distension of the stomach.

It acts to increase protein breakdown, specifically by:
- Stimulating the parietal cells of the stomach to secrete hydrochloric acid and intrinsic factor.
- Stimulating the chief cells of the stomach to secrete pepsin.
- Increasing gastric motility.
- Stimulating secretion of insulin, glucagon, and secretin.
- Relaxing the pyloric and ileocaecal sphincters.

Secretin

Secretin is secreted by the duodenum and the rest of the small intestines in response to acid secreted by the stomach. It neutralizes the acid produced by gastrin release by:
- Stimulating pancreatic and liver secretion of bicarbonate.
- Inhibiting acid secretion from the parietal cells of the stomach.
- Increasing the response to cholecystokinin (CCK).

Cholecystokinin

CCK is secreted in the duodenum in the presence of fat or amino acids. It increases the breakdown of fat by:
- Causing contraction of the gall bladder and release of bile into the duodenum.
- Stimulating pancreatic enzyme secretion (e.g. lipase).
- Causing some bicarbonate release from the pancreas.
- Producing a sensation of fullness.

Up and Coming Hormones

> The concept of hormones was first put forward by the English physiologists Bayliss and Starling in 1902. They deduced that a blood-borne signal secreted by the duodenum acted on the pancreas. They called this signal secretin and the name is still used today.

Glucose-dependent insulinotrophic peptide

Glucose-dependent insulinotrophic peptide (GIP) used to be called gastric inhibitory peptide (also GIP), but this was inaccurate and too easy to pronounce. It is secreted by the lining of the duodenum in response to fats and carbohydrates. It acts to:
- Stimulate insulin secretion if blood glucose is high.
- Inhibit gastric acid production and gastric motility.

Somatostatin

Somatostatin is secreted mainly by the δ-cells of the islets of Langerhans in the pancreas, but also by the stomach and GI tract neurons. It is also secreted by the hypothalamus where it is called growth hormone inhibiting hormone (GHIH). Secretion is stimulated by:
- Acidity and amino acids in the stomach.
- High blood glucose.
- CCK.

Somatostatin acts to slow down digestion by inhibiting:
- Secretion of all other GI tract and pancreatic hormones.
- Secretion of pancreatic enzymes and bile.
- Gastric motility.

Pancreatic polypeptide

Pancreatic polypeptide is secreted by the F-cells of the islets of Langerhans in the pancreas, in response to protein in the stomach or low blood glucose. Its actions remain unclear but it slows the absorption of food by:
- Inhibiting gall bladder contraction.
- Inhibiting pancreatic enzyme secretion.

Insulin and glucagon

Insulin and glucagon are both peptides secreted by enteroendocrine cells in the pancreas. The important functions they perform are described in Chapter 5.

Pineal gland

Structure
Macrostructure

The pineal gland secretes the hormone melatonin that regulates circadian rhythms. It is a small gland found at the posterior end of the corpus callosum, forming a section of the roof in the posterior wall of the third ventricle (Fig. 6.4).

The pineal gland begins to calcify after puberty making it a useful midline marker in X-rays and computed tomography (CT) scans.

> The French philosopher Descartes (1594–1650) believed that the pineal gland was the 'Seat of the soul' that unified the functions of body and mind.

Microstructure

The gland is composed of two types of neural cells:
- Pinealocytes—specialized secretory neurons.
- Glial support cells.

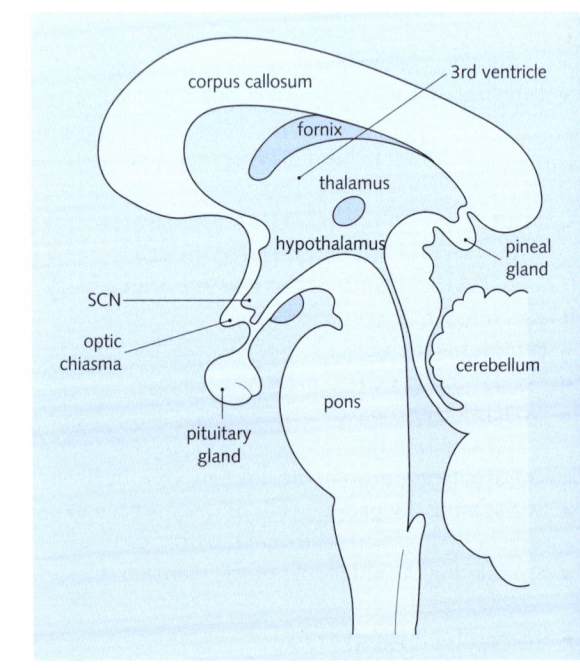

Fig. 6.4 Median section of the mid brain and brain stem showing the anatomical location of the pineal gland and suprachiasmatic nucleus (SCN).

In keeping with all endocrine organs, it has a very rich blood supply that forms a network of capillaries surrounded by the pinealocytes. It receives innervation from many parts of the brain, but the main connections are with the:
- Suprachiasmatic nucleus (SCN).
- Retina.
- Sympathetic system.
- Parasympathetic system.

Function
The pineal gland synthesizes and secretes the hormone melatonin. This is not the same as the brown skin pigment called melanin. Melatonin is a modified form of the amino acid tryptophan, which is first converted to 5-HT then to melatonin. Most modified amino acid hormones are derived from tyrosine.

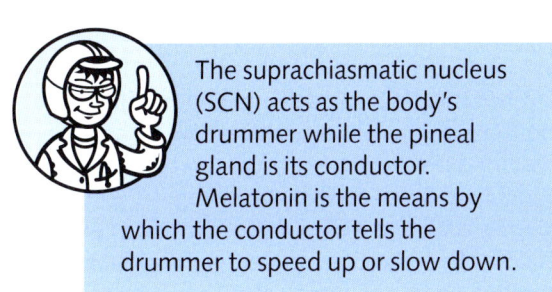

> The suprachiasmatic nucleus (SCN) acts as the body's drummer while the pineal gland is its conductor. Melatonin is the means by which the conductor tells the drummer to speed up or slow down.

Regulation
The main control of melatonin release is from the SCN. This nucleus is located in the hypothalamus, above the optic chiasma as its name suggests. It is essentially the 'body-clock' which regulates body rhythms including:
- Day and night (circadian rhythm).
- Seasonal rhythms (e.g. mating cycles).

The GI tract hormone vasoactive intestinal polypeptide (VIP) is also secreted by the cells of the SCN as a neurotransmitter.

Melatonin secretion is also regulated by light and dark. Stimulation from the retina (i.e. light) inhibits melatonin secretion, so melatonin is secreted in response to darkness. The pineal gland allows physical stimulation of light to be converted into the chemical signal melatonin.

Effects of melatonin
Melatonin has three main effects:
- It induces sleep (hypnotic effect).
- It resets the SCN.
- It influences the hypothalamus, especially the reproductive functions.

Circadian rhythms influence almost every cell in the body. Hormones secreted from the hypothalamus, pituitary gland, and gonads all respond dramatically to these rhythms—e.g. corticotrophin releasing hormone (CRH) and adrenocorticotrophic hormone (ACTH) secretion peak early in the morning. The regulation and actions of melatonin are shown in Fig. 6.5.

Jet-lag and melatonin treatment
The pineal gland has evolved to allow adaptation to changing day length (i.e. seasons). However, resetting of the SCN is best demonstrated by jet flight in the following way:
- When a person leaves their home country their SCN and pineal gland are synchronized: every night the SCN and darkness both stimulate melatonin production inducing sleep.
- If the person flies across time zones, the SCN remains set at the previous time zone so melatonin production (and, therefore, tiredness) does not change.
- The light input into the pineal gland gradually adapts melatonin secretion to fit the new time zone. It can alter secretion by a couple of hours each day.
- Melatonin receptors in the SCN allow the clock to adjust to the time zone so that the pineal gland and SCN are once again synchronized.

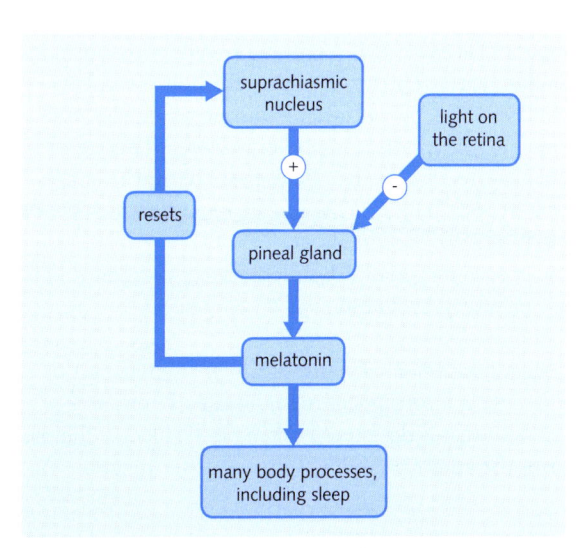

Fig. 6.5 Regulation of melatonin and its actions.

The sensation of retaining the previous time zone until resetting is complete is commonly called jet-lag. The person feels tired and sleepy at times appropriate to the previous time zone, and they may take several days to adapt. A similar effect is seen in shift work; this is especially important in medical employment. Try asking the SHO how a week of 'nights' feels.

Taking oral melatonin can shorten the period of jet lag or adaptation to a new shift. Melatonin should be taken at the times of darkness in the new time zone whilst on the plane and for several days at the destination. For shift work, the melatonin should be taken during the period of desired sleep. The SCN is reset more quickly and the body becomes resynchronized.

Melatonin is also given to some disabled children to maintain normal day and night cycles. It is given at night to induce sleep.

> Without the signals of light and dark, the body clock has a natural cycle of 25 hours. This makes adaptation to later time zones easier and accounts for jet-lag being worse when flying west to east.

Endocrine role of adipose tissue

Adipose tissue has recently been identified as an endocrine tissue that secretes a polypeptide hormone called leptin. The levels of this hormone correlate well with the percentage of adipose tissue, creating an endocrine indicator of energy stores. This signal is transported across the blood–brain barrier to act on the hypothalamus. It has two effects:
- It inhibits food intake.
- It permits gonadotrophin releasing hormone (GnRH) production.

The inhibition of food intake has prompted a lot of interest in leptin as a cure for obesity. Leptin is a protein that is digested if taken orally; like insulin, it must be administered by subcutaneous injection. At the highest doses leptin does cause some weight loss, though the effect is not impressive or effective in all patients.

Leptin resistance caused by the blood–brain barrier may prevent remarkable weight loss. It has been suggested that only a fixed amount of leptin can be transported across this barrier so that excess leptin in the blood does not increase leptin in the CNS above a fixed level. Disorders in this barrier may also account for some cases of obesity. Inherited leptin deficiency is an incredibly rare recessive disorder that causes gross obesity and infertility.

> The name leptin is derived from the Greek word *leptos* meaning thin.

The action of leptin on GnRH release (Fig. 6.6) explains the infertility of underweight women. This is a protective response to prevent pregnancy in the undernourished. Leptin is also secreted prepubertally and it plays a role in the onset of puberty. Body weight is a better predictor of the onset of menstruation than age.

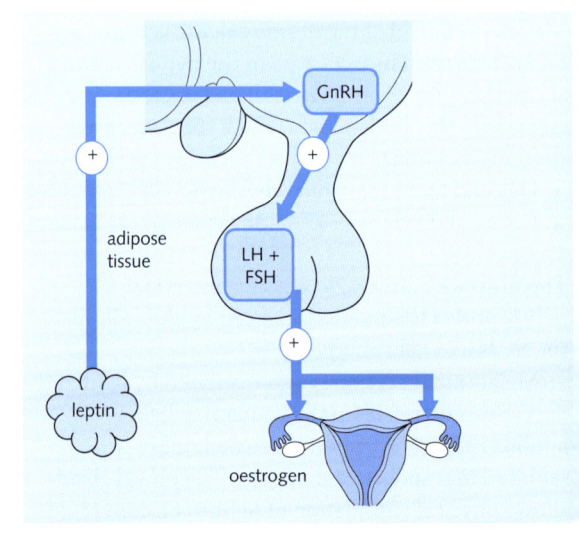

Fig. 6.6 The action of leptin on fertility. (FSH, follicle stimulating hormone; GnRH, gonadotrophin releasing hormone; LH, luteinizing hormone.)

- Define an APUD cell. List three examples with the hormones they secrete.
- Describe the actions of gastrin, cholecystokinin, and secretin.
- Describe the anatomical location of the pineal gland; what types of cells are present?
- What type of hormone is melatonin and how is it synthesized?
- What signals trigger the release of melatonin?
- Describe the function of the suprachiasmatic nucleus (SCN); how does melatonin affect these cells?
- Describe jet-lag from an endocrine point of view.
- What type of hormone is leptin?
- Why are severely underweight women infertile?
- Define leptin resistance; how is this believed to occur?

7. Endocrine Control of Fluid Balance

Water is essential for every single cell in the body, so the regulation of water levels is very important. The loss and gain of water from the body is called fluid balance. The regulation of fluid balance takes place mainly in the kidneys, where hormones control the volume and concentration of water that is excreted. The excretion of water is largely controlled by the regulation of sodium absorption. Since sodium is osmotically active, water follows it across membranes. There are many hormones involved in this process. However, the main four are:
- Antidiuretic hormone (ADH)—from the posterior pituitary gland (Fig. 7.1).
- Renin and angiotensin II—from the kidney (Fig. 7.2).
- Aldosterone—from the adrenal cortex (Fig. 7.2).

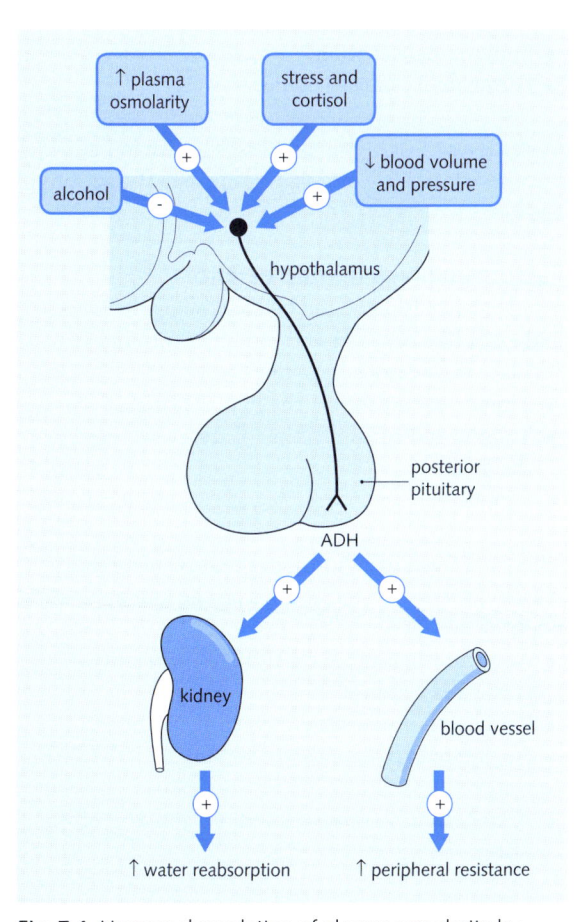

Fig. 7.1 Hormonal regulation of plasma osmolarity by antidiuretic hormone (ADH).

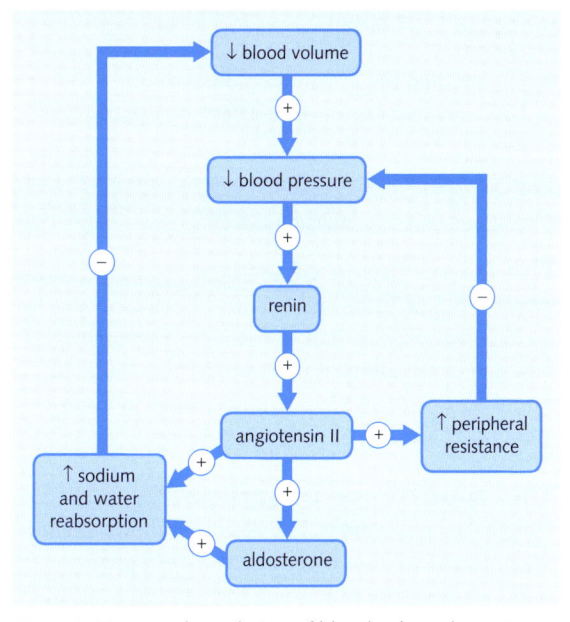

Fig. 7.2 Hormonal regulation of blood volume by renin, angiotensin II and aldosterone.

Blood pressure is also affected by fluid balance because changes in blood volume affect the pressure in arteries. Many of the hormones that regulate fluid balance also control the diameter of arteries, allowing

Important words:
Peripheral resistance: resistance created by arterioles that can be increased by reducing their diameter; blood pressure increases as resistance rises
Vasoconstriction: contraction of arterioles to raise peripheral resistance
Osmolarity: the concentration of the blood—high osmolarity means low water
Hypovolaemia: low blood volume
Diuretic: a substance that increases water loss from the kidneys
Osmotically active: the ability to attract water across a cell membrane to areas of higher concentration

Endocrine Control of Fluid Balance

blood pressure to be maintained despite loss or gain of water.

After reading this chapter you should be able to:
- Highlight the need for fluid balance.
- Explain the actions of antidiuretic hormone.
- Describe the actions of the renin–angiotensin II system.
- Understand the actions of aldosterone.
- Discuss other factors that affect fluid balance.

Fluid balance

The importance of water
Water is found inside and outside of cells and also in the blood as plasma. It is essential for the normal function of all cells so it must be tightly regulated. About 50 L water is found in the average human, accounting for 70% of body weight.

The importance of sodium
Fluid balance is intimately linked to sodium balance. Sodium ions are osmotically active (they attract water across membranes) and they are present in large quantities within the body. Water tends to passively follow movements of sodium ions. The normal plasma concentration of sodium ions is 135–145 mmol/L. In the kidney, water excretion is largely controlled by regulating sodium excretion or reabsorption.

Fluid balance and circulation
Variations in fluid balance rapidly affect blood volume and, therefore, blood pressure. Changes in fluid balance can be compensated for by variations in:
- Cardiac output (i.e. stroke volume and rate).
- Peripheral resistance (i.e. vasodilatation or vasoconstriction).

Regulation of fluid balance
Water intake
Water intake is controlled by the sensation of thirst. Thirst is determined by osmoreceptors in the hypothalamus that detect the water concentration (osmolarity) of plasma. Some hormones involved in fluid balance can also cause this feeling directly. Water is also gained from food and metabolism (Fig. 7.3).

Water excretion
Water excretion is controlled mainly by the kidney. The kidney filters the entire blood volume once every 5 minutes. Water and sodium are filtered into

Expected intake and output of water over a 24-hour period	
Water intake (mL)	Water loss (mL)
Drinking: 1500	Urine: 1500
Food: 500	Respiration: 400
Metabolism: 400	Skin evaporation: 400
	Faeces: 100
Total: 2400	Total: 2400

Fig. 7.3 Expected intake and output of water over a 24-hour period.

the kidney tubules by the glomeruli but most is reabsorbed back into the blood. Sodium ions are actively reabsorbed, while water passively follows the sodium by osmosis. Any sodium or water that is not reabsorbed is excreted in the urine.

Fluid balance can be regulated in the kidneys by altering two factors:
- Sodium reabsorption.
- Permeability of the tubules to water.

Water is also lost by the processes shown in Fig. 7.3, but these losses are less significant than the actions of the kidney.

Factors involved in fluid balance
Fluid balance is regulated by controlling the intake and excretion of water and sodium. Hormones regulate this balance by acting on:
- Thirst—stimulated in the hypothalamus.
- Volume and concentration of water excreted in the urine.
- Peripheral resistance, which affects blood pressure.

There are four main hormones that regulate fluid balance; their sites and actions on the kidney nephron are shown in Fig. 7.4:
- Antidiuretic hormone (ADH; vasopressin)—conserves water.
- Renin—stimulates angiotensin II synthesis.
- Angiotensin II—conserves sodium and, therefore, water.
- Aldosterone—conserves sodium and, therefore, water.

The actions of these four hormones are modified by other hormones, nerves and chemical factors, including:

Hormones Involved in Fluid Balance

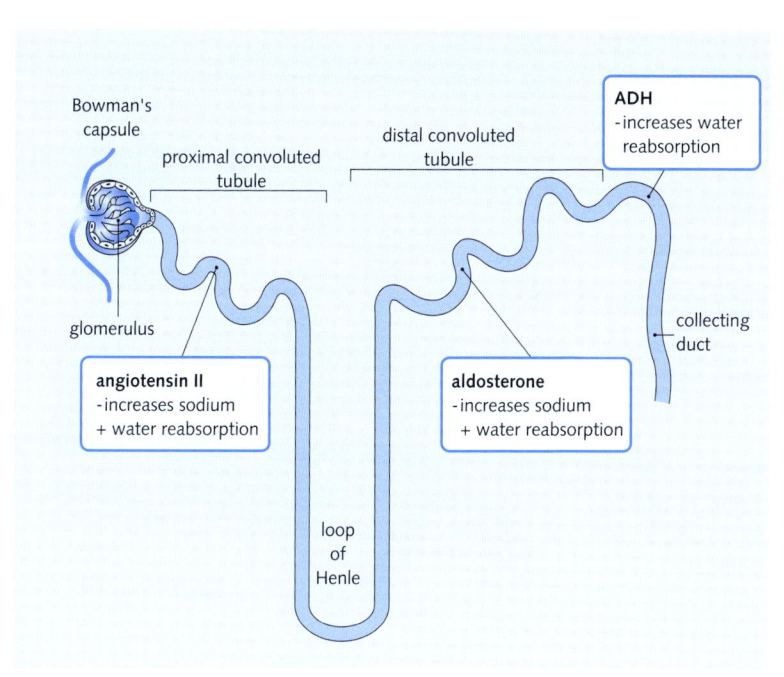

Fig. 7.4 Location and type of action by the three major hormones in the kidney nephron. (ADH, antidiuretic hormone.)

- Atrial natriuretic factor (ANF).
- Renal sympathetic nerves and catecholamines.
- Kinins.
- Prostaglandins.
- Dopamine.

Hormones involved in fluid balance

Antidiuretic hormone
ADH (or vasopressin) acts on the kidney to conserve water. It increases the permeability of the collecting ducts so that more water is reabsorbed, resulting in the production of more concentrated urine. It also causes constriction of the arterioles to raise peripheral resistance—hence vasopressin, the old name for this hormone. ADH regulates fluid balance by influencing the movement of water directly, not through sodium movement. The regulation and actions of ADH are shown in Fig. 7.1.

Synthesis and secretion
ADH is a polypeptide hormone synthesized by neurosecretory cells in the supraoptic nucleus of the hypothalamus. It is transported along their axons to the posterior pituitary gland, where it is stored in vesicles. This process is described in more detail in Chapter 2.

ADH is secreted by the posterior pituitary gland in response to:
- High plasma osmolality (concentrated blood).
- Low blood volume.

It is rapidly degraded by the liver and kidney.

> ADH secretion is inhibited by alcohol, causing large volumes of dilute urine to be excreted, resulting in dehydration the next morning.

Intracellular actions
ADH acts on G-protein linked vasopressin receptors (V receptors) found on the cell-surface of target cells. These target cells are found in two tissue types:
- Kidney—ADH acts on V_2 receptors causing pores in the membrane to open allowing water through. This process uses cAMP as a second messenger.
- Blood vessels—ADH acts on V_1 receptors causing smooth muscle contraction (i.e. vasoconstriction). Inositol triphosphate (IP_3) acts as a second messenger causing calcium levels to rise, resulting in vasoconstriction.

Effects
On the kidneys
ADH is secreted in response to high blood osmolarity. It increases the reabsorption of water in the kidney by increasing the permeability of the

collecting duct. This prevents the excretion of water whilst sodium remains unchanged, so that the blood osmolarity decreased and concentrated urine is produced.

On the blood vessels
ADH is also secreted in response to low blood volume. It causes vasoconstriction of the arterioles, which increases peripheral resistance and raises blood pressure. This action normally has little effect on blood pressure regulation, but it is important following severe haemorrhage that significantly reduces blood volume.

Deficiency and excess
ADH deficiency causes diabetes insipidus, in which excess dilute urine is produced causing fluid loss, dehydration, and blood with high osmolarity.

ADH excess is called the syndrome of inappropriate ADH secretion (SIADH). It results in water retention.

Both of these conditions are described in more detail in Chapter 2.

The renin–angiotensin II system
The actions of renin and angiotensin II are intimately linked so they are described together. The renin–angiotensin II system acts on the kidney to conserve sodium and water. Renin is produced by the kidney, and it causes the production of angiotensin II. Angiotensin II causes reabsorption of sodium in the proximal tubules of the kidney and water follows this movement. Angiotensin II also increases peripheral resistance and aldosterone release. The regulation and actions of renin and angiotensin II are shown in Fig. 7.4.

Synthesis and secretion of renin
Renin is a small peptide enzyme secreted by the cells of the juxtaglomerular complex. This structure is made of three cell types (Fig. 7.5):
- Macula densa—part of the distal tubule of the nephron that detects sodium.
- Juxtaglomerular cells—part of the afferent glomerular arteriole that releases renin.
- Extraglomerular mesangial cells—glomerular support cells that control blood flow in the afferent glomerular arteriole.

The function of this structure is complex, however it can detect both sodium in the distal convoluted tubule and blood pressure. The juxtaglomerular cells secrete renin in response to low blood pressure.

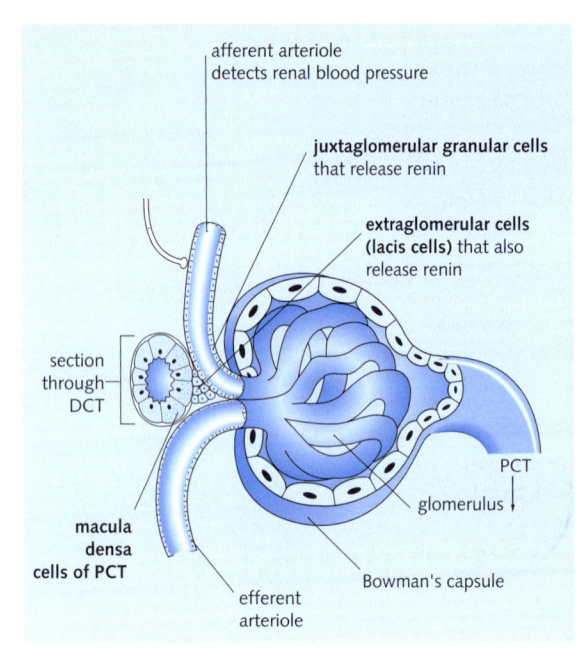

Fig. 7.5 Structure of the juxtaglomerular complex. (PCT, proximal convoluted tubule; DCT, distal convoluted tubule, lying very close to the afferent and efferent arterioles.)

Effects of renin
Renin is an enzyme that acts on a plasma protein called angiotensinogen, synthesized by the liver. It cleaves this protein to form angiotensin I. This is rapidly converted to angiotensin II by the action of angiotensin-converting enzyme (ACE) in the blood. ACE is a major site of action for antihypertensive drugs called ACE inhibitors.

> The main site of ACE action and angiotensin II synthesis is the endothelium of the capillaries in the lung.

Effects of angiotensin II
Angiotensin II has five important actions to increase blood volume and pressure:
- Stimulating sodium pumps in the proximal tubule to increase sodium reabsorption, causing water reabsorption.
- Stimulating aldosterone release from the adrenal cortex.
- Peripheral vasoconstriction to raise blood pressure.

- Inhibiting renin release (negative feedback).
- Stimulating sensation of thirst in the hypothalamus.

Thus, angiotensin II conserves sodium and water whilst raising the blood pressure.

ACE inhibitors inhibit production of angiotensin II causing lower blood volume and reduced peripheral resistance. Both actions decrease blood pressure, so these drugs are used to treat hypertension and heart failure.

> The primary function of the renin–angiotensin II system is to keep blood volume constant.

Aldosterone

Aldosterone is a mineralocorticoid steroid hormone synthesized by the zona glomerulosa cells of the adrenal cortex. The main stimulus for secretion is angiotensin II, so it is often considered part of the renin–angiotensin II system. Aldosterone causes the conservation of sodium and water. See Chapter 4 for a complete description of aldosterone and the associated disorders. The regulation and actions of aldosterone are shown in Figs 7.2 and 7.4.

Intracellular actions

Aldosterone acts on intracellular receptors in the distal convoluted tube of the kidney. It causes sodium to be reabsorbed in exchange for potassium. Water follows the movement of sodium as the internal concentration increases, and is reabsorbed.

Effects

Aldosterone is secreted in response to high potassium or reduced blood volume through the action of renin and angiotensin II. It causes sodium and water to be conserved to help rectify these changes.

Natriuretic factors

Natriuretic factors have a diuretic effect, the opposite effect of ADH, angiotensin II and aldosterone. They increase sodium excretion, so that water excretion is also increased. This causes blood volume to decrease and blood pressure is lowered. Natriuretic factors act as an 'escape mechanism' to prevent excess water retention.

Atrial natriuretic factor

ANF is a polypeptide hormone synthesized by the muscle cells of the atrium of the heart. It is secreted in response to high blood volume, which is detected by the stretching of cardiac muscle.

The main actions of ANF are:
- Decrease of sodium reabsorption by kidneys.
- Inhibition of renin secretion.
- Inhibition of aldosterone secretion.
- Vasodilatation, causing a fall in blood pressure.

Kinins

Kinins are polypeptides formed in the blood by the action of the enzyme kallikrein on plasma globulins. They act in the kidney to:
- Inhibit the action of ADH.
- Decrease sodium reabsorption.
- Stimulate prostaglandin synthesis in the kidney.
- Stimulate vasodilation in the kidney.

Renal prostaglandins

Renal prostaglandins are locally acting lipid molecules synthesized by kidney cells. They act in the kidney to:
- Inhibit the action of ADH and aldosterone.
- Stimulate vasodilatation in the kidney.

Dopamine

Dopamine is an amine synthesized in the proximal tubule cells. It acts in the kidneys to:
- Decrease sodium reabsorption.
- Induce vasodilation in the kidney.

Disorders of fluid balance

A deficiency of water is called dehydration; it may or may not be accompanied by sodium deficiency. The causes and effects of dehydration are shown in Fig. 7.6.

An excess of water is called fluid retention; it may or may not be accompanied by an excess of sodium. The causes and effects of fluid retention are shown in Fig. 7.7.

Endocrine Control of Fluid Balance

Fig. 7.6 The causes and effects of dehydration.

Causes and effects of dehydration

		Deficient sodium and water	Deficient water
Cause	Deficient input	Decreased ingestion e.g. unconscious	Unable to find water, hypothalamic thirst disorder
	Excess output	Diarrhoea, vomiting, burns, haemorrhage, aldosterone deficiency	Diabetes insipidus and mellitus
Effect	Plasma osmolarity	No change	Raised
	Symptoms	Thirst, postural dizziness, weakness, collapse, headache, apathy, confusion, coma	
	Signs	Hypotension, tachycardia, slow capillary refill, reduced skin turgor, cool peripheries, sunken eyes, dry membranes, weight loss	

Fig. 7.7 The causes and effects of water retention.

Causes and effects of fluid retention

		Excess sodium and water	Excess water
Cause	Excess input	Excess fluid transfusion	Excess drinking e.g. psychological
	Deficient output	Cardiac or renal failure	Acute renal failure or SIADH
Effect	Plasma osmolarity	No change, causes oedema	Decreased, no oedema
	Symptoms	Nausea, vomiting, anorexia, muscle weakness, headache, apathy, confusion, fits, coma	
	Signs	Hypertension, raised JVP, displaced apex beat	

- Describe the processes by which water can be lost or gained from the body. Which two processes are most important in the regulation of fluid balance?
- How do blood volume and peripheral resistance influence blood pressure?
- Explain the regulation and secretion of ADH.
- List the actions of ADH.
- Describe the regulation and secretion of renin and angiotensin II. How do ACE inhibitors act?
- Describe the actions of renin and angiotensin II.
- Explain the regulation and action of aldosterone.
- Name three peptides that affect fluid balance along with their actions.
- List the common causes of dehydration and the signs and symptoms that result.
- What are the common causes of water retention and the signs and symptoms that result?

8. Endocrine Control of Calcium Homeostasis

Calcium is essential for many important processes in the body. Bone formation, nerve depolarization, and muscle contraction, including cardiac muscle, all require calcium to function normally. Blood levels must be carefully regulated otherwise cardiac arrest can occur. Calcium regulation is not controlled by the hypothalamus or pituitary gland.

Blood calcium levels are regulated by three hormones (Fig. 8.1):
- Parathyroid hormone (PTH)—raises blood calcium levels.
- Vitamin D—raises calcium intake from the gastrointestinal tract.
- Calcitonin—raises calcium excretion and lowers blood levels.

PTH is the most important of these hormones. Low plasma calcium triggers its secretion. Three sites respond to PTH to increase calcium levels: kidneys, bones and intestines.

An excess of PTH causes hypercalcaemia, whereas PTH deficiency results in hypocalcaemia. Other causes of hypocalcaemia (e.g. chronic renal failure) cause a rise in PTH. Excess PTH always results in decalcified, weak bones.

After reading this chapter you should be able to:
- Highlight the physiological requirement for calcium.
- Understand the regulation of calcium levels.
- Explain how the three hormones control this regulation.
- Describe the common disorders of calcium metabolism.

> **Important words:**
> **Hyperparathyroidism:** an excess of parathyroid hormone
> **Osteoclast:** a cell that breaks down bone releasing calcium
> **Osteoblast:** a cell that lays down bone using calcium
> **Reabsorption:** when substances are reclaimed from the kidney tubule after being filtered out of the blood

The role of calcium

Calcium is a mineral obtained in the diet and excreted by the kidneys. The adult body contains about 1.2 kg calcium, the vast majority (99%) of which is locked in bones so it does not contribute to homeostatic control; a small reservoir of calcium is able to leave the bone. The remaining 1% of calcium is in the plasma, extracellular fluid, and inside cells.

Although only a small proportion of the body's calcium is available to cells, it performs many vital functions, especially in the neuromuscular system. The main processes that require calcium are:
- Bone formation and maintenance.
- Muscle contraction, including cardiac muscle.
- All processes that involve exocytosis, including synaptic transmission and hormone release.

The processes requiring calcium are described in Fig. 8.2.

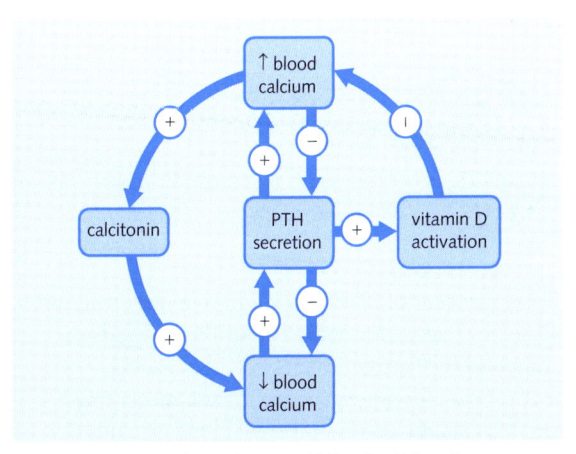

Fig. 8.1 Hormonal regulation of blood calcium by parathyroid hormone (PTH), vitamin D, and calcitonin.

Mechanisms involved in calcium homeostasis

Calcium intake

About 1 g calcium is required in the diet every day. It is found in many foods, especially dairy products. Only 30% of dietary calcium is absorbed by the

Endocrine Control of Calcium Homeostasis

Fig. 8.2 Processes that require calcium.

Processes that require calcium

System	Role of calcium
Bone formation	Calcium is a vital mineral component of bone; it makes the bone strong and rigid
Blood clotting	Many clotting factors are activated by calcium
Muscle contraction	Calcium binds to troponin, which allows myosin to bind to actin
Intracellular signalling	Calcium regulates the activity of a number of intracellular proteins in response to the second messenger IP_3
Nervous system	Calcium is essential for membrane potential and depolarization; synapses use calcium to release neurotransmitters
Endocrine system	All processes that involve exocytosis (e.g. hormone secretion) require calcium
Cardiovascular system	Calcium regulates the membrane potential and contraction of muscle cells.

upper small intestine, though this figure can vary with a number of factors:
- Diet—lactose in milk increases absorption, phytic acid in brown bread decreases absorption.
- Age—absorption is increased in the young and decreased in the elderly.
- Hormones—vitamin D increases absorption.
- Pregnancy and lactation—both increase absorption.

> Never give the antibiotic tetracycline to young children as it binds to calcium in their teeth, making them yellow. In adults milk should not be used to swallow tablets of tetracycline.

Calcium balance
Calcium balance is calculated as absorbed calcium minus excreted calcium. This balance is affected by age and some diseases:
- Children usually have a positive calcium balance; this allows bones to grow.
- In adults, input and output should be the same.
- Postmenopausal women tend to have a negative calcium balance (i.e. calcium is lost).

Calcium in the blood
Since calcium is essential, plasma levels must be tightly regulated. The normal range is 2.12–2.65 mmol/L. Dietary intake is variable, so the body must be able to adapt to raised or reduced plasma calcium. Calcium regulation involves the loss or gain of calcium from three tissues: kidneys, intestines, and bones.

The movements and distribution of calcium in the body are shown in Fig. 8.3. The movements are regulated by the three hormones discussed below.

> At any time, about 20% of bones are being broken down and rebuilt to repair numerous microfractures.

Forty per cent of plasma calcium is bound to albumin. Diseases that lower plasma albumin can increase unbound (i.e. active) calcium levels, causing symptoms of hypercalcaemia. When measuring plasma calcium levels, it is total blood calcium that is measured. Albumin levels must be taken into account to calculate unbound, corrected calcium levels.

Hormones Involved in Calcium Homeostasis

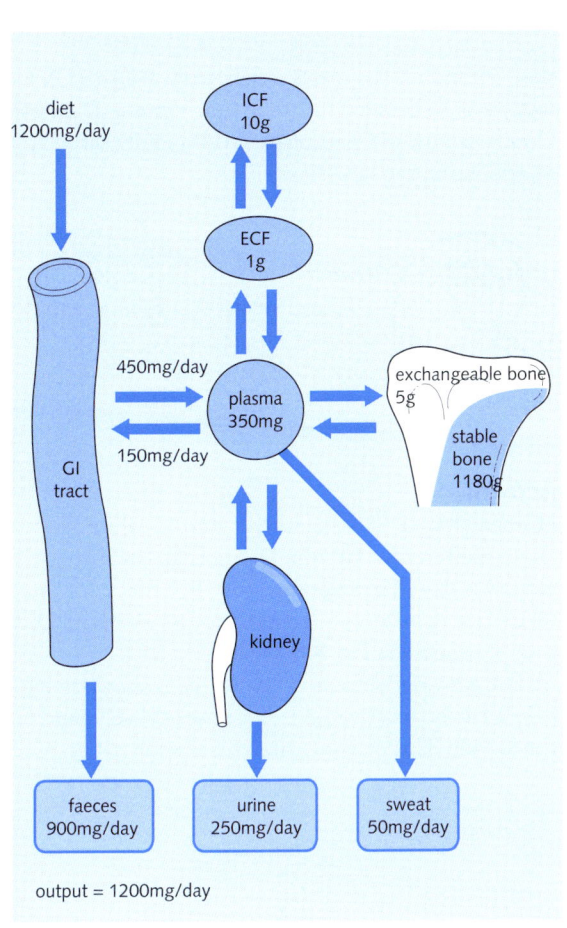

Fig. 8.3 Normal distribution and movements of calcium in the body. (ICF, intracellular fluid; ECF, extracellular fluid.)

- Calcitonin from the thyroid gland.

Their effects are summarized in Figs 8.1 and 8.4.

Parathyroid hormone (PTH)

PTH is essential to life; it is synthesized by the chief cells in the parathyroid glands. These four small glands are found behind the thyroid gland in the neck. Their development and anatomy are described in Chapter 3.

PTH is released in response to low blood calcium. Its actions are aimed at raising calcium levels back to their normal physiological concentration. Once blood calcium levels are restored, PTH production is inhibited.

Synthesis and receptors

PTH is a polypeptide hormone synthesized as pre-pro-parathyroid hormone. Two cleavage reactions result in PTH, a protein of 84 amino acid residues. This hormone is stored in cell vesicles until low plasma calcium triggers its release.

PTH acts via G-protein linked receptors on the cell surface. These receptors are found on osteoblasts, renal tubule cells, and cells in the intestinal epithelium. The receptors use cyclic AMP (cAMP) as a second messenger to regulate the phosphorylation of intracellular proteins. These proteins are either activated or deactivated as a result; this brings about the effects of PTH.

Actions

PTH is the most important regulator of blood calcium levels; it is essential to life. The actions of PTH on calcium regulation are shown in Fig. 8.5.

On the kidneys

PTH has three major effects:
- Increase of calcium ion reabsorption by stimulating active uptake in the distal convoluted tubule.

Hormones involved in calcium homeostasis

Three hormones regulate calcium levels in the blood and tissues:
- Parathyroid hormone from the parathyroid gland.
- Vitamin D (cholecalciferol) from the diet and skin.

Fig. 8.4 Summary of the actions of parathyroid hormone (PTH), calcitonin, and vitamin D on calcium regulation.

Actions of the hormones involved in calcium homeostasis

	PTH	Vitamin D	Calcitonin
Secreted/activated in response to:	Low blood calcium	PTH	High blood calcium
Kidneys	Calcium reabsorbed; vitamin D activated	Calcium reabsorbed	Calcium excreted
Bones	Calcium released	Calcium trapped	Calcium trapped
Intestines	Negligible	Calcium absorbed	Negligible

Endocrine Control of Calcium Homeostasis

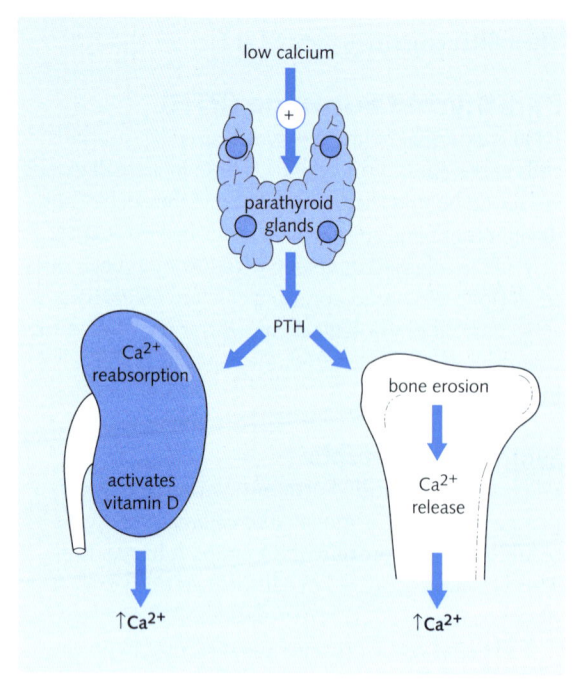

Fig. 8.5 Actions of parathyroid hormone (PTH) on the kidney and bone.

- Increase of phosphate ion excretion by inhibiting uptake in the proximal and distal convoluted tubules.
- Stimulation of 1α-hydroxylase, an enzyme that activates vitamin D.

On the bones
PTH acts only on the osteoblasts in bone. However, these cells regulate osteoclast activity via prostaglandins. In general, PTH causes the erosion of bone, releasing calcium and phosphate ions. The actions are:
- Direct inhibition of osteoblast collagen synthesis.
- Indirect stimulation of osteoclast bone erosion.
- Increased collagenase synthesis to erode the bone.
- Increased hydrogen ion release to create an acidic environment to enhance bone erosion.

On the intestines
PTH may have a direct action on calcium absorption in the upper small intestines, but this is unproven. The major effects are indirect, through the activation of vitamin D.

Vitamin D
Vitamin D is absorbed by the small intestine as part of the diet or is synthesized from cholesterol in the skin. Vitamin D synthesis requires ultraviolet (UV) light, usually derived from the sun. Vitamin D raises blood calcium and phosphate levels mainly through its actions on the intestines.

> Native Europeans probably evolved white skin to increase vitamin D production in the low levels of sunlight in the northern hemisphere.

Activation
Human vitamin D is an inactive steroid called cholecalciferol (or vitamin D_3); this fat-soluble steroid is stored in adipose tissue. Two reactions must take place in different organs to activate vitamin D; they are shown in Fig. 8.6.

The activated vitamin D molecule is shown in Fig. 8.7. It can be inactivated by 24-hydroxylase found in the kidney. This enzyme catalyses the formation of

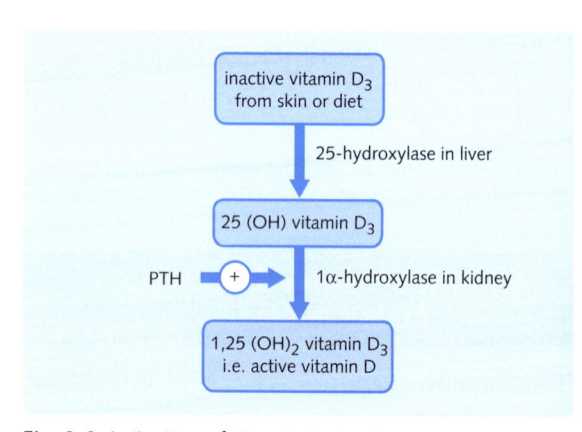

Fig. 8.6 Activation of vitamin D. (25 (OH) vitamin D_3, 25-hydroxyvitamin D_3; 1,25 (OH)$_2$ vitamin D_3, 1,25-dihydroxyvitamin D_3.)

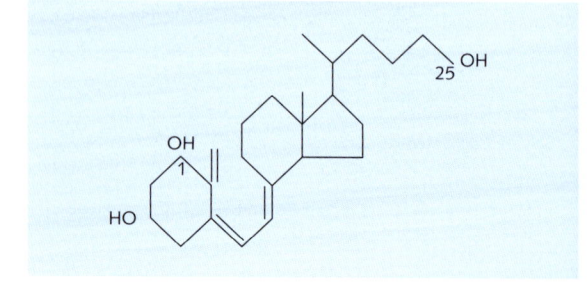

Fig. 8.7 Structure of 1,25-dihydroxyvitamin D_3, the active form of vitamin D.

1,24,25-trihydroxycholecalciferol, which is rapidly excreted.

All forms of vitamin D are transported in the blood by a specific plasma protein or within chylomicrons. Vitamin D is fat-soluble so it can cross cell membranes. It acts via specific intracellular receptors that are found in the same locations as PTH receptors.

Actions
The actions of vitamin D on calcium regulation are shown in Fig. 8.8.

On the kidney
Activated vitamin D has three effects:
- Increase of calcium reabsorption in the proximal and distal convoluted tubule.
- Increase of phosphate reabsorption in the proximal convoluted tubule.
- Inhibition of 1α-hydroxylase activity. This is a form of negative feedback.

On the bones
Vitamin D directly stimulates osteoblast activity to increase bone mass and calcification. Disorders of vitamin D absorption or activation can cause weak bones (rickets in children or osteomalacia in adults). If this is due to kidney disease, it is called renal osteodystrophy (discussed later).

On the intestines
The main action of vitamin D is stimulation of active calcium and phosphate absorption in the duodenum and jejunum. The exact mechanism is unclear, but vitamin D increases the synthesis of calcium-binding proteins in the intestinal cells. This action on the intestines takes a long time to produce an effect, so it does not raise calcium levels acutely.

Calcitonin
Calcitonin is secreted by the parafollicular cells (C cells) in the thyroid gland. The development and anatomy of this gland is described in Chapter 3. Calcitonin is secreted in response to high blood calcium, and it lowers calcium levels. It is not essential to life, but it acts to fine tune blood calcium levels.

Synthesis and receptors
Calcitonin is a polypeptide hormone that is formed by the breakdown of a larger prohormone. Calcitonin is stored in secretory vesicles, from which it is released when blood calcium levels rise.

Calcitonin acts on G-protein coupled receptors that release cAMP to bring about cellular effects.

Actions
The actions of calcitonin on calcium regulation are shown in Fig. 8.9.

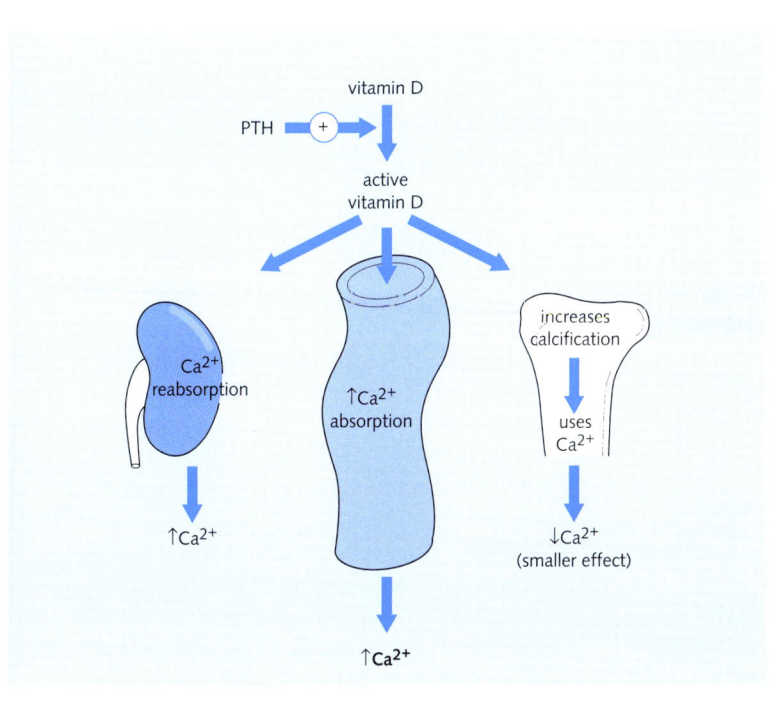

Fig. 8.8 Actions of vitamin D on the gastrointestinal tract, bone, and kidney.

Endocrine Control of Calcium Homeostasis

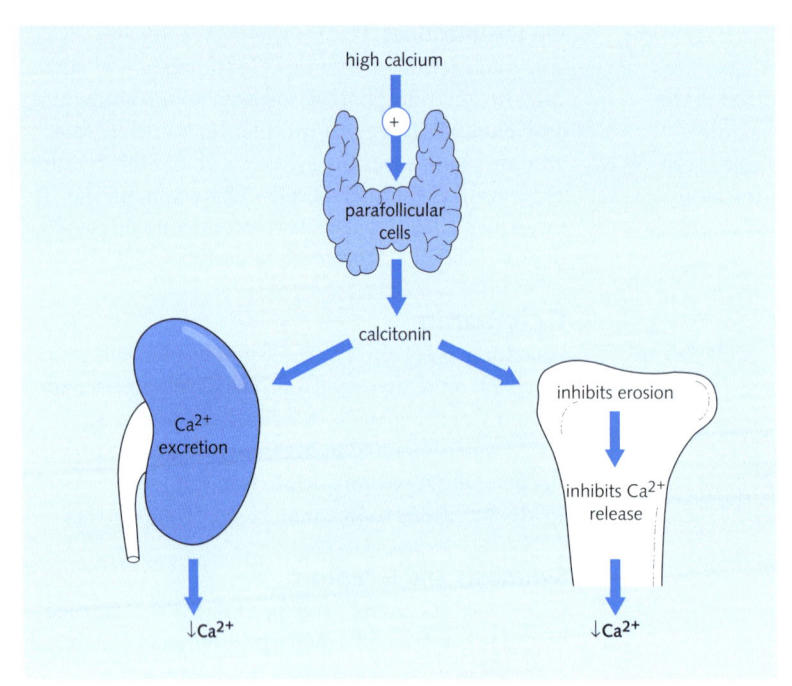

Fig. 8.9 Actions of calcitonin on the kidney and bone.

On the kidney
Calcitonin inhibits the reabsorption of calcium and phosphate. These ions are excreted as a result.

On the bones
Calcitonin acts primarily on osteoclasts in the bone. It inhibits these cells to prevent all stages of bone erosion. This prevents calcium and phosphate release into the blood, so their levels are lowered.

Calcitonin is not essential for calcium regulation, and there are no clinical consequences of calcitonin deficiency or excess. Neither removal of the thyroid gland with its parafollicular cells at thyroidectomy nor calcitonin secreting medullary cell malignancy affect calcium balance significantly.

Disorders of calcium regulation

Since PTH is the most important hormone in calcium homeostasis, disorders of this system are grouped according to their effect on PTH release. The two groups are:
- Hyperparathyroidism (excess of PTH).
- Hypoparathyroidism (deficiency of PTH).

Primary hyperparathyroidism
Primary hyperparathyroidism is excessive PTH release. All the actions of PTH raise calcium levels so hypercalcaemia (excess blood calcium) results. Hypercalcaemia and excess PTH cause the symptoms shown in Fig. 8.10. Prolonged hyperparathyroidism causes bone demineralization and softening, called osteomalacia in adults and rickets in children. The symptoms and signs of these diseases are shown in Fig. 8.11.

> When clerking a patient with suspected hypercalcaemia, remember: 'Bones, stones, abdominal groans, and psychic moans'.

Primary hyperparathyroidism is a relatively common endocrine disorder (about 1 in 1000 people), and it is especially common in postmenopausal women. The main causes are:
- Parathyroid gland adenoma (80%).
- Diffuse parathyroid gland hyperplasia (20%).

Neoplastic chief cells are not inhibited by high calcium, and consequently PTH secretion is unregulated. Malignant tumours of the parathyroid gland are very rare, but they can be associated with other endocrine tumours in multiple endocrine

Disorders of Calcium Regulation

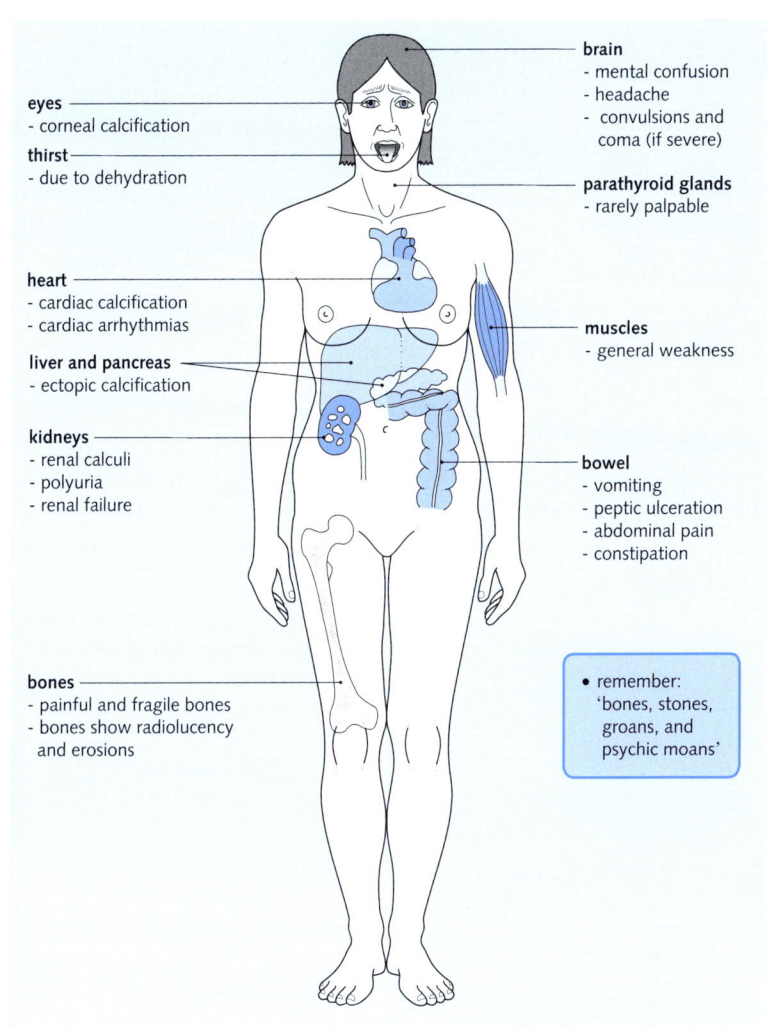

Fig. 8.10 Symptoms and signs of hypercalcaemia.

eyes
- corneal calcification

thirst
- due to dehydration

heart
- cardiac calcification
- cardiac arrhythmias

liver and pancreas
- ectopic calcification

kidneys
- renal calculi
- polyuria
- renal failure

bones
- painful and fragile bones
- bones show radiolucency and erosions

brain
- mental confusion
- headache
- convulsions and coma (if severe)

parathyroid glands
- rarely palpable

muscles
- general weakness

bowel
- vomiting
- peptic ulceration
- abdominal pain
- constipation

- remember: 'bones, stones, groans, and psychic moans'

Fig. 8.11 Symptoms and signs caused by rickets and osteomalacia.

Signs and symptoms caused by rickets and osteomalacia

Rickets (childhood)	Osteomalacia (adulthood)
'Knock-knees' or 'bow-legs' caused by bending of the long bones	Bone pain
Chest deformities, back deformities (e.g. kyphosis) and protruding forehead	Bones appear 'thin' on X-ray, with localized lucencies (called Looser's zones)
Features of hypocalcaemia	Fractures (common in the neck of the femur)
	Features of hypocalcaemia (e.g. proximal myopathy causes waddling gait)

neoplasia (MEN) syndromes. These are discussed in Chapter 10.

Diagnosis and treatment

Primary hyperparathyroidism is suspected if a patient has:
- Unexpected bone weakness.
- Hypercalcaemia symptoms and signs.
- Hypercalcaemia on a blood test with low phosphate levels.

It is investigated by:
- Measuring blood PTH.
- Radioisotope scanning of the parathyroid glands.

93

There are three treatment options:
- Restrict dietary calcium.
- Drug treatment (e.g. calcitonin).
- Surgical removal of the parathyroid gland(s).

Secondary hyperparathyroidism

Many diseases can cause hypocalcaemia (e.g. chronic renal failure), which stimulates PTH secretion as a compensatory response. If hypocalcaemia is prolonged, the parathyroid glands can enlarge by hyperplasia to secrete excess PTH. This is called secondary hyperparathyroidism.

Osteomalacia is also a feature of secondary hyperparathyroidism because of the excess PTH. Hypocalcaemia and excess PTH cause the symptoms shown in Fig. 8.12.

Plasma calcium levels determine the threshold of action potentials. Low calcium causes a low threshold, resulting in tingling sensations and muscle contractions.

Causes of secondary hyperparathyroidism

Hypocalcaemia can be caused by:
- Chronic renal failure.
- Vitamin D deficiency.

In chronic renal failure, the kidneys fail to reabsorb calcium. Renal osteodystrophy can also develop as a result of impaired 1α-hydroxylase activity. The

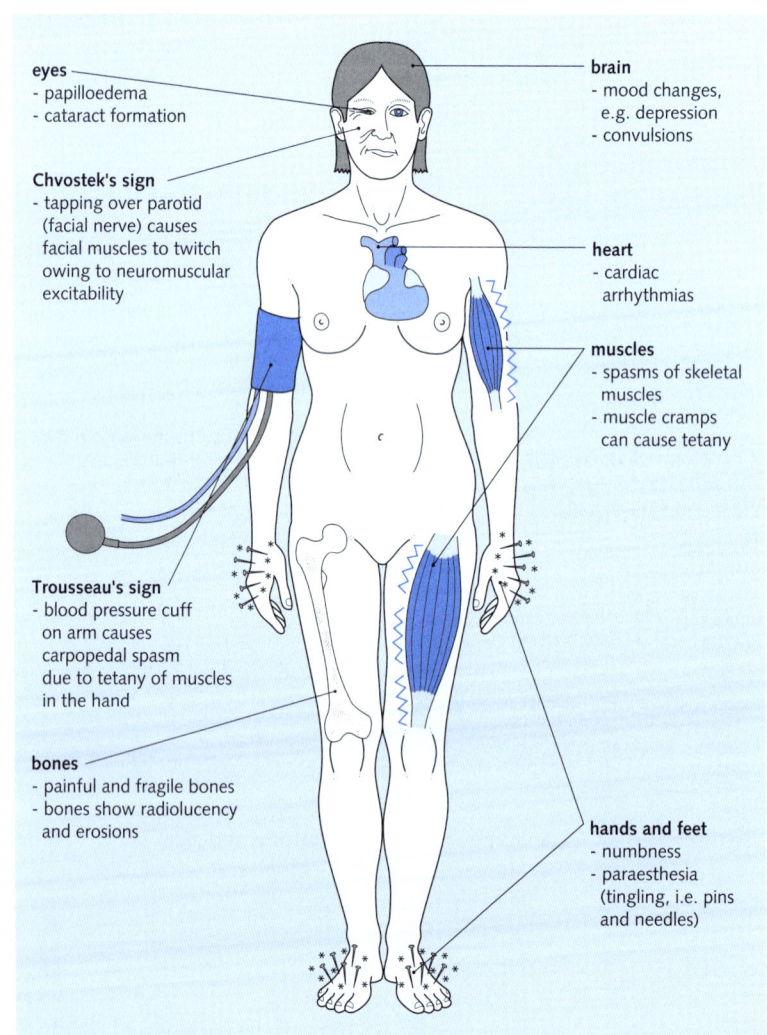

Fig. 8.12 Symptoms and signs of hypocalcaemia.

eyes
- papilloedema
- cataract formation

Chvostek's sign
- tapping over parotid (facial nerve) causes facial muscles to twitch owing to neuromuscular excitability

Trousseau's sign
- blood pressure cuff on arm causes carpopedal spasm due to tetany of muscles in the hand

bones
- painful and fragile bones
- bones show radiolucency and erosions

brain
- mood changes, e.g. depression
- convulsions

heart
- cardiac arrhythmias

muscles
- spasms of skeletal muscles
- muscle cramps can cause tetany

hands and feet
- numbness
- paraesthesia (tingling, i.e. pins and needles)

kidneys fail to activate vitamin D, consequently PTH secretion increases. The bones are demineralized, hence the name 'osteodystrophy'.

Vitamin D deficiency can occur if the diet is deficient in vitamin D or the skin does not receive sunlight (e.g. elderly people who stay indoors and only see the sun through glass, or women in cultures who cover their skin). Deficiency of activated vitamin D causes hypocalcaemia as a result of impaired calcium absorption in the intestines.

Tertiary hyperparathyroidism

Tertiary hyperparathyroidism is a complication of secondary hyperparathyroidism. Very rarely, an adenoma can develop in the hyperplastic parathyroid glands caused by prolonged hypocalcaemia. If the underlying cause of hypocalcaemia is corrected, then hypercalcaemia can develop due to excess PTH secretion. This complication is diagnosed and treated as a primary parathyroid adenoma.

Hypoparathyroidism

Hypoparathyroidism is the deficiency of PTH resulting in hypocalcaemia. It causes the usual symptoms of hypocalcaemia (Fig. 8.12) but without osteomalacia. The main causes are listed below:
- Complication of thyroid or parathyroid surgery.
- Idiopathic hypoparathyroidism—an autoimmune disorder.
- Pseudohypoparathyroidism—congenital PTH resistance.

Osteoporosis

Osteoporosis is caused by reduced osteoblast activity not raised PTH. New bone is not formed and microfractures cannot be repaired so the bones become thin and brittle. Osteoporosis is caused by a deficiency of oestrogen or testosterone. It is very common in postmenopausal women, and signs of bone degeneration are seen in 100% of 80-year-old women.

The main treatments of osteoporosis are dietary calcium and vitamin D supplements and hormone replacement therapy (HRT; see Ch. 12). Calcitonin can be used, however it is very expensive and it must be injected subcutaneously.

- Explain what happens when blood calcium levels rise.
- Explain what happens when blood calcium levels fall.
- Describe the action of parathyroid hormone on the kidneys, bones and intestines.
- Describe the action of vitamin D on the kidneys, bones and intestines.
- How is vitamin D activated and which hormone controls this process?
- Describe the action of calcitonin on the kidneys, bones and intestines.
- What is the difference between primary, secondary and tertiary hyperparathyroidism? What disorders commonly cause these conditions?
- List the symptoms caused by primary hyperparathyroidism; do calcium levels rise or fall?
- List the symptoms caused by secondary hyperparathyroidism; do calcium levels rise or fall?
- List the symptoms and causes of hypoparathyroidism.

9. Endocrine Control of Growth

Growth hormone (GH) is often described as a pituitary hormone that acts directly on tissues instead of stimulating peripheral endocrine tissues like other anterior pituitary hormones. This view has been challenged by the discovery of insulin-like growth factors (IGFs) secreted by the liver in response to GH. The regulation of growth, therefore, follows the conventional pattern starting in the hypothalamus (described in Ch. 2).

Growth is a process that takes place at many levels. It can be defined as an increase in:
- Anabolism (e.g. protein synthesis).
- Cell size and number.
- Cell maturation and maintenance.
- Organ size.
- Body size or weight.

> **Important words:**
> **Anabolism:** the process of building large molecules from smaller ones
> **Epiphysis:** the end of a long bone (plural, epiphyses)
> **Epiphyseal growth plate:** an area of cartilage between the epiphysis and shaft of the bone that proliferates during childhood, resulting in elongation of the bone
> **Growth factor:** any chemical that stimulates cellular growth
> **Cell maturation:** when a cell differentiates to reach its final form

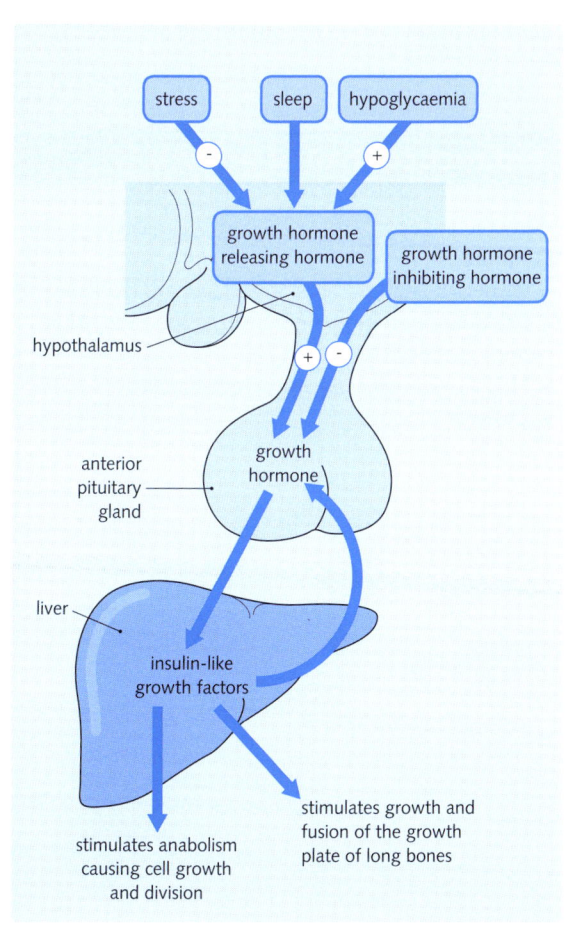

Acting through IGFs, GH stimulates all the processes listed above. By promoting anabolic processes, the cell increases in size. This promotes cell division and maturation causing the organ to grow. The cellular actions of GH begin before birth and continue throughout life, though the rate varies. The fastest rate of growth is in the fetus and neonate, however a growth spurt also occurs during puberty.

The growth of the body is limited by the epiphyses (growth plates) at the ends of the long bones. GH stimulates these plates to grow, causing the bones to lengthen and body height to increase. It also stimulates fusion of these growth plates preventing further growth.

After reading this chapter you should be able to:
- Understand the regulation of growth hormone.
- List the actions of insulin-like growth factors.
- Discuss other factors that affect growth.
- Explain how height is determined.
- Describe the disorders of excess or deficient growth hormone.

Fig. 9.1 Hormonal regulation of growth hormone.

Direct control of growth

Growth hormone (GH)

GH (also called somatotrophin) is a polypeptide that is secreted by the somatotroph cells in the anterior pituitary gland. Like many pituitary hormones, it is

97

Endocrine Control of Growth

synthesized as a precursor molecule (pre-progrowth hormone). Two cleavages release the active hormone. For more information about the anterior pituitary see Chapter 2.

Regulation of secretion
GH secretion is regulated by two hypothalamic releasing factors:
- Growth hormone releasing hormone (GHRH).
- Somatostatin (also called growth hormone inhibiting hormone or GHIH).

GHRH is released in a pulsatile manner, especially during deep sleep or hypoglycaemia, and GH release follows this pattern. Secretion of GH from the anterior pituitary gland is also regulated by the negative feedback of IGF-1 and other growth factors (Fig. 9.1).

Effects
GH promotes the growth and maintenance of most cells. It has an anabolic effect:
- Stimulates the uptake of amino acids.
- Stimulates the synthesis of proteins.

The majority of actions associated with GH are actually mediated by IGFs. GH promotes their synthesis, mainly in the liver but also in other tissues.

GH is transported in the blood bound to GH-binding protein. It acts via G-protein and Janus kinase (JAK) receptors on the cell surface of target cells.

Insulin-like growth factors
Insulin-like growth factors (IGFs or somatomedins) are polypeptide hormones that exist in two forms: IGF-1 and IGF-2. They resemble insulin in structure and they act through similar receptors. IGF-1 is more important as a stimulator of growth. IGFs are transported in the blood by a number of IGF-binding proteins.

Metabolic actions
Both IGF hormones have some insulin-like actions, e.g. increasing amino acid uptake and protein synthesis. However, they also oppose the actions of insulin on glucose by preventing glucose uptake and causing glycogen breakdown to raise blood glucose.

Growth actions
The increase in protein synthesis caused by the metabolic effects of IGF hormones causes cells to grow. This stimulates cell division and maturation causing organs and soft tissues to enlarge.

The growth of the long bones depends on the state of the epiphyseal growth plate. This plate is a layer of chondrocytes (cartilage cells) located between the end (epiphysis) and shaft (diaphysis) of the bone (Fig. 9.2). Before puberty, IGFs stimulate these chondrocytes to grow, divide, and mature into osteocytes (bone cells), allowing the bone to lengthen whilst maintaining a population of chondrocytes in the plate for further growth. During puberty, IGFs and sex steroids stimulate the chondrocytes within the plate to mature into osteocytes so that the epiphysis and diaphysis become fused together. The bone is no longer able to lengthen with further IGF stimulation so final adult height is reached.

Other growth factors
Growth in specific tissues is also stimulated by a number of growth factors, many of which are small peptides that act in a paracrine (local) manner. Their relationship to GH is not known. The actions and secretion of several such peptides are described in Fig. 9.3.

Indirect control of growth

Many factors apart from GH control growth, including:
- Genetics—tall parents often have tall children.
- Adequate nutrition—however, excess nutrition does not increase height.
- Health—chronic disease affects height.
- Other hormones.

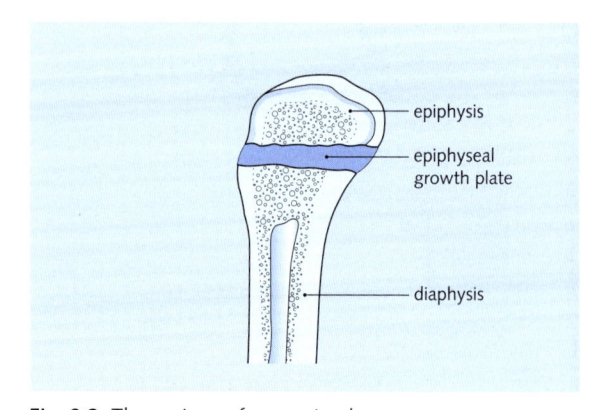

Fig. 9.2 The regions of a growing bone.

Indirect Control of Growth

The secretion of growth factors and their effects

Growth factor	Mode of delivery	Action on growth and development	Method and control of secretion
Nerve growth factor (NGF)	Paracrine	Induces neuron growth and helps to guide growing sympathetic nerves to organs they will innervate (may also act on the brain and aid memory retention)	Secreted by cells in path of growing axon; regulation of secretion not yet understood
Epidermal growth factor (EGF)	Paracrine and endocrine	Promotes cell proliferation in the epidermis, maturation of lung epithelium, and skin keratinization	Secreted by many cell types i.e. not only epidermal cells (EGF is also found in breast milk); regulation of secretion not yet understood
Transforming growth factors (TGF-α, TGF-β)	Paracrine	Stimulate growth of fibroblast cells; TGF-α acts similarly to EGF; TGF-β especially affects chondrocytes, osteoblasts, and osteoclasts	Secreted by most cell types but especially platelets and cells in placenta and bone; regulation of secretion not yet understood
Fibroblast growth factor (FGF)	Paracrine	Mitogenic effect in several cell types; may induce angiogenesis (formation of new blood vessels), which is essential for growth and wound healing	Secreted by most cell types; regulation of secretion not yet understood
Platelet-derived growth factor (PDGF)	Paracrine	Potent cell-growth promoter; chemotactic factor (involved in inflammatory response)	Secreted by activated blood platelets during blood vessel injury
Erythropoietin	Endocrine	Stimulates the production of erythrocyte precursor cells	Secreted by the kidney in response to falling tissue oxygen concentration
Interleukins (IL) (8 known)	Autocrine and paracrine	IL-1 stimulates B-cell proliferation and helper T cells to produce IL-2; IL-2 autoactivates helper T cells and activates cytotoxic T cells	IL-1 is secreted by activated macrophages; IL-2 is secreted by activated helper T cells

Fig. 9.3 The secretion of growth factors and their effects.

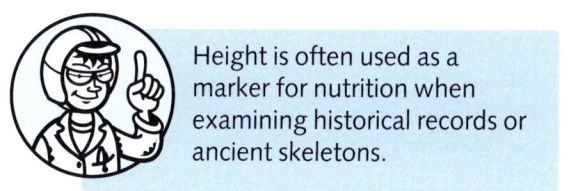

Height is often used as a marker for nutrition when examining historical records or ancient skeletons.

Other hormones
Growth problems can be caused by the abnormal secretion of a number of hormones, including:
- Insulin.
- Antidiuretic hormone (ADH).
- Parathyroid hormone and vitamin D.
- Cortisol.
- Sex steroids.
- Thyroid hormones.

Thyroid hormone is described in Chapter 3; it stimulates cell metabolism promoting cell growth and division especially in the skeleton and developing central nervous system (CNS). It also stimulates GH secretion from the pituitary.

Cortisol is described in Chapter 4; it inhibits pituitary GH secretion, so chronic ill health or stress can suppress growth.

Normal growth can only occur if the internal environment and nutrition are suitable.

Fetal growth
In the fetus, a hormone called placental lactogen is secreted from the placenta. It stimulates fetal cartilage development, and it acts in a similar manner to prolactin on the maternal mammary glands.

Thyroid hormones are essential for the development of the skeleton and CNS. A deficiency in the fetus or neonate results in cretinism.

Puberty
Sexual maturation and the pubertal growth spurt are described in Chapter 11. The main hormones involved are:

Endocrine Control of Growth

- Gonadotrophins—luteinizing hormone (LH) and follicle stimulating hormone (FSH).
- Sex steroids—oestrogen and testosterone.

> Tall stature is most often caused by tall parents; an excess of growth hormone is very rare.

Determination of height

A person's final height is determined simply by the rate and duration of growth. While the rate is determined by the growth hormones and factors already described, the duration is determined by their action on the bones.

During puberty, the epiphyseal growing plates at the end of the long bones begin to fuse. This fusion prevents further growth and, therefore, further height gain. Total fusion occurs between 18 and 20 years of age in males, and earlier in females.

The fusion of the epiphyseal plates is stimulated primarily by GH and sex steroids, however thyroid hormones also promote this effect. A simple increase in GH during puberty is not sufficient to increase final height since the bones simply mature faster and stop growing.

Only the bones that grow in this manner are prevented from responding to further GH. The jaw and skull can continue to grow past puberty; this effect is seen in GH excess. Ultimately height is determined by multiple genetic factors.

Disorders of growth

Excess of growth hormone

The rare excess of GH in children is called gigantism. The secretion occurs before puberty and fusion of the epiphyses, so the child is very tall for their age. Since the epiphyses also fuse at an earlier age, the child may have an unremarkable height in adulthood. Diabetes is very common in this group because of the opposing actions of GH and insulin on blood glucose.

An excess of GH is slightly more common in adults, in whom it is called acromegaly. The signs and symptoms are shown in Fig. 9.4. See also Fig. 17.7. The long bones can no longer lengthen, so there is no increase in height. However, the soft tissues and other bones can still grow, causing the distinctive features of this condition.

The excess of GH in any age group is usually caused by a somatotroph adenoma of the anterior pituitary gland. This can also cause other symptoms by compressing surrounding structures (see Ch. 2).

Diagnosis and treatment

Excess GH can be diagnosed by high IGF-1 levels, but the best test is to measure GH levels following an oral glucose tolerance test. GH levels should fall with the rise in glucose. Computed tomography (CT) or magnetic resonance imaging (MRI) scans can be used to confirm the presence of a functional pituitary adenoma.

Somatotroph adenomas are removed surgically. Following the operation, the patient must be routinely monitored for GH levels and other pituitary hormones throughout life. If surgery is not appropriate then bromocriptine or octreotide (a GHIH analogue) can be used.

Deficiency of growth hormone

The deficiency of GH in children is called dwarfism. It is detected by short stature along with either:

- Dropping between growth chart centiles (i.e. not following expected course).
- Being significantly shorter than mean parental height (MPH).

MPH is the average of the parents' height plus 7 cm in boys or minus 7 cm in girls. Final height is usually within 10 cm in either direction of the MPH.

The most common cause of dwarfism is a deficiency of GHRH from the hypothalamus; craniopharyngiomas can also be responsible. See Chapter 2 for more details.

Diagnosis and treatment

GH deficiency is diagnosed using a stimulation test. GH levels are measured after exercise or a dose of clonidine, both of which should raise GH levels. Insulin-induced hypoglycaemia is no longer routinely used in children due to the potential risk of severe hypoglycaemia. GH deficiency is treated with subcutaneous injections of synthetic GH before bedtime every night.

Disorders of Growth

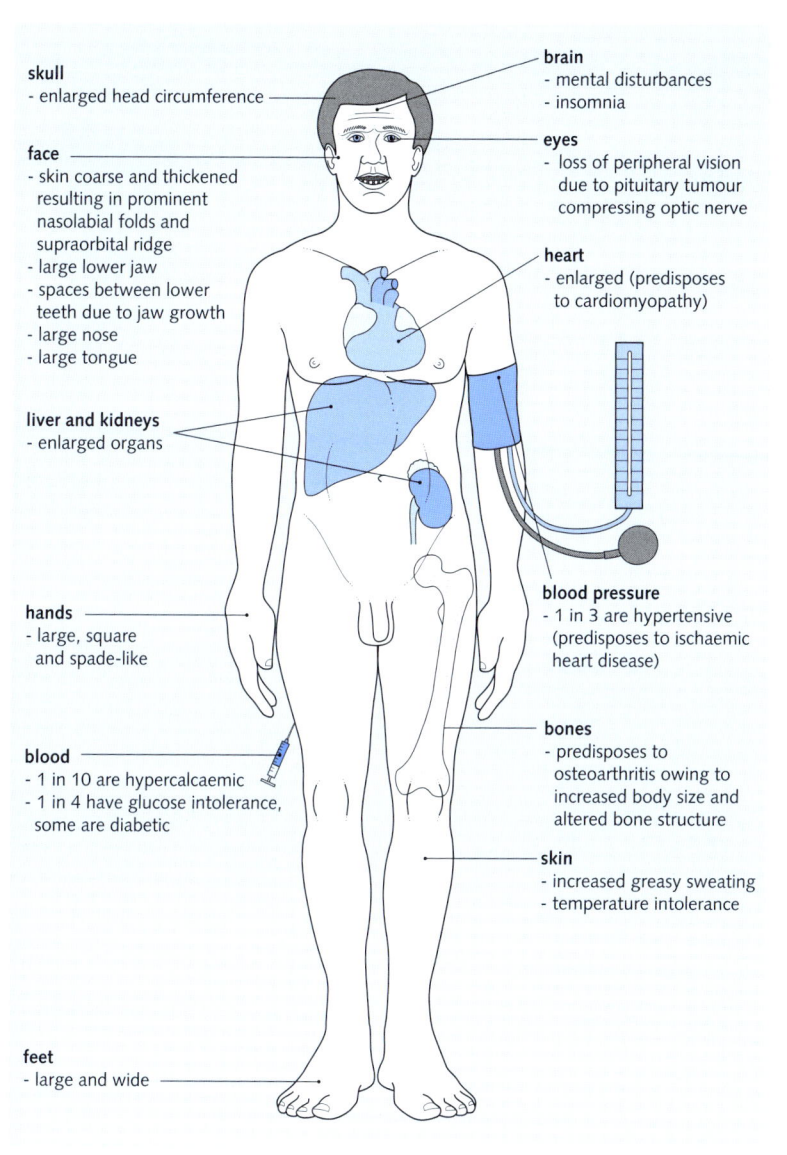

Fig. 9.4 The symptoms and signs of acromegaly.

skull
- enlarged head circumference

face
- skin coarse and thickened resulting in prominent nasolabial folds and supraorbital ridge
- large lower jaw
- spaces between lower teeth due to jaw growth
- large nose
- large tongue

liver and kidneys
- enlarged organs

hands
- large, square and spade-like

blood
- 1 in 10 are hypercalcaemic
- 1 in 4 have glucose intolerance, some are diabetic

feet
- large and wide

brain
- mental disturbances
- insomnia

eyes
- loss of peripheral vision due to pituitary tumour compressing optic nerve

heart
- enlarged (predisposes to cardiomyopathy)

blood pressure
- 1 in 3 are hypertensive (predisposes to ischaemic heart disease)

bones
- predisposes to osteoarthritis owing to increased body size and altered bone structure

skin
- increased greasy sweating
- temperature intolerance

- Describe the regulation of growth hormone and IGF secretion.
- What factors stimulate the secretion of growth hormone?
- Describe the actions of IGFs on cells.
- Describe the actions of IGFs on the growth plate of bones.
- What other factors affect growth?
- Name four peptide growth factors, the tissue that secretes them, and where they have an effect.
- How is height determined?
- How is the mean parental height calculated, and what are the boundaries of 'normal' height from the MPH?
- List the symptoms of acromegaly.
- How is acromegaly diagnosed?

10. Endocrine Disorders of Neoplastic Origin

The 'disorders' sections of several chapters have mentioned several multiple endocrine neoplasia (MEN) syndromes and ectopic hormones. This chapter describes these conditions in more detail.

MEN syndromes are endocrine tumours that originate in multiple sites. There are three patterns called MEN-I, MEN-IIa, MEN-IIb. These syndromes are rare, but they can cause tumours in young adults. They are usually inherited.

The term 'ectopic hormone' is used to describe hormone secretion from tissues that do not usually secrete that specific hormone. There are many tumours that secrete ectopic hormones, often in tissues that are significant components of the endocrine system.

After reading this chapter you should be able to:
- Discuss the theories of MEN aetiology.
- List the tumours caused by the three MEN syndromes.
- List the most common ectopic hormones.
- Describe the theories of ectopic hormone production.

MEN syndromes are rare, but they may be life threatening.

Multiple endocrine neoplasia syndromes

MEN syndromes are patterns of endocrine tumours that often occur at the same time. The tumours are rare, usually aggressive, and arise in multiple tissues; they occur earlier than single sporadic tumours. The underlying cause is probably genetic since these syndromes have a very strong (but not complete) autosomal dominant family history.

It is not known why MEN tumours commonly occur at the same time, but there are three theories:
- The affected tissues may share a common embryological origin that is affected by an abnormality.
- A circulating abnormal growth factor induces excess cell division in a number of endocrine tissues.
- A genetic abnormality is expressed in certain endocrine cells, resulting in tumour growth.

Three patterns of MEN have been described: MEN-I, MEN-IIa, and MEN-IIb (Fig. 10.1).

MEN-I (Wermer's syndrome)
The most common tumours are:
- Parathyroid hyperplasia.
- Pancreatic islet-cell tumours.
- Pituitary adenoma.

The pancreatic islet-cell tumours may secrete ectopic hormones—e.g. gastrin in Zollinger–Ellison syndrome (see Ch. 5). Thirty per cent of the very rare Zollinger–Ellison tumours are caused by MEN-I. Less commonly MEN-I is associated with:
- Parathyroid adenoma.
- Hyperplasia of thyroid parafollicular cells.
- Adrenal cortical hyperplasia.

MEN-IIa (Sipple's syndrome)
The main tumours of the MEN-II syndrome are:
- Phaeochromocytoma (often bilateral).
- Medullary cell carcinoma of the thyroid (often multi-focal).

Occasionally parathyroid hyperplasia can develop.

MEN-IIb
The very rare MEN-IIb is sometimes called MEN-III. However, the tumours that develop are similar to MEN-IIa. Two other types of tumour develop in the skin and submucosa throughout the body:
- Neuromas (tumours of neurons).
- Ganglioneuromas (tumours of neuronal ganglia).

Ectopic hormone syndromes

Ectopic hormones
Ectopic means out of place. The term ectopic hormone is used when a tissue secretes a hormone

103

Endocrine Disorders of Neoplastic Origin

Tumours and hormones associated with the three MEN syndromes

Syndrome	Associated tumours	Hormones secreted
MEN-I	Parathyroid hyperplasia, pancreatic islet-cell, pituitary adenomas	PTH, insulin, prolactin
MEN-IIa	Medullary carcinoma of the thyroid, phaeochromocytomas	Calcitonin, adrenaline
MEN-IIb	Medullary carcinoma of the thyroid, phaeochromocytomas, neuromas	Calcitonin, adrenaline

Fig. 10.1 The principal tumours and hormones secreted in the three multiple endocrine neoplasia (MEN) syndromes.

that it does not normally secrete. The hormone is released in an uncontrolled manner by a tumour (benign or malignant). The tumour can be:
- Endocrine tissue secreting unusual hormones.
- Non-endocrine tissue secreting any hormone.

Symptoms are usually caused by the excess of the ectopic hormone while the tumour is still small. Examples of syndromes caused by ectopic hormone secretion are listed in Fig. 10.2.

Examples of syndromes caused by ectopic hormone secretion

Syndrome	Hormone secreted by tumour cells	Tumour
Hypercalcaemia	Parathyroid hormone (PTH) or PTH-like peptide	Squamous cell carcinoma of the lung, breast carcinoma
Hyponatraemia	ADH	Oat cell carcinoma of the bronchus, some intestinal tumours
Hypokalaemia (symptoms of Cushing's syndrome caused by ACTH excess take longer to develop)	ACTH and ACTH-like peptides	Oat cell carcinoma of the bronchus, medullary carcinoma of the thyroid, thymic carcinoma, islet cell tumours
Gynaecomastia	Human placental lactogen	Carcinoma of the bronchus, liver, or kidney
Galactorrhoea	Prolactin	Carcinoma of the bronchus, hypernephroma
Polycythaemia	Erythropoietin	Hypernephroma, carcinoma of the uterus
Hypoglycaemia	Insulin (rare)	Hepatomas, large mesenchymal tumours
No syndrome	Calcitonin	Oat cell carcinoma of the lung

Fig. 10.2 Examples of syndromes caused by ectopic hormones. (ACTH, adrenocorticotrophic hormone; ADH, antidiuretic hormone; PTH, parathyroid hormone.)

Treatment

The tumours are treated in a similar manner to any symptomatic tumour, i.e.:
- Surgical removal.
- Irradiation.
- Chemotherapy.

Aetiology of ectopic hormones

The exact mechanism behind ectopic hormone release is not fully understood, and it may vary between tumours; there are two main theories:
- The tumour originates from cells that normally secrete small amounts of hormones—e.g. cells of the bronchial mucosa normally secrete adrenocorticotrophic hormone (ACTH) and anaplastic carcinoma of the lung secretes ectopic ACTH.
- Mutations associated with the transformation to neoplasia activate dormant genes resulting in ectopic hormone production.

Types of ectopic hormones

Ectopic hormones are almost always peptide hormones because their synthesis requires expression of only a single gene. Steroid hormones synthesis requires the expression of a complicated series of enzymes.

The ectopic hormone is often not an exact version of a normal hormone, e.g. breast cancer cells secrete PTH-like peptide. This would fit the second theory of aetiology since the normal processing enzymes may not be present.

- Describe the theories of MEN tumour formation.
- List the common tumours associated with each of the MEN syndromes.
- What features of an endocrine tumour suggest a MEN syndrome?
- Describe the theories behind ectopic hormone secretion.
- List four tumours known to secrete ectopic hormones, along with the relevant hormone.

105

11. Development of the Reproductive System

The development of the reproductive system begins 4 weeks after conception and continues through to puberty. The majority of the structures are derived from the middle embryological layer called the mesoderm; however, the cells that will give rise to the gametes (sperm and oocytes) are derived from the endodermal yolk sac.

Between weeks four and seven there are no clear differences between the male and female development. Differences begin to appear by the 7th week when the indifferent genitalia develop to form distinct reproductive systems for each sex.

Sexual development is arrested soon after birth until puberty. At puberty the gonads are reactivated by luteinizing hormone (LH) and follicle stimulating hormone (FSH) secretion, and sex steroids (e.g. oestrogen and testosterone) are produced. These cause secondary sexual development and the attainment of sexual maturity.

After reading this chapter you should be able to:
- Discuss how gender is determined.
- Visualize the embryological development of the male and female reproductive systems.
- Describe the changes that occur in males and females at puberty.

> **Important words:**
> **Ectoderm:** outer embryological layer that forms the skin and nervous system
> **Mesoderm:** middle embryological layer that forms many organs and the cardiovascular system
> **Endoderm:** inner embryological layer that forms the intestines and the germ cells
> **Mesenchyme:** support tissue derived from the mesoderm
> **Gonads:** the ovaries or testes

Embryological development of gender

Genetic determination of gender

The gender of a fetus is determined at conception by the sex chromosome in the sperm that fertilizes the oocyte (chromosome 23, which can be either X or Y). A single gene on the Y chromosome, called the sex-determining region (SRY), is responsible for the male (XY) phenotype. The protein transcribed by this is called testis-determining factor (TDF). The absence of this gene and protein results in a female (XX) phenotype.

Early indifferent development
Gonads

The SRY gene is not activated until the 7th week of gestation. For the first 6 weeks, development is identical in both sexes. The gonads begin to form during the 5th week. Three types of cells form the gonads (Fig. 11.1):
- Mesenchymal cells, developing support tissue from the mesoderm.
- Mesothelial cells, a type of mesenchyme that forms the lining of body cavities, e.g. peritoneal lining.
- Primordial germ cells, the developing gamete-producing cells.

The mesothelium and mesenchyme proliferate to form a bulge called the gonadal or genital ridge. This is found at the back of the developing abdominal cavity embedded in the mesonephros; these are ridges of tissue that act as primitive kidneys until the permanent kidneys develop.

The mesenchyme forms an inner medulla while the mesothelium forms an outer cortex. The cortex has finger-like projections that reach into the

> The mesonephros are evolutionary remnants of primitive kidneys similar to those found in amphibians. They regress when the permanent kidneys form.

107

Development of the Reproductive System

Fig. 11.1 The origins and fates of the cells that form the gonads.

The origins and fates of the cells that form the gonads

Cells	Origins	Structure at 6 weeks	Adult structure
Mesothelial cells	Mesodermal lining of the peritoneum	Cortex of the gonadal ridge and primary sex cords	Ovarian follicles or seminiferous tubules
Mesenchymal cells	Surrounding mesoderm	Medulla of the gonadal ridge	Leydig cells in the testes, and supporting stroma in the ovaries
Primordial germ cells	Migrate from the endodermal lining of the yolk sac	Primary sex cords	Gamete-producing male spermatogonia and female oocytes

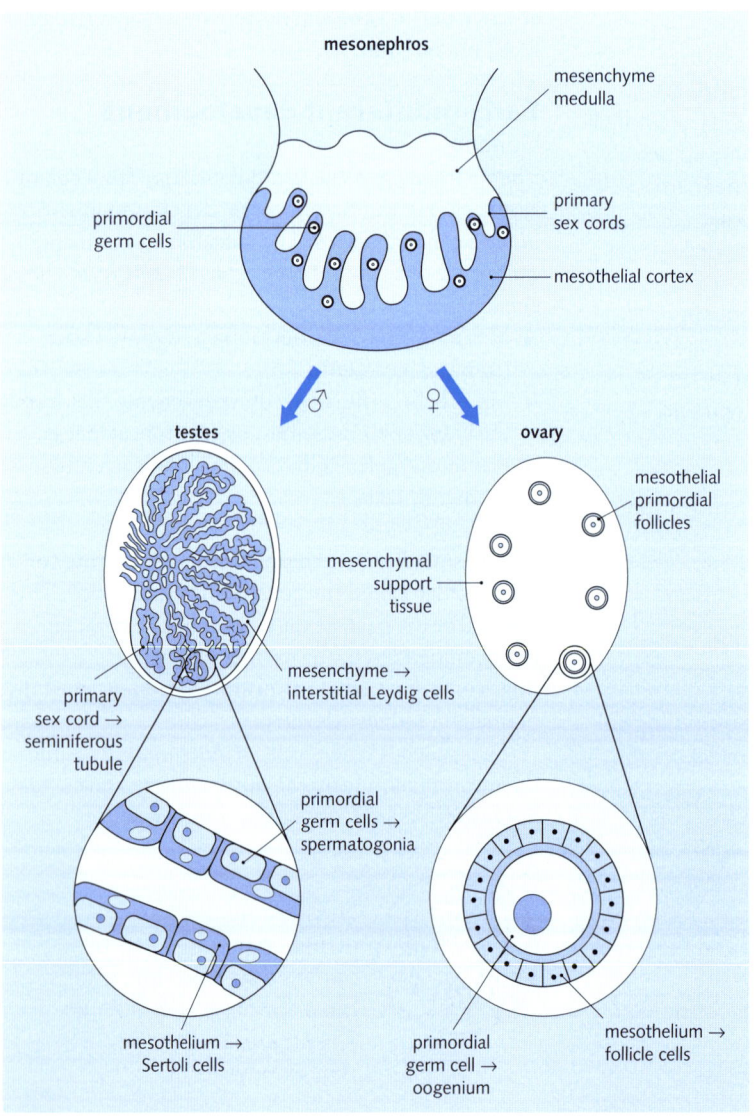

Fig. 11.2 The development of the male and female gonads.

108

Embryological Development of Gender

medulla; these are called primary sex cords. The primordial germ cells enter the genital ridge and join the primary sex cords. The indifferent gonads are now complete (Fig. 11.2).

Genital ducts
While the indifferent gonads are developing, two genital ducts are formed from the mesoderm (Fig. 11.3):
- Mesonephric (Wolffian) duct—this duct drains the urine from the mesonephros; it forms the male genital ducts, e.g. epididymis, vas deferens.
- Paramesonephric (Müllerian) duct—this funnel-ended duct lies laterally to the mesonephric duct; it forms the female genital ducts, e.g. fallopian tubes, uterus.

External genitalia
The external genitalia also begin to develop around the 4th week. Initially five mesenchymal swellings covered with ectoderm develop round the cloacal membrane; this membrane covers the end of the hindgut and urethra, both of which are endodermal structures. These five swellings are also shown in Fig. 11.4:
- One genital tubercle.
- Two urogenital folds.
- Two labioscrotal folds.

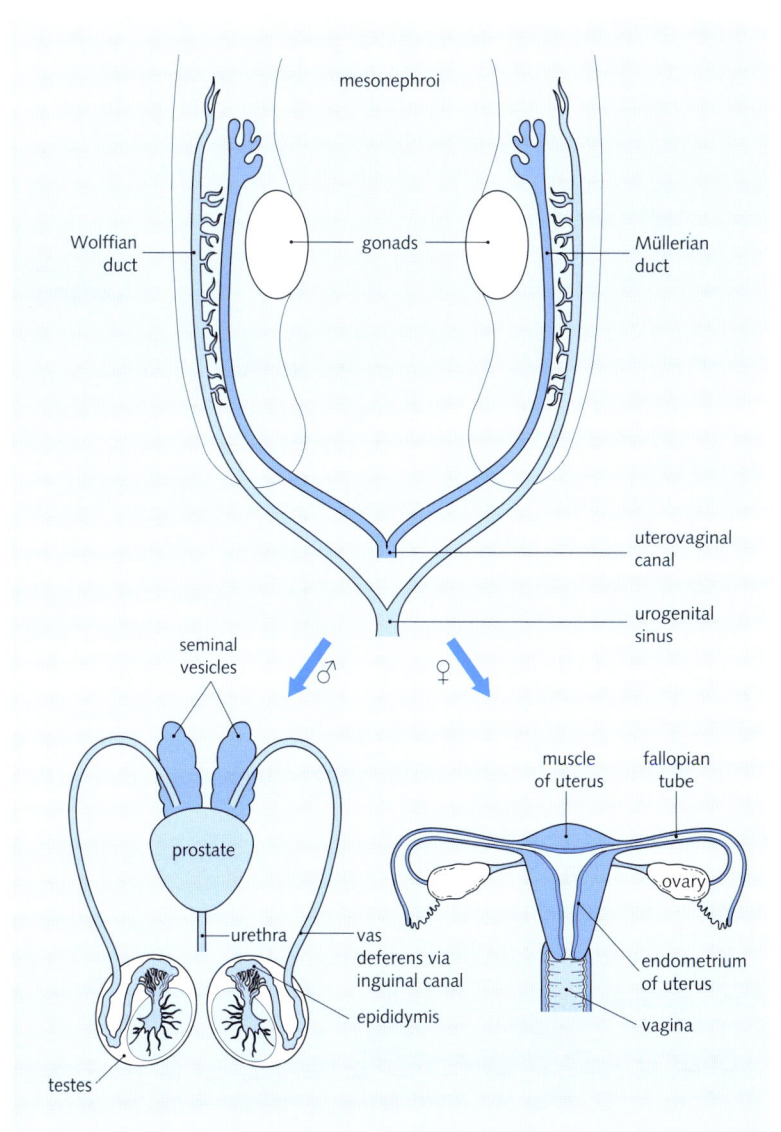

Fig. 11.3 The development of the male and female internal genitalia.

Development of the Reproductive System

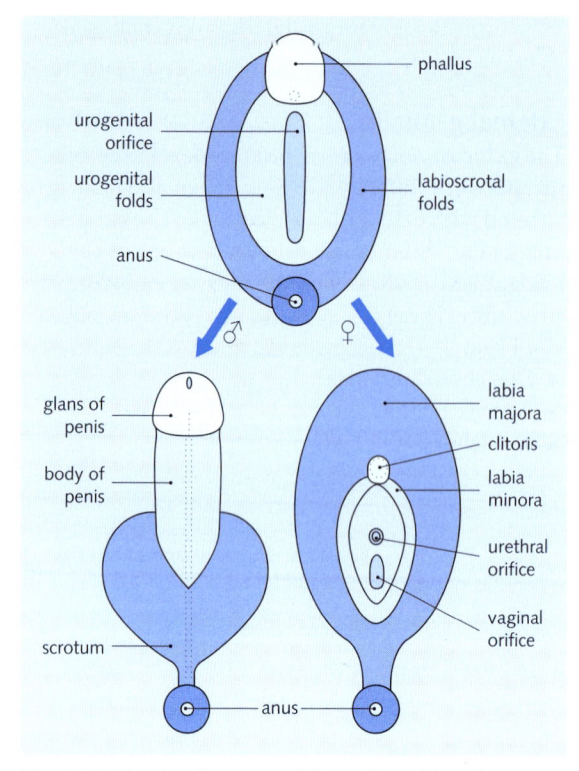

Fig. 11.4 The development of the male and female external genitalia.

The genital tubercle enlarges to form the phallus. The cloacal membrane divides into two and then ruptures to form:
- Urogenital orifice, though the vagina remains covered by the hymen.
- Anus.

A ligament forms between the indifferent gonad and the labioscrotal swellings through the inguinal canal. It is called the gubernaculum. It guides the descent of the testes into the scrotum and forms the round ligaments of the uterus and ovaries.

Male development
The testes
As the 6th week ends the SRY gene is transcribed and testis-determining factor (TDF) is produced. This factor acts on the mesothelial primary sex cords, which differentiate into the seminiferous cords. These separate from the surrounding mesenchyme to form the seminiferous tubules. The fate of the three cell types in the testes is shown in Fig. 11.2:
- Mesenchyme gives rise to interstitial (Leydig) cells.
- Mesothelium forms Sertoli cells.
- Primordial germ cells form spermatogonia.

The testes enlarge and separate from the mesonephros. They follow the path of the gubernaculum to reach the scrotum via the inguinal canal. The layers of the abdominal wall travel ahead of the testis into the inguinal canal, passing into the scrotum to form the layers of the scrotal wall and spermatic cord. A thin fold of peritoneum also descends, however, its connection with the abdomen (called the processus vaginalis) is lost. The small peritoneal sack remains in the scrotum as the tunica vaginalis. The testes finish their descent around the 7th month; they retain their abdominal blood vessels and lymphatic drainage. This migration and the formation of the scrotum is shown in Fig. 11.5.

Internal genitalia
The Leydig cells of the testes begin to secrete androgens (e.g. DHT, testosterone) from the 8th week. These androgens stimulate the further development of the Wolffian ducts, which differentiate into the:
- Epididymis.
- Ductus deferens.

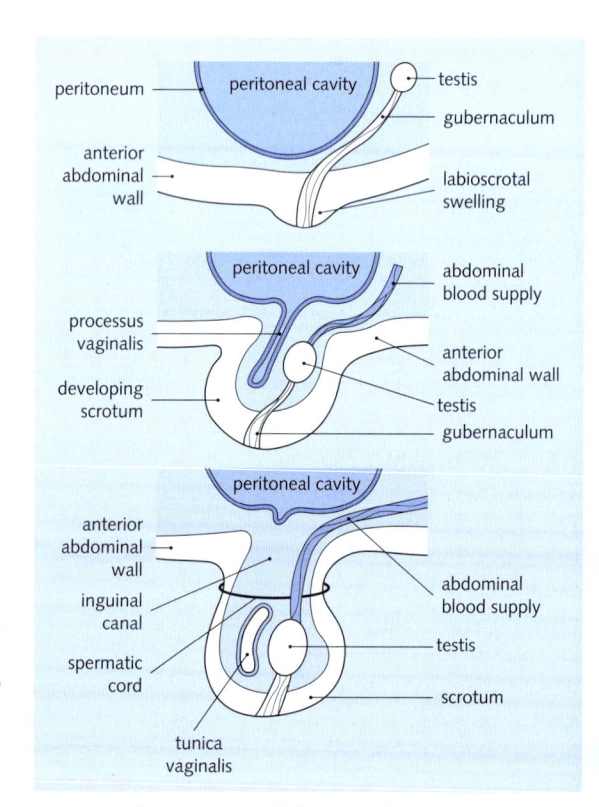

Fig. 11.5 The migration of the testes through the anterior abdominal wall.

- Seminal vesicles.
- Ejaculatory ducts.

The Sertoli cells also secrete Müllerian inhibiting substance (MIS)—a hormone that causes the Müllerian ducts to regress (see Fig. 11.3).

The prostate develops as endodermal outgrowths from the urethra surrounded by mesenchyme.

External genitalia

Testosterone secreted by the Leydig cells is also responsible for the development of the male external genitalia shown in Fig. 11.4:
- Phallus enlarges to form the glans (distal end) of the penis.
- Urogenital folds fuse ventrally to form the body of the penis.
- Labioscrotal folds fuse to form the scrotum.

All of these structures remain covered in ectoderm that forms the skin covering the penis. A clear ventral line called the scrotal and penile raphe remains from the fusion process. The ectoderm over the glans breaks down to form the foreskin (or prepuce), which remains attached to the glans preventing retraction of the foreskin until late infancy.

The urethra is an endodermal structure that is enclosed by the urogenital fusion. Failure of this process results in epispadias (see Ch. 13).

> Remember that in the anatomical position the penis is erect and pointing upwards. Ventral is underneath and dorsal is on top.

Female development
The ovaries

Female sex is determined by a number of genes on the X chromosome. SRY overrides these genes (e.g. the XXY genotype in Klinefelter syndrome is phenotypically male). The genes of both X chromosomes are needed for normal female development, therefore the XO genotype in Turner syndrome usually results in infertility and ovarian degeneration. The signals involved in female development have not been determined.

The mesothelial primary sex cords degenerate and secondary sex cords develop from the mesothelium.

The primordial germ cells migrate into these new cords, which then break up to form primordial follicles. The mesenchymal medulla forms the connective tissue stroma that supports these follicles. The primordial germ cells develop into oogonia, which undergo mitotic division to increase the number of germ cells. They enter the first prophase of meiosis before birth, after which further mitosis is not possible; there is no stem cell system equivalent to that found in males. The cells are called oocytes once the meiotic division has begun.

A primordial follicle is composed of:
- A single oocyte from the primordial germ cell.
- A single layer of follicular cells from the mesothelium, which surround the oocyte.

At birth about 750 000 primordial follicles are present; their meiotic division will only be completed many years later (see Fig. 11.2).

The ovaries separate from the mesonephros and become suspended in the pelvis by their mesentery.

> The ovum from which you were conceived was formed about 7 months before your mother's date of birth.

Internal genitalia

The internal genitalia develop due to the absence of testosterone and MIS; the unstimulated Wolffian duct regresses, while the Müllerian ducts develop. The funnel-shaped end nearest the ovary forms the:
- Fallopian (uterine) tubes.
- Uterine endometrium.

The other two layers (including the muscle) of the uterus are formed from surrounding mesenchyme (see Fig. 11.3).

The vaginal epithelium is derived from the endodermal urethra, and the other two layers develop from the surrounding mesenchyme. The hymen is formed from the cloacal membrane covering the urogenital orifice that the vagina develops from.

External genitalia

The external genitalia also develop due to the absence of testosterone (see Fig. 11.4):

- Phallus forms the clitoris.
- Urogenital folds do not fuse completely, and they form the labia minora.
- Labioscrotal folds do not fuse completely, and they form the labia majora.

The urogenital sinus is formed by the incomplete fusion of these two folds. The vagina and urethra open into this sinus. Skin covers the clitoris, labia majora, and labia minora. There is no breakdown of the skin covering the clitoris (called the prepuce) comparable to the development of the foreskin.

Development of the breast

The mammary glands (breasts) develop from apocrine sweat glands (i.e. those associated with hair follicles) in the mesenchymal layer directly beneath the skin; this accounts for their very superficial nature. Development is identical in males and females until puberty.

At the 4th week of development, a line of thickened ectoderm (skin) develops from the inguinal region to the axilla; this is called the mammary ridge. In humans, this ridge regresses except in the pectoral region, however failure of regression can cause extra nipples to form.

In the 6th week, single mammary buds develop as downgrowths into the mammary ridge on either side. Under the influence of placental hormones (e.g. human placental lactogen), they branch to form 15–20 lactiferous ducts. The surrounding mesenchyme develops into the fat and connective tissue of the breast. The functional glandular components of the lactiferous ducts in females are called mammary glands; they develop under the influence of oestrogen at puberty.

The nipple is formed by depression of the skin before birth. Shortly after birth the skin surrounding the nipple pit begins to grow, raising the nipple to form the usual shape.

Postnatal development

Soon after birth, the anterior pituitary gland begins to secrete gonadotrophins (LH and FSH) at roughly adult levels. Within 2 years, secretion declines rapidly to very low levels that are maintained until puberty. Sexual maturation is halted and the reproductive organs cease to develop.

Adrenarche

At about 8 years of age, the zona reticularis of the adrenal cortex reaches maturity. It begins to secrete adrenal androgens (see Ch. 4). This event is called adrenarche. These weak androgens contribute to the growth of pubic and axillary hair at puberty, especially in females. They do not cause puberty or the growth spurt.

Puberty

Puberty is the reactivation of gonadotrophin (LH and FSH) release after the dormancy of childhood. The age of pubertal onset varies widely between individuals (females 8–13 years; males 9–14 years). Puberty is characterized by a number of processes:

- Pubertal growth spurt.
- Development of secondary sexual characteristics.
- Achievement of fertility.
- Psychological and social development.

Gonadarche and the initiation of puberty

From an endocrine perspective, puberty is marked by the onset of pulsatile gonadotrophin release from the anterior pituitary gland during the night. Gonadotrophins stimulate the production of sex steroids (i.e. testosterone and oestrogen) from the gonads; the activation of the gonads is called gonadarche.

The onset of puberty is not fully understood, however, the CNS integrates a number of signals. A reduction in hypothalamic sensitivity to the negative feedback of the sex steroids causes the hypothalamus to secrete higher levels of gonadotrophin releasing hormone (GnRH) in a pulsatile manner. Secretion of growth hormone (GH), thyroid stimulating hormone (TSH) and adrenocorticotrophic hormone (ACTH) are also increased.

Body weight and puberty

Over the last few decades the onset of puberty has occurred at an increasingly young age. This change is often attributed to improved nutrition and rising body weight. In fact, achieving a body weight of 47 kg is a better predictor of the start of periods than age.

In recent years, a possible mechanism for this effect has been found. The hormone leptin is secreted by adipose tissue, and it is a hormonal indicator of body fat: higher levels of leptin are present with increasing body fat. Puberty cannot begin without leptin, but evidence suggests that it is only one of a number of factors.

The pubertal growth spurt
The earliest developmental event in puberty is an increase in growth velocity called the growth spurt. It occurs about 2 years earlier in females, giving a temporary height advantage. The initial rise in growth velocity is slight, so growth of the breasts or testes is usually noticed first.

The increase in growth rate is caused by increased GH and sex steroid secretion. Sex steroids also cause bone maturation. As the bones mature, the growing plates (epiphyses) fuse, preventing further growth. This fusion occurs 2 years earlier in females, giving males an extra 2 years of growth. This largely accounts for the increased height of adult males.

Puberty in the male
Puberty usually occurs between 9 and 14 years of age in boys, however, it is considered normal if it occurs between 9 and 16 years of age. Once the testes have developed, male pubertal changes are brought about by the secretion of androgens such as testosterone.

Testes development and early puberty
Growth of the testes is often the first sign of puberty noticed in boys. The increase is mainly due to proliferation of the seminiferous tubules under the influence of FSH. LH stimulates the interstitial Leydig cells to secrete testosterone. The scrotum becomes larger, thicker, and pigmented; pubic hair growth follows.

Spermatogenesis begins once the testes have enlarged and matured. The first ejaculation is often around 13–14 years of age.

> Clinically, male puberty has begun when the testes reach 4 mL. This is measured with an orchidometer.

Penile development and late puberty
The penis begins to enlarge after the testes about the same time the growth spurt is noticed. The penis doubles in size during puberty to reach an average size of 9.5 cm flaccid or 13.2 cm erect.

Facial and axillary hair growth usually start at about 15 years of age. The sebaceous glands in the skin are also activated, often causing acne.

The breaking of the voice is also a late feature. The larynx, cricothyroid cartilage, and laryngeal muscles enlarge to give an Adam's apple.

Puberty in the female
Puberty usually occurs between 8 and 13 years of age in girls, however, it is considered normal between 8 and 15 years of age. The changes caused by oestrogens and progesterone are shown in Fig. 11.6.

Breast development and early puberty
The development of breast buds is often the first sign of puberty noticed in girls. The breast then continues to grow under the influence of oestrogen while the ductal system develops (see Fig. 11.6). Pubic hair begins to grow about 6 months later.

Menarche and late puberty
The uterus begins to enlarge after the development of pubic hair. The onset of menstruation (periods) is called menarche. The mean age of menarche is

The changes caused by oestrogen and progesterone during female puberty

Oestrogen-mediated changes	Progesterone-mediated changes
Fat deposition and proliferation of the ductal system in the breasts, causing growth	Proliferation of the secretory lobules and acini in the breast
Growth of the vagina and maturation of the epithelium	Contribution to vaginal and uterine growth
Growth of the clitoris	Initiation of cyclical changes in endometrium and ovary

Fig. 11.6 The changes caused by oestrogen and progesterone during female puberty.

13 years, making it a late feature of puberty. In the ovary, follicular development begins and the first ovulation occurs 10 months after menarche, on average.

The menstrual cycle

The menstrual cycle is regulated by the interactions of a number of hormones. Pituitary FSH stimulates a group of ovarian follicles to develop, and these follicles release oestrogen. When oestrogen levels reach a threshold, they stimulate a surge of LH from the pituitary gland. This surge causes ovulation and development of the corpus luteum, which secretes progesterone. In the absence of pregnancy, the corpus luteum degenerates resulting in oestrogen and progesterone levels falling; this allows FSH secretion to rise restarting the cycle. This sequence of events is described in more detail in Chapter 12. The cycle may take some time to become regular.

- How is gender determined from a genetic and endocrine perspective?
- Describe the early development of the gonads along with the fate of the three types of cell.
- Describe the two ducts that form the male and female internal genital tracts.
- List the structures formed by these ducts in the male and female.
- Name the five swellings that form the external genitalia.
- What structures do these swellings form in the male and female?
- Describe the development of the breasts, especially the line along which they develop.
- Describe the endocrine changes that characterize puberty.
- What are the changes that occur during male puberty?
- What are the changes that occur during female puberty?

12. The Female Reproductive System

The female reproductive system must perform five main functions:
- Oogenesis and ovulation—production and release of oocytes (female gametes).
- Fertilization—allowing the sperm and oocyte to meet and fuse.
- Pregnancy—providing a suitable environment for the fetus to grow.
- Parturition—expelling the fetus with minimal trauma to the mother and baby.
- Lactation—providing the baby with nutrition.

After menarche (the start of periods) the female body prepares for pregnancy every month until menopause. This process is regulated by four main hormones (Fig. 12.1):
- Follicle stimulating hormone (FSH).
- Luteinizing hormone (LH).
- Oestrogen.
- Progesterone.

These hormones regulate all of the process described above. In the absence of pregnancy, the hormone levels rise and fall in the same pattern every month. These fluctuations, and the changes they cause, are called the menstrual cycle. Menstrual cycles continue until the menopause (p.136) when they cease.

Important words:
Oocyte: the female gamete that is released at ovulation and fertilized by sperm
Progestogen: a synthetic version of progesterone
Menstruation: vaginal bleeding at the beginning of the menstrual cycle also called a 'period'
Amenorrhoea: absence of menstruation
Menorrhagia: excessively heavy periods
Dysplasia: cellular changes that suggest malignancy

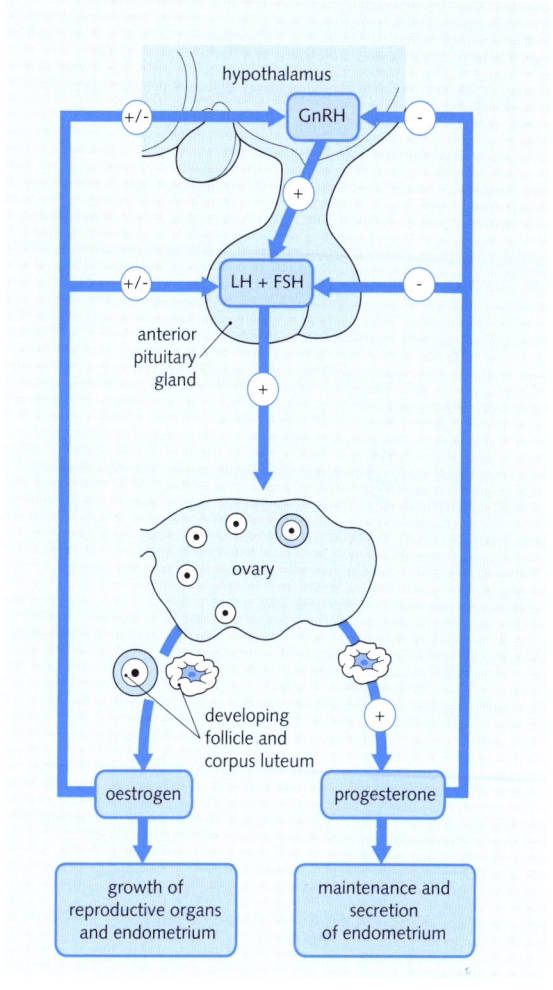

Fig. 12.1 Hormonal regulation of the female reproductive system. (FSH, follicle stimulating hormone; GnRH, gonadotrophin releasing hormone; LH, luteinizing hormone.)

After reading this chapter you should be able to:
- Visualize the anatomical structure of the female reproductive system.
- Describe the hormones that regulate the female reproductive system.
- Explain the sequence of events during the menstrual cycle.
- Discuss the common disorders of the female reproductive system.

The Female Reproductive System

Organization

The female reproductive system consists of six main components:
- **Ovaries**—produce oocytes and female sex steroids (e.g. oestrogens).
- **Fallopian tubes**—connect the ovaries to the uterus; they are the normal site of fertilization.
- **Uterus**—supports the implantation and development of the fetus.
- **Vagina**—normal site for the deposition of sperm.
- **Vulva**—the structures surrounding the introitus (external orifice of the vagina).
- **Breasts**—provide milk for the baby.

The ovaries lie inside the peritoneal cavity while all the other components lie outside; the ovarian end of the fallopian tubes open into this cavity. The peritoneum covers the uterus and fallopian tubes to form a fold called the broad ligament. Each of these components is discussed individually in the following sections. Their anatomical locations are shown in Fig. 12.2, and their blood supply, lymphatics and innervation are shown in Fig.12.3.

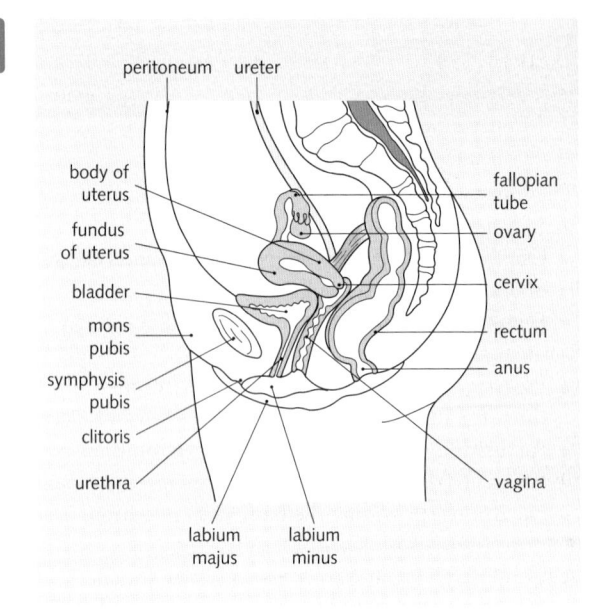

Fig. 12.2 Sagittal section of the female pelvis showing the locations of the reproductive organs.

Ovaries

The ovaries are two oval organs that produce oocytes (female gametes) and sex steroid hormones in

Blood supply, lymphatics and innervation of the female reproductive organs

Organ	Arterial supply	Venous drainage	Innervation	Lymphatic drainage
Ovaries	The ovarian arteries from the aorta via the suspensory ligaments	Forms the pampiniform plexus that drains into the ovarian veins in the suspensory ligaments	Autonomic nerves via the suspensory ligaments	Para-aortic lymph nodes
Fallopian tubes	Uterine and ovarian arteries	Uterine and ovarian veins	From the uterovaginal plexus and suspensory ligaments	Iliac, sacral, and aortic lymph nodes
Uterus	Uterine arteries, branches of the internal iliac arteries	Forms a plexus in the broad ligament that drains into the uterine veins	Uterovaginal plexus in the broad ligament	Iliac, sacral, aortic (and inguinal) lymph nodes
Vagina	Uterine arteries from the internal iliac arteries	Vaginal venous plexus that drains into the internal iliac veins	Uterovaginal plexus in the broad ligament	Iliac and superficial inguinal lymph nodes
External genitalia	Pudendal arteries	Pudendal veins	Pudendal and ilioinguinal nerves, S2–S4	Superficial inguinal lymph nodes
Breasts	Internal thoracic, lateral thoracic, and intercostal arteries	Axillary and internal thoracic veins	Intercostal nerves, mainly T4	Axillary and parasternal lymph nodes and the contralateral breast

Fig. 12.3 Blood supply, lymphatics and innervation of the female reproductive organs.

Organization

response to pituitary gonadotrophins (LH and FSH). The position of the ovaries is variable, but they usually lie lateral to the uterus, fixed to the posterior of the broad ligament. The opening of the fallopian tubes (infundibula) lies laterally to the ovaries, allowing oocytes to enter the infundibula at ovulation.

The ovaries are held on the surface of the broad ligament by a fold of peritoneum called the mesovarium, which is continuous with the germinal epithelium that forms their outer surface. Ovarian nerves, arteries, and veins enter the hilum of the ovary from the mesovarium. The relationship of the ovaries to the uterus, fallopian tubes and ligaments is shown in Fig. 12.4.

Two peritoneal ligaments attach to the ovary:
- Suspensory ligament of the ovary—from the mesovarium to the pelvic wall; it contains the blood vessels and nerves.
- Round ligament of the ovary—from the ovary to the fundus (top) of the uterus; it is a remnant of the upper section of the gubernaculum.

Microstructure

The ovary has three components (Fig. 12.5):
- Surface—simple cuboidal epithelium called the germinal epithelium.
- Cortex—composed of connective tissue stroma supporting thousands of follicles. Every month a group of preovulatory primary follicles begin to enlarge and synthesize steroid hormones. One of these follicles will eventually ovulate and form a postovulatory corpus luteum whilst the others will regress. Therefore, the cortex supports preovulatory, postovulatory and degenerating follicles.
- Medulla—composed of supporting stroma, and it contains a rich network of vessels and nerves that enter the ovary from the mesovarium.

The fallopian (uterine) tubes

The fallopian tubes are two 'J' shaped tubes lying in the upper border of the broad ligament. The tubes extend from their opening into the peritoneal cavity near the ovaries to the fundus (top) of the uterus where they open into the uterine cavity. As a result, there is a connection between the peritoneal cavity and the external reproductive tract. Infection can travel up this route, however the cervix acts as a barrier. The fallopian tubes are described in four parts, from lateral to medial:
- Infundibulum—the funnel shaped opening of the tube that is closely related to the ovary; it collects the oocytes from the surface of the ovary using ciliated, finger-like fimbriae.
- Ampulla—the widest section is where fertilization normally occurs.
- Isthmus—connects the ampulla to the uterus.
- Uterine section—the section as the tube penetrates the uterine muscle.

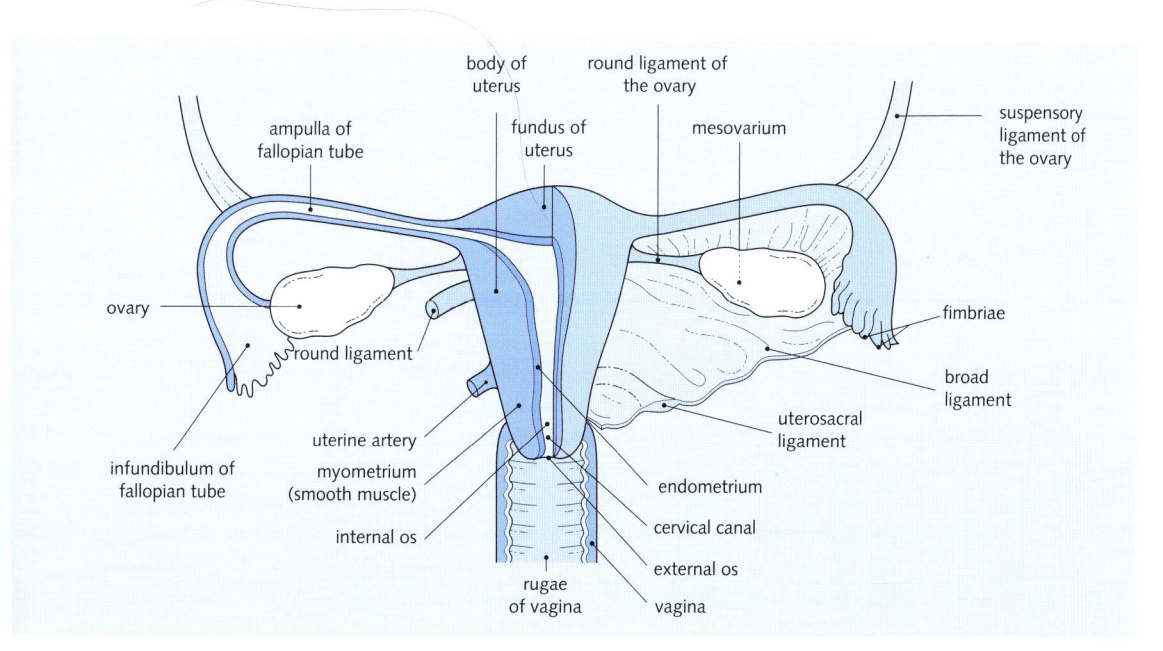

Fig. 12.4 Structure of the ovaries, fallopian tubes and uterus.

The Female Reproductive System

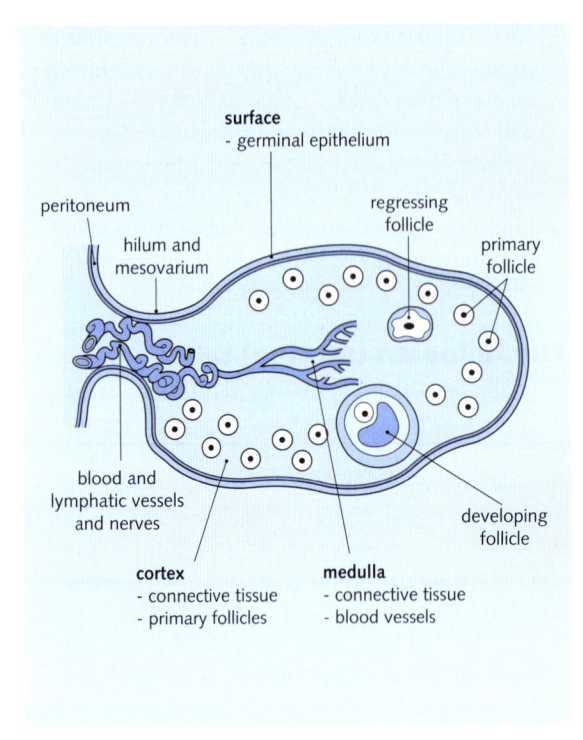

Fig. 12.5 Microstructure of an ovary.

The fallopian tubes are lined by ciliated and secretory cells that waft the oocyte towards the uterus and supply it with nutrients. Two layers of spiral muscle surround this lining and help move the oocyte and sperm by peristalsis. These muscles are sensitive to sex steroids (e.g. oestrogen) so that motility is most rapid when sex steroid levels are highest. The 'morning after pill' uses oestrogen to increase this motility so that the oocyte is ejected before the uterus is ready for implantation, thus preventing pregnancy. If motility is slow there is a risk of ectopic pregnancy in which implantation occurs in the fallopian tube. The relationship of the fallopian tubes to the uterus, ovaries and broad ligament is shown in Fig. 12.4.

> At ovulation the oocyte briefly enters the peritoneal cavity by breaking through the germinal epithelium. The opening of the fallopian tubes is in this cavity, so there is a direct connection to the outside world via the uterus and vagina.

Uterus and cervix

The uterus is a pear-shaped, muscular organ that can enlarge greatly to accommodate the growing fetus. It is lined by a specialized epithelium called the endometrium. The fallopian tubes join superiorly, and the vagina is inferior. These relationships are shown in Fig. 12.4.

The uterus is described in three sections:
- Fundus—above the entry point of the fallopian tubes.
- Body—this is the usual site of implantation.
- Cervix—the lower third; links the uterus and vagina.

The cylindrical cervix is structurally and functionally distinct from the rest of the uterus. The junction of the cervix with body is called the internal os and that with the vagina, the external os. The passage between these two junctions is called the endocervical canal. The ureters pass 1 cm lateral to the internal os on either side; this relationship is important when considering cervical carcinoma. The external os can be visualized in the conscious patient using a speculum, and the surrounding cells are sampled in a smear test.

The cervix and all structures superior are sterile areas. This sterility is maintained by the frequent shedding of the endometrium, thick cervical mucus and the narrow external os of the cervix.

Support and ligaments of the uterus

A number of structures hold the uterus in position and prevent prolapse into the vagina. The main support is derived from the pelvic floor formed by the levator ani, coccygeus and pubococcygeus muscles. The other structures are called 'ligaments' though most of them are not actually ligaments. The functions, relations and locations of these structures are common exam questions:

- Broad ligament—a double layer of peritoneum that surrounds the uterus with the fallopian tubes forming its superior border; it does not support the uterus.
- Round ligament—from the uterus body (anteroinferiorly to the insertion point of the fallopian tubes) to the labia majora through the inguinal canal; it is the remnant of the lower sections of the gubernaculum. It holds the uterus in an anteverted position. (The fundus lies superior and anterior to the cervix.)
- Uterosacral ligaments—from the cervix either side of the rectum to the piriformis muscle

over the sacrum; these structures **support** the uterus.
- Transverse cervical (cardinal) ligaments—from the cervix and superior vagina to the lateral pelvic walls; these structures support the uterus.
- Pubocervical ligaments—from the cervix to the pubis bone; they do not support the uterus.

Microstructure of the uterus
The body and fundus are composed of three tissue layers (Fig. 12.6):
- Serosa—the peritoneal covering.
- Myometrium—the thick smooth muscle layer; it is sensitive to hormones, e.g. oxytocin.
- Endometrium—the inner lining of the uterus that varies through the menstrual cycle; it is sensitive to hormones, e.g. oestrogen and progestogen.

The endometrium is further divided into two layers:
- Deep basal layer—this changes little through the menstrual cycle and is not shed at menstruation.
- Superficial functional layer—a hormone-sensitive layer that proliferates in response to oestrogen and becomes secretory in response to progesterone; it is shed at the end of the menstrual cycle and regenerates from cells in the basal layer.

The arterioles of the superficial endometrial layer lie alongside the glands of the endometrium. They have a characteristic spiral appearance. As progesterone levels fall at the end of the menstrual cycle they respond by constriction. The superficial layer becomes ischaemic and undergoes necrosis; this causes the shedding and haemorrhage that is menstruation.

Microstructure of the cervix
The cervix consists mainly of collagen and small amounts of smooth muscle. The columnar epithelium lining the endocervical canal secretes mucus that changes in consistency during the menstrual cycle. Oestrogen promotes a watery mucus that allows sperm to pass. Progesterone causes the production of a viscous mucus that is hostile to sperm.

The cervix is divided into two sections:
- Endocervix—superior, related to the uterus body.
- Ectocervix—inferior, related to the vagina.

The anatomical definition of this division is different from the histological definition (Fig. 12.7). The anatomical division is generally located superiorly to the histological division.

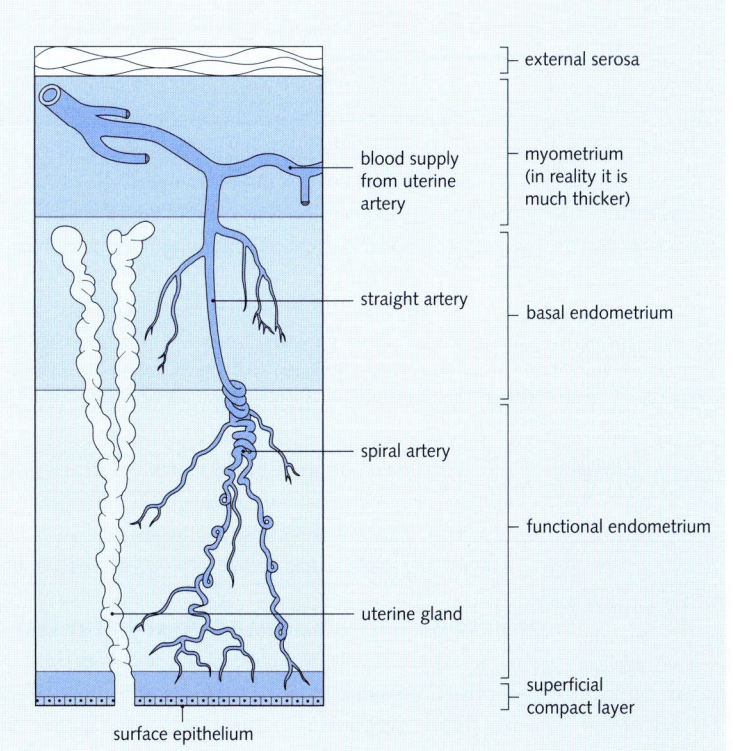

Fig. 12.6 Microstructure of the uterus.

	Endocervix	Ectocervix
Anatomy	Above the internal os; covered by peritoneum anteriorly	Below the internal os; not covered by peritoneum anteriorly
Histology	Columnar endometrial epithelium; does not menstruate	Stratified squamous vaginal epithelium; does not menstruate

Fig. 12.7 Comparison of the endocervix and ectocervix.

During puberty, high oestrogen levels cause the columnar epithelium of the cervix to extend beyond the external os into the vagina; this is called cervical ectopy. The acidic vaginal pH induces squamous metaplasia (changing to squamous epithelium) of the columnar epithelium. The cells that change are within an area called the transformation zone, and they are susceptible to dysplasia (pre-cancerous changes). Cervical smears take samples of these cells to detect early signs of dysplasia allowing curative treatment (see p. 138).

Vagina

The vagina is a 9 cm long muscular tube that runs upwards and backwards from the external genitalia (vulva) to the cervix. The urethra and bladder are anterior and the rectum posterior. In most women the cervix is inserted into the anterior wall of the vagina at an angle of 90°; this is called the anteverted position of the uterus.

The cervix projects into the vagina creating a small dome with the external os at the centre. The vaginal lumen around the cervix is divided into anterior, posterior, and two lateral fornices. The posterior fornix is the deepest and it is covered by peritoneum on its internal surface. This is an important relationship since an object that penetrates this area (e.g. during a 'backstreet' abortion) will enter the abdominal cavity, potentially causing an infection.

The word vagina is Latin for 'scabbard'.

Microstructure

The structure of the vaginal wall allows expansion during intercourse and childbirth. The wall of the vagina is composed of four layers:

- Stratified squamous epithelial lining for protection.
- Elastic lamina propria.
- Fibromuscular layer (two layers of smooth muscle).
- Fibroelastic adventitia.

The wall contains few sensory fibres and no glands; the lining epithelium is lubricated by cervical mucus. During sexual arousal the vagina is further lubricated by the secretions from Bartholin's glands next to the introitus (external vaginal orifice) and the transudation of fluid across the vaginal epithelium.

Unlike the cervix, uterus and fallopian tubes, the vagina is a non-sterile area; the main organisms present are the commensal *Lactobacillus vaginalis*. After puberty, oestrogen stimulates the cells of the vaginal epithelium to secrete glycogen. The lactobacilli digest this glycogen to release lactic acid and thus lower the pH of the vagina below 4.5; this prevents infection by other organisms. Other commensal organisms present include *Candida* and *Escherichia coli*. As in the gastrointestinal tract, antibiotics can disrupt the flora to cause overgrowth and infections such as candidiasis (thrush). Low oestrogen levels can also result in infection.

External genitalia (vulva)

The appearance of the vulva is shown in Fig. 12.8. The introitus or external vaginal orifice opens into the vestibule, which is the area between the labia. The short urethra opens anterior to the introitus within the vestibule. The labia minora are two hairless folds of skin that surround the vestibule; they fuse anterior to the urethral opening to form the prepuce (hood) of the clitoris. Beneath the prepuce lies the clitoris; this is composed of two erectile corpora cavernosa that become engorged with blood upon sexual stimulation.

The labia minora lie within two larger, hair-bearing skin folds called the labia majora; the labia majora fuse posteriorly and extend anteriorly to the mons pubis. The mons pubis is a fat pad covered in pubic hair at the anterior of the vulva.

Two pairs of small glands are located either side of the introitus called the vestibular and Bartholin's glands. During sexual arousal they secrete a lubricating mucus into the vestibule via small ducts.

Female breasts (mammary glands)

The breasts lie in the superficial fascia over the pectoral muscles on the anterior of the chest. They are very superficial structures composed largely of fat; the size and shape varies between individuals.

Oogenesis

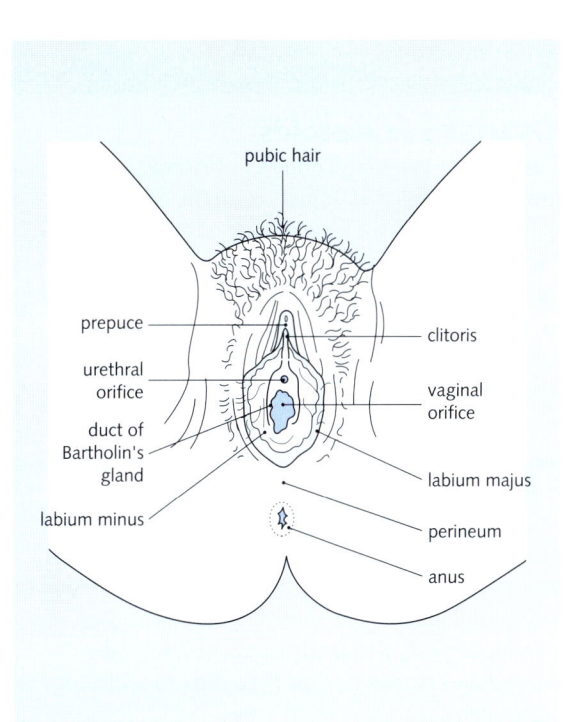

Fig. 12.8 Structure of the vulva.

They extend towards the axilla and this axillary tail must be checked on examination. The pigmented skin around the nipple is called the areola. In white-skinned women it permanently changes from pink to brown during pregnancy.

Microstructure
Embedded in the fatty tissue of the breast there are 15–20 independent glandular lobules with openings on the nipple. Each opening is a lactiferous duct that forms a lactiferous sinus beneath the areola. Within the breast, the lactiferous ducts branch extensively to end in secretory acini potentially capable of secreting milk. The ducts and lobules are surrounded by myoepithelial cells that contract in response to oxytocin and expel the milk on stimulation of the nipple. The internal structure of the breast is shown in Fig. 12.9.

Oogenesis

Meiosis in the female fetus
Oogenesis is the process by which haploid (23 chromosome) **oocytes** are formed from diploid (46 chromosome) stem cells called **oogonia**. This process requires a meiotic division that is begun before birth and ends when the oocyte is fertilized.

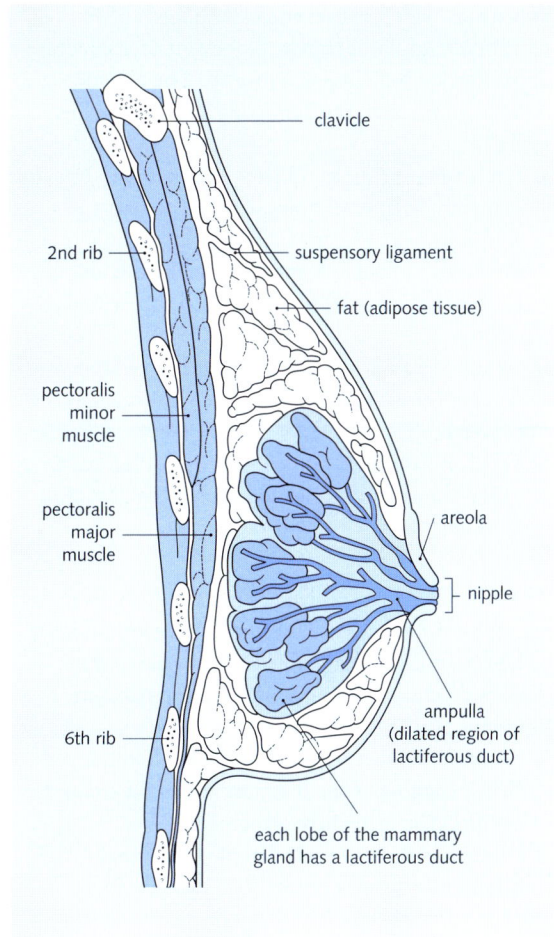

Fig. 12.9 Internal structure of the breast.

Oogonia are ovarian stem cells derived from the primordial germ cells in the yolk sac (see Ch. 11); initially they divide by mitosis to increase their numbers. In the second trimester of pregnancy all the oogonia develop into **primary oocytes** and enter the prophase of the first meiotic division. The meiotic division is arrested at this stage until menarche. Since no oogonia remain, further mitotic divisions to increase the number of oocytes are not possible.

A single layer of flat granulosa cells surround each primary oocyte to form a **primordial follicle**. The follicles are located within the ovarian cortex.

Meiosis and follicle development in the menstrual cycle
The primordial follicles and primary oocytes remain unchanged until puberty and menarche. Once the

121

menstrual cycle has become established, FSH secretion stimulates the development of a selection of primordial follicles each month. The chosen follicles go through the following stages (See Fig. 12.13):

- Unilaminar primary follicle—the primary oocyte and a single layer of granulosa cells enlarge in size.
- Multilaminar primary follicle—the granulosa cells divide to form layers and the zona pellucida (glycoprotein shell) forms around the primary oocyte.
- Secondary follicle—the surrounding ovarian cortex forms the secretory theca interna and theca externa.
- Graafian (tertiary) follicle—a fluid filled cavity called the antrum develops within the granulosa cell layer.

Once the Graafian follicle has matured, the primary oocyte completes the first meiotic division under the influence of LH. Unlike most cell divisions this process produces two cells of differing size and function:

- Secondary oocyte—the mature haploid oocyte that is capable of fertilization; it has the majority of the cytoplasm and organelles.
- First polar body—a small haploid cell that degenerates; it has virtually no cytoplasm.

The secondary oocyte begins the second meiotic division immediately however it is arrested at metaphase II until fertilization.

The secondary oocyte is released at ovulation surrounded by two layers:

- Zona pellucida—the glycoprotein layer.
- Corona radiata (also called the cumulus oophorus)—a covering of granulosa cells from the follicle.

Meiosis at fertilization

The second meiotic division is not completed unless the secondary oocyte is fertilized. The calcium influx caused by the fusion of the sperm and secondary oocyte stimulates the completion of this division. Two haploid cells are produced:

- Female pronucleus—the functional gamete; it has the majority of the cytoplasm and fuses with the male pronucleus (see Ch. 14).
- Second polar body—another small haploid cell that degenerates; it has virtually no cytoplasm.

Hormones

Ovarian sex steroids

The ovaries produce a number of steroid hormones in response to gonadotrophins from the anterior pituitary. The main hormones produced are:

- Oestrogens, e.g. oestradiol.
- Progestogens, e.g. progesterone.
- Androgens, e.g. androstenedione.

Oestrogens

Oestrogens are secreted at the start of the menstrual cycle in response to LH and FSH. Their synthesis takes place in the developing ovarian follicle, requiring both the thecal and granulosa cells. The theca interna secretes androgens in response to LH. LH activates the enzyme that converts cholesterol to pregnenolone (i.e. the first step in steroid production), however the thecal cells lack the aromatase enzyme necessary to convert androgens to oestrogens.

The majority of androgens cross the basement membrane into the granulosa cells. FSH activates the aromatase enzyme produced by the granulosa cells allowing the thecal androgens to be converted to oestrogens (mainly oestradiol-17β). The process of oestrogen synthesis is shown in Fig. 12.10. After ovulation, oestrogens are produced by the corpus luteum formed from the follicle.

Oestrogens are transported in the blood bound to sex-hormone-binding globulin (SHBG) and albumin. They act via intracellular receptors in the target cells. Oestrogens act on the anterior pituitary and hypothalamus to provide feedback (usually inhibition), which regulates the system. The actions of oestrogens are shown in Fig. 12.11; the main actions are:

- Development of the reproductive organs and secondary sexual characteristics.
- Proliferation of the functional layer of uterus endometrium.

Progestogens

Progestogens are secreted in the second half of the menstrual cycle by the corpus luteum. This structure is formed by the transformation of the granulosa cells in the follicle after ovulation; LH maintains the secretory activity of these cells. The main progestogen is progesterone, which is synthesized from cholesterol in just two steps. During pregnancy, progesterone production is taken over by the placenta.

Hormones

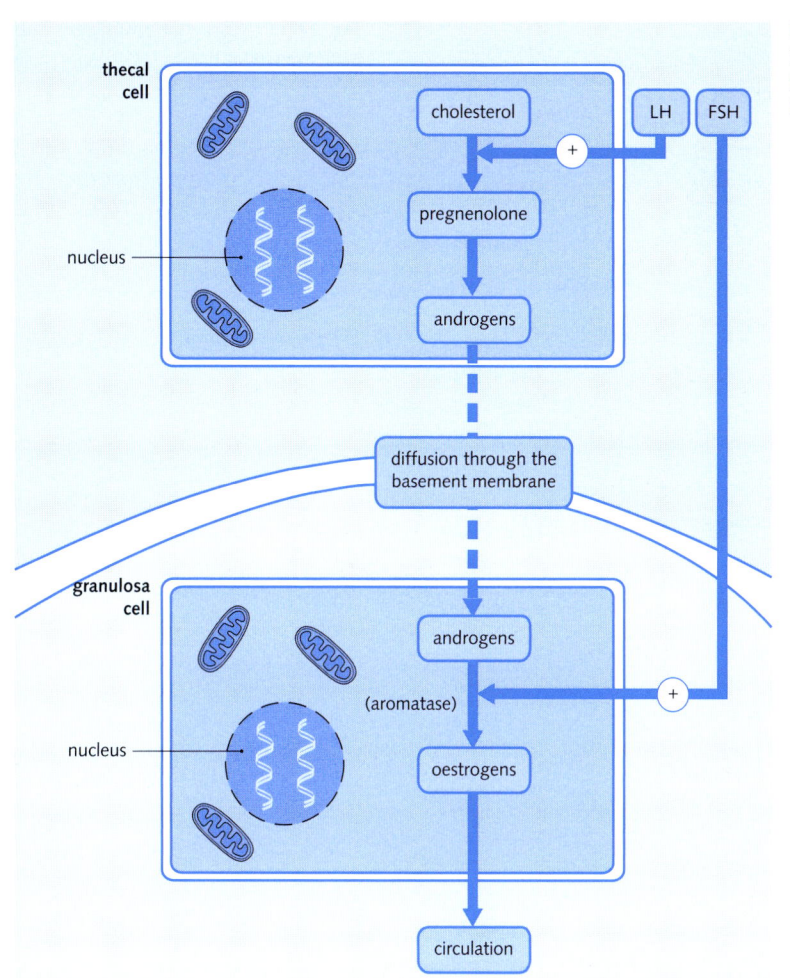

Fig. 12.10 Synthesis of oestrogens by the developing follicle. (FSH, follicle stimulating hormone; LH, luteinizing hormone.)

Progesterone is transported in the blood bound to corticosteroid-binding globulin (CBG) and albumin. It acts via intracellular receptors in the target cells. Progesterone acts on the anterior pituitary and the hypothalamus to provide negative feedback. The actions of progesterone are shown in Fig. 12.11; its main actions are:
- Maintenance of the uterine endometrium.
- Stimulation of uterine secretions.

Androgens
The androgens are precursors of oestrogens, however small quantities are released systemically. They act with adrenal androgen to promote pubic and axillary hair growth during puberty.

Control of ovarian steroid production
Ovarian steroids are regulated in a similar manner to many other major hormones (as shown in Fig. 12.1).

Gonadotrophin releasing factor (GnRH) is synthesized by the hypothalamus and transported to the anterior pituitary gland in the portal veins. Here, it acts on gonadotroph cells to stimulate the release of gonadotrophins (i.e. LH and FSH). This process is described in more detail in Chapter 2.

Gonadotrophins reach the ovaries in the blood and stimulate the release of the ovarian sex steroids. Both LH and FSH stimulate enzymes involved in oestrogen synthesis. LH also allows the formation and maintenance of the corpus luteum that synthesizes progestogens.

Oestrogens and progestogens feedback to the anterior hypothalamus to regulate their release. This feedback is usually inhibitory and it prevents excess secretion. Before ovulation the feedback becomes positive triggering the surge in LH and FSH release that causes ovulation.

123

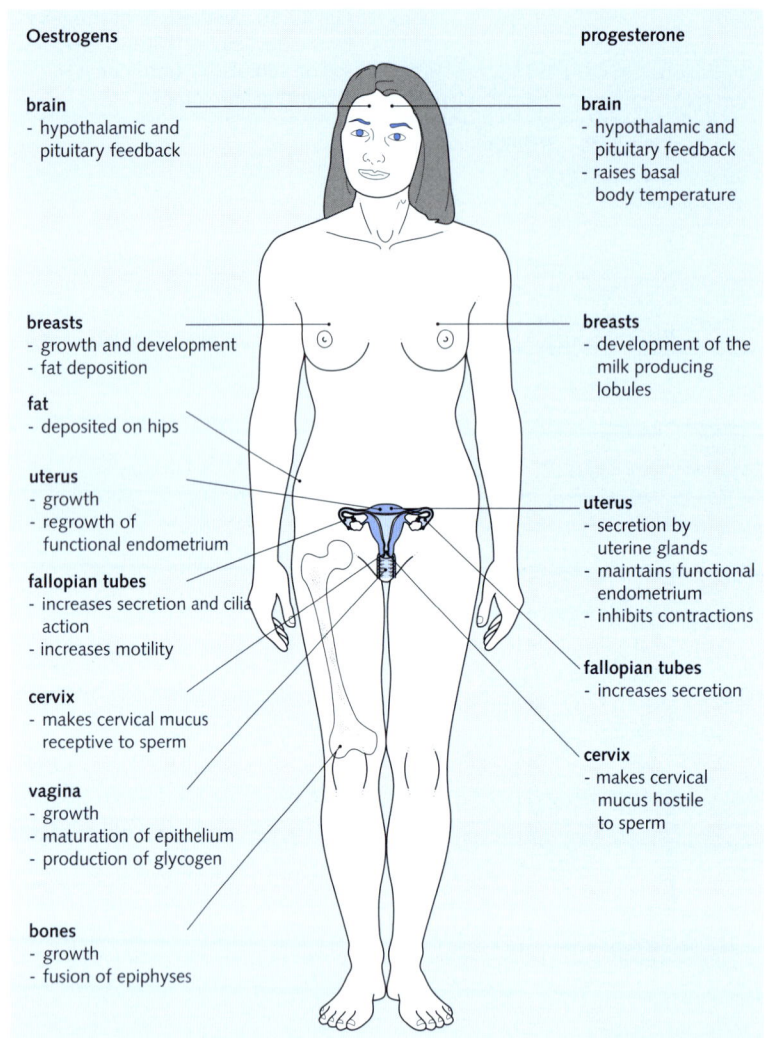

Fig. 12.11 Actions of oestrogens and progesterone.

Oestrogens

brain
- hypothalamic and pituitary feedback

breasts
- growth and development
- fat deposition

fat
- deposited on hips

uterus
- growth
- regrowth of functional endometrium

fallopian tubes
- increases secretion and cilia action
- increases motility

cervix
- makes cervical mucus receptive to sperm

vagina
- growth
- maturation of epithelium
- production of glycogen

bones
- growth
- fusion of epiphyses

progesterone

brain
- hypothalamic and pituitary feedback
- raises basal body temperature

breasts
- development of the milk producing lobules

uterus
- secretion by uterine glands
- maintains functional endometrium
- inhibits contractions

fallopian tubes
- increases secretion

cervix
- makes cervical mucus hostile to sperm

Other ovarian hormones
Inhibin and activin
Inhibin and activin are polypeptide hormones secreted by the granulosa cells of the ovarian follicles. Inhibin stimulates androgen synthesis but inhibits conversion to oestrogens while activin inhibits androgen production but stimulates conversion to oestrogens. Together they regulate local sex steroid levels and the balance between oestrogens and androgens.

Relaxin
This is a polypeptide hormone secreted by the corpus luteum and placenta. It prepares the body for childbirth by causing cervical softening and relaxation of pelvic ligaments.

The menstrual cycle

The menstrual cycle is the process by which the female prepares for possible fertilization of the oocyte. The cycle lasts 28–32 days and begins on the first day of menstruation (also called a 'period'). A number of changes occur in the ovaries and endometrium; these are regulated by hormones. The hormonal, ovarian and endometrial changes are shown in Fig. 12.12.

The cycle is divided into two stages, each lasting about 14 days. Between these stages (about the 14th day) ovulation occurs.

The first half of the cycle
The first half of the cycle begins on the first day of menstruation and lasts until ovulation. The length of

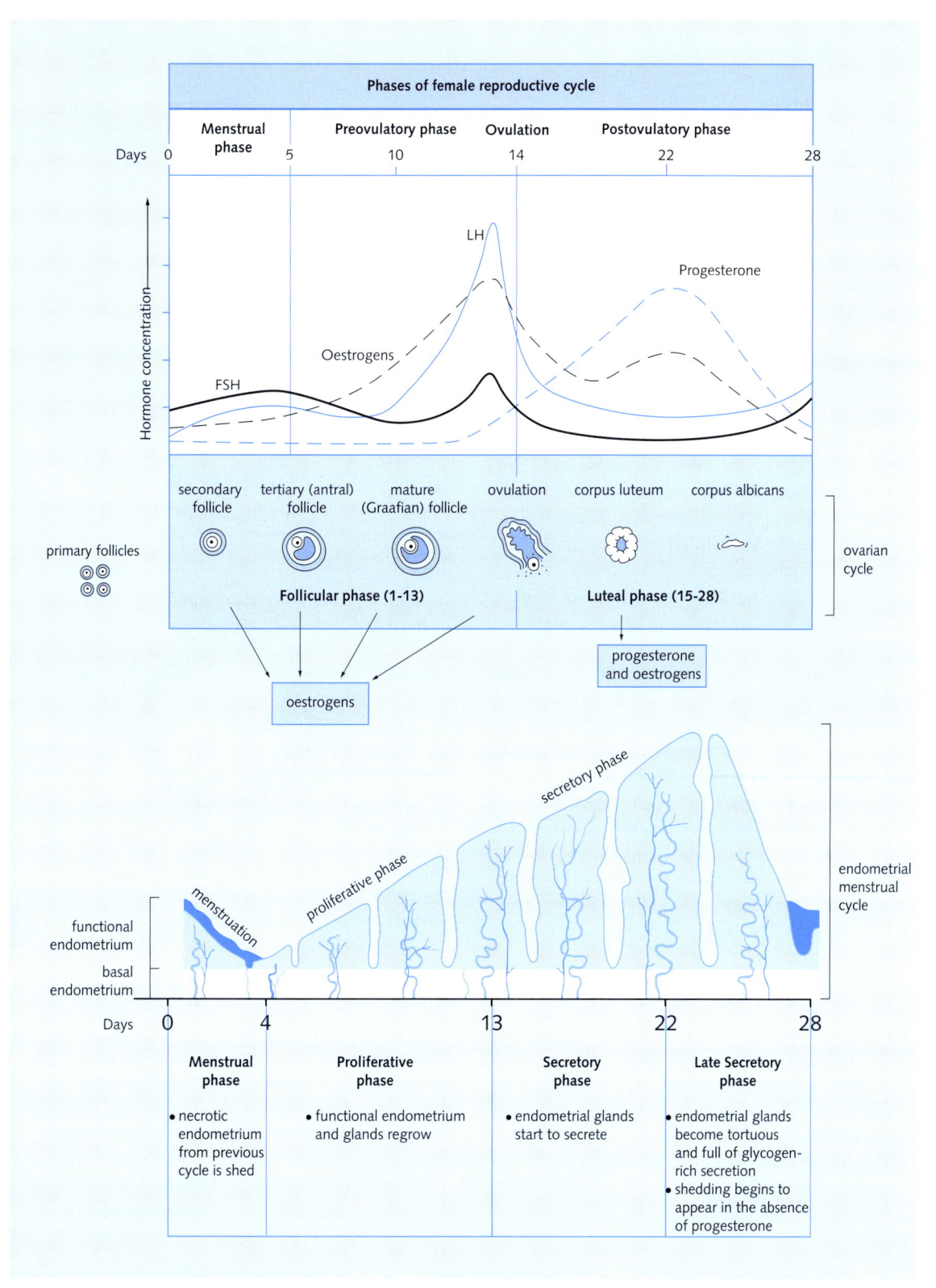

Fig. 12.12 The hormonal, ovarian and endometrial changes during the menstrual cycle. (FSH, follicle stimulating hormone; LH, luteinizing hormone.)

125

14 days is variable; if a woman has a long cycle it is the first stage that is prolonged.

Ovarian changes

This stage of the cycle is called the follicular stage in the ovary.

During menstruation LH and FSH levels rise. As its name implies, FSH stimulates several primary follicles to mature into secondary follicles. This involves proliferation of the granulosa cells while stromal cells surrounding the follicle line up to form thecal cells. These two components allow oestrogen production to begin and for the next 12 days oestrogen levels rise exponentially.

A few days later, fluid begins to collect between the granulosa cells forming a cavity called the antrum. The follicle is now called a tertiary or antral follicle. The oestrogens stimulate synthesis of FSH and LH receptors in the granulosa cells and growth accelerates. The oestrogens also have a negative feedback effect on the pituitary gland to cause a drop in FSH and LH levels. Usually only one follicle will maintain growth as the levels of FSH fall. The other follicles regress (a process called atresia).

Oral contraceptives: Administration of synthetic oestrogen and/or progestogen through the first half of the menstrual cycle prevents FSH secretion. This prevents follicular growth so that ovulation cannot occur.

Further growth results in the formation of a mature Graafian follicle with a diameter of about 2.5 cm just before ovulation. The development of a follicle is shown in Fig. 12.13; they are composed of seven layers, from the inside out they are as follows:

- **Oocyte**—the female gamete, arrested in first meiotic prophase.
- **Zona pellucida**—a glycoprotein layer that surrounds the oocyte like an egg shell.
- **Granulosa cells**—cuboidal cells surrounding the oocyte; they secrete oestrogens.
- **Antrum**—fluid filled cavity within the granulosa cells.
- **Basement membrane/lamina**.
- **Theca interna**—a layer of stromal cells that secrete androgens.
- **Theca externa**—a non-secretory stromal cell layer.

Endometrial changes

The first half of the cycle is separated into two phases:

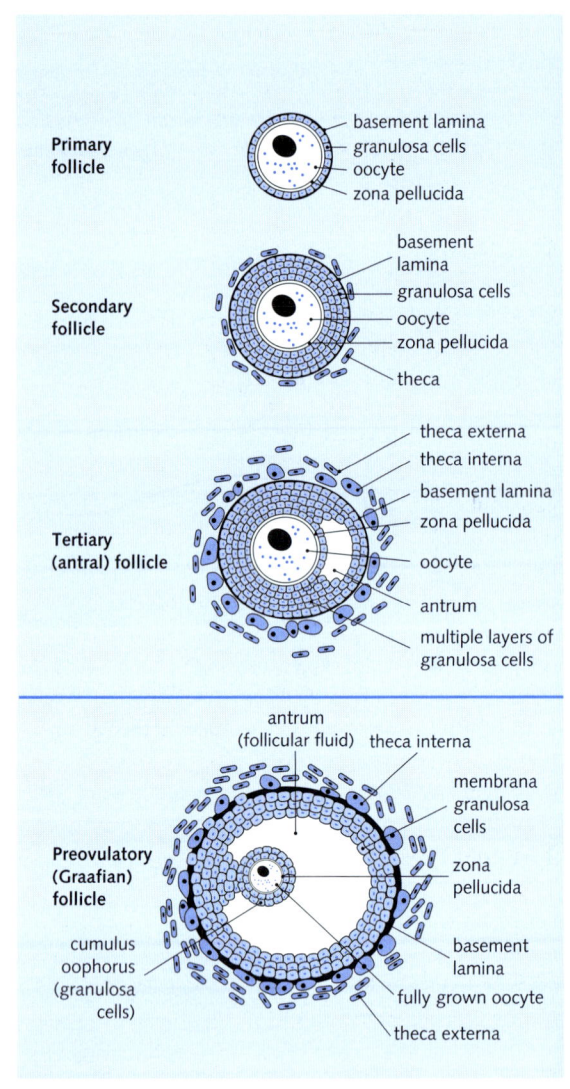

Fig. 12.13 Development of an ovarian follicle through the follicular phase of the menstrual cycle.

- Menstrual phase.
- Proliferative phase.

During the menstrual phase (days 1–4) the ischaemic and necrotic functional layer of the endometrium is lost. This dead tissue passes out of the vagina along with blood from the degenerating spiral arteries.

The proliferative phase (days 4–13) is caused by the rising oestrogen levels. These stimulate cells in the basal layer of the endometrium to proliferate and form a new functional layer. Glands are formed in this layer but they are not yet active.

The rising oestrogen also stimulates secretion of a clear, watery cervical mucus that facilitates sperm transport across the cervix. At other times the mucus is scant and thick.

Oral contraceptives: Administration of progestogen through the first half of the cycle causes the cervical mucus to remain thick. This forms a barrier that prevents the passage of sperm.

Ovulation

At the end of the follicular stage, the mature Graafian follicle secretes such large quantities of oestrogen that the feedback on the pituitary gland changes. The feedback turns from negative to positive and the very high oestrogen levels cause a dramatic surge in the release of LH and FSH. LH causes the Graafian follicle to rupture through the germinal epithelium—a process called ovulation. The oocyte and its first polar body are released into the peritoneal cavity; they are surrounded by the zona pellucida and a few granulosa cells. The ovulated oocyte is swept into the fallopian tubes by the wafting action of the cilia of the fimbriae.

Oral contraceptives: Administration of oestrogen and/or progestogen through the first half of the menstrual cycle can prevent the preovulatory surge of LH that stimulates ovulation.

The second half of the cycle

The second half of the cycle is the time between ovulation and menstruation; the average length is 14 days and this remains constant despite changes in cycle length. The length is determined by the lifespan of the corpus luteum (about 10 days).

Ovarian changes

This stage of the cycle is called the luteal stage in the ovary.

The LH surge continues to act on the granulosa and theca cells in the empty follicle once ovulation has occurred. The cells divide and become yellow. They are now called lutein cells (lutein means yellow, hence the name 'luteinizing' hormone) and the ruptured follicle is called the corpus luteum.

Over the next 10 days these cells secrete high levels of progesterone and oestrogens, but they then spontaneously involute (shrink) and lose their secretory ability.

The progesterone and oestrogen secreted by the corpus luteum inhibit LH and FSH release from the pituitary gland. The falling LH levels fail to maintain the corpus luteum so it undergoes involution. As a result progesterone and oestrogen levels fall dramatically so their negative feedback to the pituitary gland is lost. FSH and LH secretion rise causing ovarian follicles to grow, thus starting the next cycle.

Oral contraceptives: The use of oestrogen and/or progestogen in oral contraceptives aims to mimic the early stages of the second half of the menstrual cycle.

Endometrial changes

After ovulation the progesterone secretion by the corpus luteum activates the endometrium. A number of changes occur:

- Nutrients are stored in the cells.
- Glands become tortuous (irregular shaped) in preparation for secretion.

About 5 days after ovulation, the glands begin to secrete a glycogen rich 'milk' in preparation for a potential embryo; as a result the changes to the endometrium during the second half of the menstrual cycle are called the secretory phase.

As progesterone and oestrogen levels fall, the spiral arteries supplying the functional endometrium begin to coil and constrict causing ischaemia and necrosis. Blood leaks from the damaged vessels into the endometrium before the whole functional endometrium is shed. Menstruation occurs and this marks the first day of the next cycle.

Disorders of the ovaries and fallopian tubes

Pelvic inflammatory disease

Inflammation of the ovaries, fallopian tubes or uterus is called pelvic inflammatory disease (PID), whereas inflammation specific to the fallopian tubes is called salpingitis. PID can run an acute or chronic course and it is usually caused by the following organisms:

- Sexually transmitted diseases ascending from the vagina, e.g. *Chlamydia trachomatis* (60% of all PID) and *Neisseria gonorrhoeae* (30%).
- Direct infection following childbirth, surgery or the insertion of a coil.
- Infection from adjacent organs, e.g. from appendicitis.
- Blood-borne infection, e.g. tuberculosis.

These last three infections are usually caused by staphylococci, streptococci, *E. coli* or anaerobes; together they account for only 10% of PID.

Acute pelvic inflammatory disease

In acute PID the lining of the internal genital tract becomes inflamed and swollen; excess mucus is secreted along with a fibrinous exudate (pus). Acute PID can be asymptomatic, but moderate infection causes the following symptoms:

- Severely painful and tender lower abdomen.
- Fever, often with rigors.
- Vaginal discharge.
- Painful intercourse (dyspareunia).

On examination there is often abdominal guarding and vaginal examination will cause extreme pain. Acute PID must be treated with antibiotics since failure of treatment causes damage to the fallopian tubes in 10% of cases. Recurrent asymptomatic infection with *Chlamydia trachomatis* can also cause damage and potentially infertility.

Chronic pelvic inflammatory disease

A failure to treat acute PID can also result in chronic PID. The fallopian tubes can become sealed by the pus resulting in the following complications:

- Hydrosalpinx—severe swelling due to outflow obstruction.
- Pyosalpinx—an abscess develops in the fallopian tube and adhesions form to surrounding structures, especially the ovaries.
- Infertility.

The patient may complain of menorrhagia and intermittent pelvic pain, often worse before menstruation. On vaginal examination, a tender swelling may be felt and further investigation by laparoscopy may be needed. It is usually treated by surgical removal of the affected organs, especially the fallopian tubes.

Polycystic ovarian syndrome

Polycystic ovarian syndrome (PCOS) is a common but very complicated disease characterized by several endocrine abnormalities and multiple cysts in the ovaries. The 'cysts' are actually multiple immature follicles that develop in the ovaries and are visible on ultrasound examination. Twenty per cent of women have polycystic ovaries (PCO), however only a fraction develop symptoms. The symptoms are caused by the endocrine abnormalities, of which the most important are:

- Excess of LH and deficiency of FSH secretion from the anterior pituitary.
- High insulin levels and insulin resistance, though this is linked to weight gain.
- Excess testosterone.

These endocrine abnormalities cause the following symptoms:

- Amenorrhoea or oligomenorrhoea (no or infrequent periods) with infertility.
- Hirsutism (male pattern hair growth).
- Acne.
- Weight gain.

PCOS is diagnosed from the history, ultrasound examination and raised LH:FSH ratio. The syndrome is linked to obesity and the best treatment is weight loss. Further treatment is aimed at treating the symptoms:

- Infertility is treated by raising pituitary FSH secretion, often by using the anti-oestrogen clomiphene.
- Insulin resistance can be treated by the diabetic medication, metformin, and this may also help infertility.
- Anti-androgen drugs are given with a combined oral contraceptive pill to improve the hirsutism though cosmetic treatment is often better.
- In severe cases, destruction of the follicles by laparoscopic ovarian 'drilling' will relieve symptoms for about a year.

Benign ovarian tumours and cysts

Benign ovarian masses are common during the reproductive years, however it is often impossible to distinguish them from malignant ovarian masses from the history and examination alone. Cysts develop in the ovary during the normal menstrual cycle (i.e. the follicle and corpus luteum). These cysts reach 2–2.5 cm diameter at which stage they are visible on transvaginal ultrasound examination but usually impalpable on pelvic examination; they should regress within 1 month. Persistent, large or abnormal cysts merit further investigation.

There are many types of ovarian masses, which can make this topic difficult to remember. The masses described in the next section are arranged according to the part of the ovary from which they derive (Fig. 12.14). These masses are usually benign, however many have the potential to become malignant.

Benign masses derived from the follicles

Abnormal development of the ovarian follicles can result in functional ovarian cysts that secrete

Fig. 12.14 Common types of ovarian neoplasia.

Common types of ovarian neoplasia

Tumour origin	Name	Frequency	Description
Epithelial cell	Serous cystadenoma	30%	Benign, clear-fluid filled cyst
	Serous cystadenocarcinoma	5%	Malignant, clear-fluid filled cyst
	Mucinous cystadenoma	10%	Benign, mucin filled cyst
	Mucinous cystadenocarcinoma	0.5%	Malignant, mucin filled cyst
	Endometrioid	8%	Benign, solid and brown
Germ cell	Benign teratomas	20%	Benign cyst with several tissue types
	Immature teratomas	0.1%	Malignant cyst with several tissue types

hormones; they are not neoplastic. They are larger than normal follicles but normally shrink without treatment. They are the most common cause of ovarian masses during the reproductive years. There are two types:
- Follicular cysts—an unruptured and persistent follicle that secretes oestrogen often causing menorrhagia (heavy periods).
- Luteal cysts—a persistent corpus luteum secreting progesterone causing irregular bleeding and severe premenstual syndrome (PMS).

Benign masses derived from the epithelium
Cystadenomas
These are benign tumours of the ovarian epithelium that secrete fluid to form a cyst; they can grow to massive sizes. They are the most common cause of neoplasia during the reproductive years. There are two types:
- Serous—secretes a thin, watery substance.
- Mucinous—secretes protein-rich fluid called mucin.

Endometriotic cysts
Endometriosis (see p. 131) can result in functional endometrial tissue being deposited on the ovaries. This tissue is stimulated by the hormonal changes of the menstrual cycle, which results in periodic bleeding. With time the blood becomes dark brown and thick; these cysts are, therefore, called chocolate cysts.

Brenner tumours
These are very rare, benign and solid tumours that resemble the transitional epithelium of the urinary tract. They are a type of fibroma.

Benign masses derived from the germ cells
Benign cystic teratomas
These tumours (also called dermoid cysts) are a common type of ovarian tumour. The tumour is composed of cells from at least two germ layers (i.e. ectoderm, mesoderm, and endoderm); they are often lined by skin with hairs and sebaceous glands. Other recognizable structures and tissues may be present (e.g. teeth, bone, muscle and neural tissue). The teratoma is benign if the tissues are mature.

Struma ovarii
These are very rare types of benign cystic teratoma composed mainly of thyroid tissue; they may present with hyperthyroidism.

Benign masses derived from the stroma
Malignant tumours in the ovarian stroma are exceedingly rare, however the stromal mesenchyme of the ovarian cortex and medulla can become neoplastic. These rare benign tumours are often associated with hormone production:
- Thecomas—develop from the thecal cells; they often secrete oestrogens causing endometrial hyperplasia and a higher risk of endometrial carcinoma.
- Granulosa cell tumours—develop from the granulosa cells; like thecomas they often secrete oestrogens with similar effects.
- Androblastoma—the tumour resembles testicular cells (e.g. Sertoli and Leydig cell); they secrete androgens.
- Fibromas—solid, white tumour made of fibrous tissue.

Malignant ovarian disease

Ovarian carcinoma is a relatively rare form of carcinoma, but the 5-year survival is poor because of late detection. It usually affects postmenopausal women. Risk is increased by the *BRCA1* gene and low parity (i.e. few children); the contraceptive pill has a protective effect. All large, abnormal or persistent ovarian masses should be considered malignant until proven otherwise.

As with benign ovarian masses there are many types, they are described according to their tissue of origin (see Fig. 12.14).

Malignant masses derived from the epithelium
Cystadenocarcinomas
Benign cystadenomas may undergo malignant change to form malignant cystadenocarcinomas. The two types remain:
- Serous—the most common type of ovarian malignancy (about 50%).
- Mucinous—accounts for 10% of ovarian malignancy.

Endometrioid tumours
These are primary tumours that resemble adenocarcinoma of the endometrium. They usually arise spontaneously, though rarely they develop from endometrioid cysts. Uterine adenocarcinoma is sometimes present.

Clear cell tumours
These are a less common type of endometrioid tumour with pale, glycogen-rich cytoplasm that resembles the secretory phase endometrium.

Malignant masses derived from the germ cells
Dysgerminomas
These are the most common malignant germ cell tumour; the cells resemble seminomas found in the male testes. They occur mainly in adolescents and young women; they are highly malignant.

Yolk sac tumours
Derived from the endoderm of the yolk sac, these are highly malignant tumours that secrete α-fetoprotein (AFP), which can be detected in the blood. They occur mainly in adolescents and young women.

Solid teratoma
Teratomas that contain embryonal tissues are highly malignant. They occur mainly in adolescents.

Choriocarcinoma
These are highly malignant tumours that secrete human chorionic gonadotrophin (hCG); they are derived from trophoblastic tissue found in teratomas.

Metastatic ovarian tumours
Tumours may metastasize to the ovaries, especially from:
- Endometrium.
- Breast.
- Stomach (Krukenberg tumour).
- Colon.

Diagnosis and treatment of ovarian masses
Symptoms
Ovarian masses are frequently asymptomatic. Pelvic or abdominal pain may occur; large masses can cause noticeable increases in abdominal girth. Advanced malignancy may cause appetite and weight loss, tiredness, and general malaise. Ovarian masses are most commonly detected through pelvic examination or ultrasound scans.

Investigations
The definitive diagnosis of an ovarian mass can only be made from a biopsy taken during laparotomy. Initial investigations aim to determine which masses require surgical exploration. These investigations include:
- Pregnancy test—to eliminate the risk of ectopic pregnancy.
- CA-125—a blood-borne tumour marker.
- Ultrasound scan—determines the location, size and nature of the mass.

If a malignancy is suspected, further investigations and imaging aim to determine the stage of the carcinoma to allow appropriate treatment.

Unsuspicious (benign) ovarian masses can be followed up using repeated pelvic examinations and ultrasound scans. They usually do not require treatment.

Staging
The staging of ovarian carcinoma is shown in Fig. 12.15.

Treatment
Ovarian carcinoma is usually treated at the initial investigative laparotomy. Suspicious masses are removed along with both ovaries, the uterus and the

Disorders of the Endometrium and Myometrium

Staging and prognosis of ovarian carcinoma

Stage	Description	Five-year survival (%)
I	Limited to one or both ovaries	80–100
II	Other pelvic sites involved	80–100
III	Sites involved above the pelvic brim within the peritoneal cavity	15–20
IV	Distant metastases	5

Fig. 12.15 Staging and prognosis of ovarian carcinoma.

omentum. These tissues are sent for histological analysis and diagnosis. The surgeon will also explore the abdomen and pelvis for signs of metastases. Surgery is followed by chemotherapy unless the carcinoma is stage I.

Tumours of the fallopian tubes

Tumours of the fallopian tubes are extremely rare. Benign adenomatoid tumours can form in the superior border of the broad ligament (mesosalpinx). Primary adenocarcinoma of the epithelium rarely occurs in postmenopausal women, and it has an extremely poor prognosis due to late presentation.

Disorders of the endometrium and myometrium

Inflammation of the endometrium (endometritis)

Inflammation of the endometrium is called endometritis; it can follow an acute or chronic course. Acute endometritis is a bacterial infection, often following trauma (e.g. childbirth, surgical termination, cervical surgery or insertion of the coil). The endometrium usually avoids infection by frequent shedding (menstruation) and the thick cervical mucus. The main causative organisms are staphylococci, streptococci, clostridia and anaerobes.

Endometritis presents in a similar manner to PID with lower abdominal pain, tenderness and fever. In severe cases, cervical obstruction can occur so the uterus fills with pus (pyometra). This is treated with antibiotics following cervical swabs to determine the organism and its antibiotic sensitivity.

Untreated acute endometritis or PID can cause chronic endometritis. Women develop menstrual irregularities, particularly heavy periods, and the infection may spread to affect other reproductive organs.

Adenomyosis

Adenomyosis is when the basal endometrium penetrates the myometrium (muscular layer of the uterus). Cells from the basal layer of the endometrium form small deposits within the smooth muscle that grow and stimulate proliferation of the muscle. The uterus develops a tumour-like mass or enlarges diffusely often causing menstrual pain and heavy periods. Symptomatic adenomyosis is usually treated by hysterectomy since the basal cell layer is insensitive to hormones.

Endometriosis

Endometriosis is the presence of functioning endometrial tissue outside the uterus. It is a very common gynaecological disorder occurring in about 5% of women, however many are undetected. The main locations of the ectopic endometrial cells are shown in Fig. 12.16. The two most common sites are the:
- Ovaries.
- Ligaments of the uterus.

The ectopic tissue still responds to hormonal stimuli, so cyclic proliferation and bleeding occurs. The bleeding often forms a cyst that enlarges every month. The size of the cyst is limited by rupture; this may cause adhesions.

The exact cause of endometriosis is still under debate, however there are three theories:
- Retrograde menstruation—menstrual debris enters the peritoneal cavity via the fallopian tube in most women. A lack of immune activity could cause implantation and disease.
- Metaplasia of peritoneal epithelium—an unknown stimulus causes the epithelium to transform into endometrial tissue.
- Metastatic spread—emboli of endometrial tissue may travel via blood and lymph vessels to reach ectopic sites.

In 25% of women endometriosis is asymptomatic. Women who do have symptoms may have:
- Lower abdominal pain during menstruation (75%).
- Constant pain if adhesions are present.

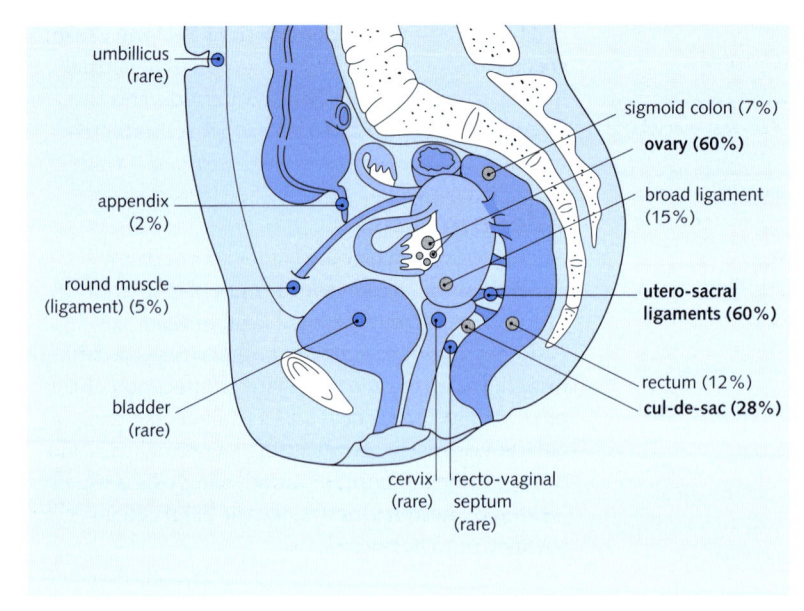

Fig. 12.16 Locations of endometriosis with the relative frequencies (multiple sites are common).

- Menstrual irregularities (60%).
- Deep dyspareunia (pain on intercourse; 30%).
- Infertility (30%).

Endometriosis is diagnosed from the history and by laparoscopy, which shows red spots. Since endometriosis is common, it is important to know if it is causing the presenting symptoms. This can be achieved by a short course (up to 6 months) of continuous GnRH analogues that stop cyclical sex steroid changes and endometrial symptoms (they downregulate the GnRH receptors). This investigation also acts as a treatment that persists after the GnRH analogue is discontinued.

Endometriosis can be further treated by:
- Continuous use of combined oral contraceptives.
- Medroxyprogesterone acetate (a progestogen).
- Laparoscopic ablation of the endometrial deposits and adhesions, often used if fertility is desired.
- Danazol—an antioestrogen and antiprogesterone with androgenic activity (up to 6 months).

Severe endometriosis may require hysterectomy and bilateral salpingo-oophorectomy (removal of the uterus, fallopian tubes and both ovaries).

Endometrial hyperplasia

Endometrial hyperplasia is caused by an excess of oestrogens (e.g. from medication, oestrogen-secreting tumours or anovulatory cycles). The proliferation of the functional endometrium during the first half of the menstrual cycle is enhanced, causing irregular and heavy menstruation. Endometrial hyperplasia is important because it carries a higher risk of developing into endometrial carcinoma. A spectrum of malignant change is seen:
- Simple hyperplasia—diffuse enlargement, dilated glands; low risk of carcinoma.
- Complex hyperplasia—focal areas of severe hyperplasia with irregular glands.
- Complex atypical hyperplasia—like complex hyperplasia but cells show atypical malignant changes; there is a high risk of carcinoma.

Endometrial hyperplasia is investigated by biopsy of the endometrium. If there is simple hyperplasia in patients who wish to remain fertile, conservative treatment with cyclical progesterone may be sufficient. In severe and atypical cases hysterectomy is recommended.

Functional endometrial disorders
Anovulatory cycles
At the extremes of reproductive age (i.e. around menarche and menopause) menstruation is frequently irregular. This is due to a failure of ovulation followed by excessive oestrogen secretion. The endometrial glands proliferate as a result.

Inadequate luteal phase
A failure in progesterone secretion from the corpus luteum causes inadequate endometrial secretion resulting in infertility.

Oral contraceptives

Starting or changing the 'Pill' can cause breakthrough bleeding (bleeding in the middle of the cycle). It usually settles down within a few months, and it can be reduced by taking the pill at the same time every day. Higher doses of oestrogen may be needed if it does not settle. Prolonged use of oral contraceptives reduces the thickness of the endometrium and inactivates the glands, which reduces the amount of menstrual discharge. There is also a 50% lower risk of endometrial and ovarian cancer.

Menopausal changes

At menopause, ovulation and menstruation become irregular followed by complete cessation of the cycles and menstruation within a few months or years. The endometrium reverts to the prepubertal state and the columnar epithelium may undergo metaplasia to form squamous epithelium; cysts can also develop.

Neoplastic disorders

The most common types of uterine neoplasia are shown in Fig. 12.17.

Benign endometrial polyps

Endometrial polyps are very common around the menopause. They are benign tumours caused by the over-proliferation of endometrial glands in response to oestrogen. The polyps are usually 1–3 cm in size, and they form smooth, firm nodules within the endometrium. They cause menstrual pain and irregularities and can be removed using forceps and a speculum.

Benign leiomyomas (fibroids)

Fibroids are benign tumours of the myometrium (muscle layer of the uterus); they are the most common tumours in the genital tract and affect 20% of menopausal women. The tumours are round, well-defined growths of the smooth muscle cells that often occur in several locations at the same time (Fig. 12.18). Fibroids are oestrogen dependent so they enlarge during pregnancy and with the use of oral contraceptive, but regress after the menopause.

The majority of fibroids are asymptomatic. In the remainder symptoms are caused by the presence of a uterine mass and the extra endometrium required to cover it. Symptoms include:
- Menorrhagia—periods are heavy and prolonged.
- Pelvic pain—torsion of the fibroid can cause ischaemia and pain.
- Pelvic mass—large fibroids can be felt in the abdomen and compress surrounding structures.
- Infertility—interference with embryo implantation or causing recurrent spontaneous abortion.

Treatment is only needed if the fibroids are symptomatic and troublesome. Continuous GnRH agonists may cause the fibroid to regress, however women who have completed their families often choose hysterectomy. The fibroids can be removed sparing the unaffected uterus (myomectomy), but adhesions are a common complication.

Fig. 12.17 Types of uterine neoplasia.

Types of uterine neoplasia

Uterine layer	Name	Frequency	Description
Endometrium	Endometrial hyperplasia	Common	Benign overgrowth of endometrium
	Oestrogen sensitive endometrial carcinoma	Less common	Malignant tumour derived from endometrial hyperplasia
	Oestrogen insensitive endometrial carcinoma	Less common	Malignant tumour that originates spontaneously
	Endometrial polyps	Common	Benign enlargement of endometrial glands
Myometrium	Leiomyoma	Very common	Benign smooth muscle tumour
	Leiomyosarcoma	Very rare	Malignant smooth muscle tumour

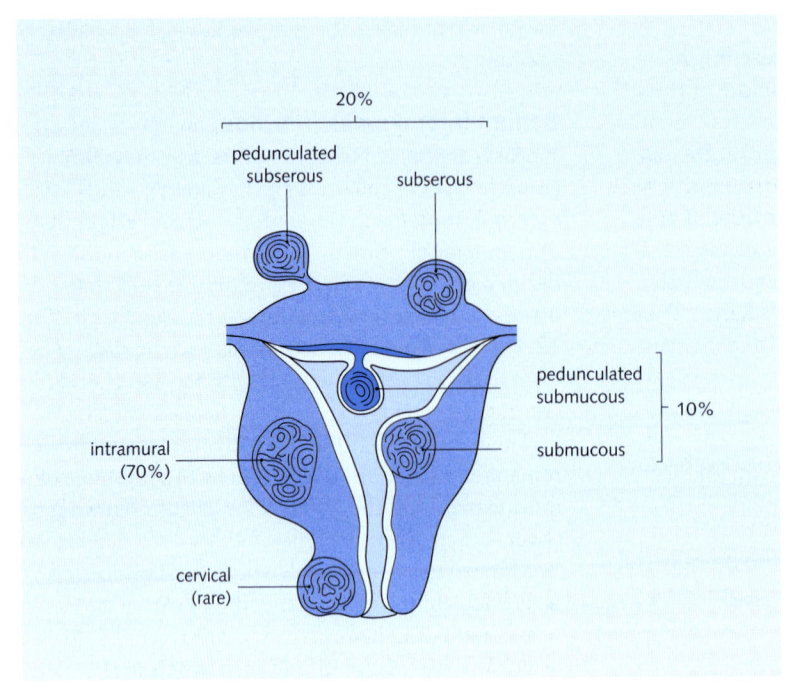

Fig. 12.10 Locations of fibroids with relative frequencies.

Endometrial carcinoma

Endometrial carcinoma is the commonest malignancy of the female genital tract. It is usually found during or after the menopause; two patterns are seen:
- Following endometrial hyperplasia—this affects menopausal women; the tumour is an oestrogen-dependent adenocarcinoma with a good prognosis.
- Independent of oestrogen—this affects post-menopausal women; it can follow endometrial squamous metaplasia and it has a poor prognosis.

Endometrial carcinoma spreads mainly by local invasion. Initially, this affects the myometrium but the bladder and rectum may become involved with time. It presents with postmenopausal bleeding that becomes progressively more severe. This symptom must be investigated by hysteroscopy (viewing the endometrium through an endoscope inserted through the cervix) and endometrial biopsy.

The staging of endometrial carcinoma is shown in Fig. 12.19. Early stage carcinoma can be cured by hysterectomy and bilateral salpingo-oophorectomy (removal of the uterus, fallopian tubes and both ovaries) followed by radiotherapy. Patients with inoperable carcinoma may benefit from high doses of progestogens and/or radiotherapy.

Other uterine carcinomas

Rarely, malignancy may develop from the stroma of the endometrium, usually with a poor prognosis.

Staging and prognosis of endometrial carcinoma

Stage	Description	Five-year survival (%)
I	Limited to the endometrium and myometrium, but not serosa	75–100
II	Limited to the uterus and cervix, but not serosa	60
III	Involvement of serosa and/or metastases to other pelvic organs	50
IV	Distant metastases	20

Fig. 12.19 Staging and prognosis of endometrial carcinoma.

They present in a similar manner to endometrial carcinoma but an epithelial component is often present. There are three main types:
- Endometrial stromal sarcoma—consisting of stromal spindle cells.
- Adenosarcoma—malignant stromal and benign epithelial components.
- Carcinosarcoma—malignant stromal and epithelial components, may contain non-uterine tissues.

Leiomyosarcomas are extremely rare, malignant tumours of the myometrium that tend to occur after the menopause. They often metastasize by vascular spread to the lungs.

Menstrual disorders and the menopause

Abnormal frequency, duration and volume of menstruation are common presentations of many gynaecological diseases. Normal menstruation:
- Occurs once every 22–35 days.
- Lasts less than 7 days.
- Less than 80 mL of fluid is discharged.

Amenorrhoea

If menstruation has not occurred within 35 days of the start of the last cycle it is called oligomenorrhoea. An absence of menstruation for 70 days or more is called amenorrhoea; this is normal before puberty, during pregnancy, and after the menopause. Pathological amenorrhoea is divided into primary and secondary causes.

Primary amenorrhoea

This is the failure to start menstruating by the age of 16. It is a relatively rare condition. The most common cause is delayed puberty that will resolve spontaneously, however a number of diseases and conditions can cause primary amenorrhoea:
- Low body fat, e.g. anorexia nervosa, malnutrition.
- Congenital malformations of the vagina, uterus or ovaries.
- Imperforate hymen.
- Turner syndrome—lack of an X chromosome (i.e. XO).
- Testicular feminization syndrome—XY but insensitive to testosterone.

Secondary amenorrhoea

This is when a woman who has begun to menstruate fails to have a period for 70 days or more. It is a relatively common disorder that affects about 1% of women of reproductive age. The most common causes of secondary amenorrhoea are shown in Fig. 12.20.

Investigation and treatment

Amenorrhoea is investigated with hormone tests (prolactin, FSH, LH and thyroid function tests) and an ultrasound scan of the pelvis. Treatment depends on the cause of the amenorrhoea.

Menorrhagia

Menorrhagia is excessive (>80 mL) menstrual bleeding; it may also be prolonged. Severe menorrhagia may result in iron deficiency anaemia. The main causes are:
- Dysfunctional uterine bleeding (80%).
- Endometrial tumours and polyps that distort the endometrium.
- Adenomyosis.
- Chronic pelvic inflammatory disease.
- Hypothyroidism.

Menorrhagia should be investigated if the woman is over 40 years of age, has intermenstrual bleeding or postcoital bleeding. Hysteroscopy, endometrial biopsy, transvaginal ultrasound, thyroid function tests and clotting studies are used to exclude pathology. In the majority of patients with menorrhagia no underlying pathology is found and the disease is termed **dysfunctional uterine bleeding**; this condition can also cause irregular bleeding. This is more common at the extremes of reproductive age and is due to an excess of endometrial prostaglandin synthesis that may cause excessive uterine contractions and abnormal blood clotting.

Fig. 12.20 Common causes of secondary amenorrhoea.

Common causes of secondary amenorrhoea		
Cause	Aetiology	Frequency (%)
Weight loss	Deficiency of leptin from fat prevents GnRH release	35
Polycystic ovaries	FSH deficiency and multiple endocrine abnormalities prevent follicle development	25
Pituitary insensitivity	Following the 'Pill', stress or illness	15
Hyperprolactinaemia	Microadenoma of the pituitary gland (see Ch. 2)	15
Primary ovarian failure	Premature menopause, possibly with autoimmune involvement	5

If an underlying cause is found it should be treated; otherwise dysfunctional uterine bleeding can be assumed. Menorrhagia can be treated with:
- Oral contraceptives.
- Mirena® (progestogen releasing intrauterine system).
- Mefenamic acid (a non-steroidal anti-inflammatory drug).
- Tranexamic acid (fibrinolysis inhibitor).
- Endometrial ablation or hysterectomy if the bleeding is severe.

Dysmenorrhoea
Dysmenorrhoea means painful menstrual periods; it has either primary or secondary causes:

Primary dysmenorrhoea
This is caused by an imbalance in prostaglandin synthesis that results in ischaemia and hyperexcitability of the myometrium. Uterine spasms result, causing cramping pains before and at the start of menstruation. This disorder is very common soon after menarche when it affects about 75% of girls. It decreases with age but NSAIDs or oral contraceptives usually help.

Secondary dysmenorrhoea
This affects older women and is usually due to endometriosis or PID. Cramping pains are felt before menstruation, however they persist and worsen through the period. The underlying pathology should be treated.

Premenstrual syndrome
Premenstrual syndrome (PMS) describes a negative mood and several physical symptoms that can occur in the luteal phase (days 14–28) of the menstrual cycle. Mild PMS is very common, but 5–15% of women suffer from life-disrupting symptoms regularly. The symptoms are shown in Fig. 12.21.

The cause remains unknown, however fluctuating oestrogen levels may be responsible, possibly mediated via decreased levels of serotonin (5-hydroxytryptamine; 5-HT) in the CNS. Diagnosis is made from the history and can be confirmed by keeping a diary of symptoms; these must correspond to the menstrual cycle. Treatment can be symptomatic (e.g. analgesia) or aimed at preventing oestrogen fluctuations (e.g. oral contraceptives or oestrogen patches).

> Extensive research amongst female medical students at Nottingham University has revealed the following 'treatments' for PMS: bananas, chocolate, vitamin B_6, sage, fennel, chocolate, low fat diet, evening primrose oil and chocolate.

Menopause
The menopause is the cessation of menstruation and ovulation that usually occurs between the ages of 45 and 55 years (average 51). The term 'climacteric' includes the time before and after the menopause during which 'menopausal' symptoms are noticed.

The ovaries gradually become less sensitive to FSH and LH from about the age of 40 years because of the loss of follicles and receptors. This causes anovulatory cycles and a progressive decrease in oestrogen production. As oestrogen levels fall, FSH and LH secretion increases because of the lack of negative feedback. The ovaries resist this increase and the woman enters a period of oligomenorrhoea followed by amenorrhoea. After 6 months of

Fig. 12.21 Symptoms of premenstrual syndrome.

Symptoms of premenstrual syndrome		
Psychological	Behavioural	Physical
Anxiety Depression Increased appetite Irritability Loss of libido Sleep disturbance Tension	Anger Impulsiveness and accident-prone behaviour Poor concentration Poor tolerance to stress	Acne Weight gain Breast tenderness and swelling Abdominal bloating Change in bowel habit Headache Pelvic pain

amenorrhoea the woman is said to have reached menopause. With time, the FSH and LH levels begin to decline along with oestrogen levels.

Other tissues are capable of secreting oestrogens independently, e.g. adipose tissue and the adrenal cortex. The oestrogens produced do not equal the premenopausal levels so women become oestrogen deficient. The lack of oestrogen predisposes women to three main complications (Fig. 12.22):

- Osteoporosis.
- Heart disease.
- Collagen breakdown.

The climacteric period is symptomatic in 75% of women and severe in 40%. The climacteric symptoms are listed below (those caused by low oestrogen levels are in bold):

- **Hot flushes**; usually at night, often with sweating.
- **Dry, burning vagina** with dyspareunia (pain on intercourse).
- Painful joints.
- Headaches.
- Depression, anxiety, irritability and dizziness.
- Palpitations.
- Urinary incontinence and infection.

The menopause is often said to reduce sexual desire. This is true in 20% of women, but 20% report an increase in sexual desire.

Hormone replacement therapy (HRT) is the replacement of oestrogens via a tablet, implant or skin patch. It is often used to treat climacteric symptoms and prevent the long-term complications of oestrogen deficiency. There are three types of HRT:

- Cyclical combined HRT—continuous oestrogens, 12/28 days of progestogens; they cause regular withdrawal bleeds.
- Continuous combined HRT—continuous oestrogens and progestogens; they can only be used 12 months after the last menstrual period.
- Oestrogen only—continuous oestrogens; they can only be used if the woman has had a hysterectomy.

Progestogens must be included if the woman has not had a hysterectomy to prevent endometrial hyperplasia and carcinoma. Cyclical progestogens cause withdrawal bleeding (similar to using the Pill); after one year the woman may use a continuous combined preparation in which progestogens are taken constantly to prevent withdrawal bleeds.

Long-term use of HRT has a life-saving effect by reducing cardiovascular disease and osteoporosis; there is a slight increase in the risk of breast cancer and deep vein thrombosis.

The contra-indications of HRT are common exam questions; they include:

- Oestrogen-dependent cancer (including breast cancer).
- Thromboembolic disorders.
- Liver disease with abnormal liver function tests (LFTs).
- Undiagnosed vaginal bleeding.
- Pregnancy or breastfeeding.

There is no male equivalent of the menopause. Men continue to produce testosterone and spermatozoa well into their 80s, but the amount and quality decline with age. The evolutionary basis for this difference is not fully understood.

Fig. 12.22 Long-term complications of oestrogen deficiency following the menopause.

Symptoms/disease	Cause	Consequence
Osteoporosis	Accelerated bone loss	Increased risk of bone fractures, especially the femoral neck at the hip and crush fractures of the vertebrae
Cardiovascular disease	Oestrogens have a beneficial effect on the type of lipid in the blood and in doing so protect against cardiovascular disease	Increased risk of coronary artery disease, myocardial infarction and strokes
Loss of collagen	Weakening in the pelvic ligaments, joints, and muscles, and loss of elasticity in the skin	Predisposes to uterovaginal prolapse, immobility, muscle weakness and causes skin wrinkling

Disorders of the cervix

Cervical ectopy
Cervical ectopy is a normal physiological finding that has been called cervical ectropion or erosion in the past. It describes the extension of the columnar epithelium of the endocervix beyond the external os under the influence of oestrogen. Any process that raises oestrogen levels can temporarily result in a cervical ectopy, including puberty, pregnancy and the first months of using the 'Pill'. In ectopy, the columnar epithelium appears as a red ring around the external os compared with the pink squamous epithelium. With time, metaplasia occurs and the columnar epithelium converts to stratified squamous epithelium. The presence of cervical ectopy may account for small amounts of postcoital or intermenstrual bleeding; it may raise susceptibility to sexually transmitted infection.

Inflammation of the cervix
Cervicitis
Cervicitis is inflammation and infection of the cervix. It is usually asymptomatic, although vaginal discharge, postcoital bleeding, dyspareunia and pelvic pain may be present; on examination it may be inflamed and tender. The infection is often caused by:
- *Chlamydia trachomatis* (very common, see below).
- *Neisseria gonorrhoeae* (common, see below).
- *Trichomonas vaginalis* (see p. 141).
- Herpes simplex (see p. 141).

The infection is diagnosed using swabs (cervical, *Chlamydia* cervical, and high vaginal) along with a wet mounted cervical smear for *Trichomonas*. Asymptomatic infection should be treated aggressively because of the risk of PID and infertility. The partner often requires treatment to prevent reinfection.

Chlamydia trachomatis This intracellular bacterium is sexually transmitted. It is thought to cause asymptomatic infection in about 5% of young, sexually active women. Infection is usually asymptomatic, however it can cause urethritis (infection of the urethra), cervicitis and pelvic inflammatory disease (PID, see p. 127). It is usually diagnosed by direct fluorescent antibody tests (DFA) that require a special culture medium; it is treated using the antibiotics doxycycline or erythromycin.

Neisseria gonorrhoeae This Gram-negative diplococcus is an intracellular bacterium that is often called gonococcus; it is a sexually transmitted infection. Like *Chlamydia* it usually causes asymptomatic urethritis, cervicitis and pelvic inflammatory disease. It is diagnosed by Gram stain and culture of the swab sample; it is treated with the antibiotics cefixime and ceftriaxone. *Chlamydia* is also present in 50% of gonococcus infections, so doxycycline is often given as well.

Neoplasia of the cervix
During puberty the vagina becomes more acidic due to the presence of glycogen in the vaginal walls and the action of lactobacilli that colonize the vagina. The columnar epithelium of the endocervix reacts to the acid environment by transforming into stratified squamous epithelium—a process called metaplasia. This results in a transformation zone between the ectocervix and endocervix which is susceptible to dysplasia (pre-cancerous changes) in a similar manner to the squamocolumnar transformation zone found in Barrett's oesophagus. The 'smear test' is used to identify and treat these dysplastic changes before they progress to cervical carcinoma.

Cervical intraepithelial neoplasia
The dysplastic changes leading to cervical carcinoma are called cervical intra-epithelial neoplasia (CIN); they are graded from I–III according to the severity and depth of the changes. The risk and rate of progression to cervical carcinoma increases with each grade, however all stages are treatable. All stages are asymptomatic and undetectable by simply looking at the cervix.

Risk factors
CIN is strongly associated with certain strains of human papilloma virus (HPV); CIN is, therefore, a sexually transmitted disease. Any factor associated with exposure to HPV increases the risk of CIN, including multiple sexual partners and early age of first intercourse. Cigarette smoking is also a risk factor.

The smear test
Women between the ages of 20 and 64 years are offered free Pap smears (smear tests) every 3 years by their doctor. The test involves scraping cells from the cervix using a speculum to open the vagina and wooden spatula to take the sample. Between 2 and 5% of smears are reported as abnormal and 10% have an insufficient sample for estimation (a repeat smear is required).

Since the sample is taken from the surface layer, the depth of epithelial involvement cannot be

Disorders of the Cervix

Fig. 12.23 Classification, natural history and treatment of cervical intra-epithelial neoplasia (CIN).

Classification, natural history, and treatment of cervical intra-epithelial neoplasia

Grade	CIN I	CIN II	CIN III
Pap smear classification	Mild dysplasia	Moderate dysplasia	Severe dysplasia / carcinoma *in situ*
Extent of epithelium involved on biopsy	Third nearest the basement membrane	Two thirds nearest the basement membrane	Full thickness
Percentage who progress to carcinoma if untreated	1%	8%	20%
Treatment	Follow up in 6 months or refer to colposcopy	Refer to colposcopy	Refer to colposcopy

measured directly. The number and severity of dysplastic cells are used to estimate the grade of CIN (Fig. 12.23). This estimate is not entirely accurate; about 30% of normal smears are false negatives and 4% of abnormal smears are false positives. Estimation of borderline smears can be improved by testing for the presence of oncogenic HPV strains.

Management of positive smears

Receiving a diagnosis of a positive (abnormal) smear test is commonly misinterpreted as a diagnosis of cervical cancer; the patient will often be afraid and anxious. It is important to explain that it is not a diagnosis of cancer but that follow-up smears/treatment is very important. The management is shown in Fig. 12.23.

Colposcopy

Abnormal smears are often followed up using colposcopy; this is usually performed at a hospital outpatient clinic. A colposcope looks like a pair of binoculars; it magnifies the cervix by 5–20 times. Acetic acid is applied to the cervix so that abnormal areas of cervical epithelium turn white. These areas are inspected and biopsied so that the depth of involvement can be assessed to accurately diagnose the grade of CIN. It is often treated immediately.

Treatment

Moderate to severe (II and III) CIN is treated by removal or destruction of the abnormal epithelium; this is usually performed in colposcopy using local anaesthetic. There are a number of methods of treatment:

- Large loop excision of the transformation zone (LLETZ).
- Laser therapy.
- Cryotherapy.
- Cone biopsy (grade III, may raise miscarriage risk).

Vaginal bleeding and discharge is common for about 2 weeks after treatment; sexual intercourse and the use of tampons should be avoided for 4 weeks after treatment. The treatment is followed up with a Pap smear test 6 months later and yearly smear tests for 5 years. Ninety per cent of women are cured; 10% require repeated treatment.

Cervical carcinoma

Cervical carcinoma is usually a squamous cell carcinoma arising in the transformation zone between the ectocervix and endocervix. There is a clear progression from CIN to carcinoma and both diseases have the same risk factors, including HPV infection.

Cervical carcinoma usually affects women after the menopause, but the incidence in younger women is high enough to warrant population screening from the age of 20 years. The early stages are frequently asymptomatic; by the time of presentation advanced disease with poor 5-year survival is often present. Symptoms include abnormal vaginal bleeding (classically postcoital), vaginal discharge or renal failure in advanced stages; it can be detected through abnormal smear tests. The staging and 5-year survival figures are shown in Fig. 12.24. Invasion of the ureters (which pass 1 cm lateral to the internal os) in stage IIb is a poor prognostic feature.

Suspected cervical carcinoma is investigated by colposcopy with biopsy; if advanced disease is

Fig. 12.24 Staging and prognosis of cervical carcinoma.

Staging and prognosis of cervical carcinoma

Stage	Description	Five-year survival (%)
Ia	Carcinoma only in cervix, <5 mm	95
Ib	Carcinoma only in cervix, >5 mm	85
IIa	Invasion outside the cervix but not the parametrium (surrounding tissue, including ureters)	75
IIb	Invasion outside the cervix including the parametrium (+/− ureters)	55
III	Invasion of ureters, lower third of the vagina, or pelvic walls	30
IV	Invasion of bladder, rectum, or outside the pelvis	10

suspected then imaging techniques are used to identify the extent of invasion and metastases:
- Stage Ia—cone biopsy or simple hysterectomy.
- Stage Ib/IIa—radical hysterectomy (includes parametrium and pelvic lymph nodes) or pelvic radiotherapy.
- Stage IIb—combination of radical surgery, chemotherapy and radiotherapy.

A smear test should be performed once every 3 years on sexually active women aged 20–64 years to detect cervical intra-epithelial neoplasia.

Cervical polyps and benign cervical tumours
Cervical polyps are small, round, benign growths of the cervix that often protrude through the external os. They are very common, affecting about 5% of women; they may cause irregular vaginal bleeding or vaginal discharge. They are treated by surgical excision that is often performed in outpatients.

Other benign tumours of the cervix are uncommon though leiomyomas (smooth muscle tumours) can occur.

Disorders of the vagina and vulva

Infections
The female genital tract is susceptible to many infections, including a number that are sexually transmitted. A summary of the most common female sexually transmitted infections is shown in Fig. 12.25.

Infections of the vagina
Vaginal discharge is a common symptom that can be caused by infections of the cervix (e.g. *Chlamydia*, discussed on p. 138) and vagina (e.g. thrush). The most common causes of vaginal discharge are not sexually transmitted, instead they are caused by overgrowth of normal vaginal flora. This is caused by a rise in vaginal pH (as occurs in pregnancy and diabetes) or loss of the lactobacilli (as occurs when taking antibiotics). These diseases are:
- Bacterial vaginosis.
- Candidiasis (thrush).

Nonetheless, women presenting with vaginal discharge are investigated for a number of sexually transmitted diseases (STDs) due to the high prevalence of sexually transmitted infection. The main vaginal infections are described below.

Bacterial vaginosis
This disease is usually caused by the overgrowth of the anaerobic bacteria *Gardnerella vaginalis*, although other anaerobes may be responsible. It is classically associated with a smooth, white vaginal discharge with a distinctive 'fishy' smell. It is diagnosed if the discharge has:
- pH >5.5.
- Ammonia smell when mixed with potassium hydroxide.
- Clue cells on microscopic examination.

It is treated using the antibiotic metronidazole.

Disorders of the Vagina and Vulva

Common sexually transmitted infections in women

Infection	Site of infection	Organism	Symptoms	Treatment
Pelvic inflammatory disease	Upper genital tract	*Chlamydia trachomatis* (atypical bacteria) or *Neisseria gonorrhoeae*	Abdominal pain and tenderness, dyspareunia, pus discharge	Doxycycline or ceftriaxone
Gonorrhoea	Cervix	*Neisseria gonorrhoeae*	Urinary frequency, dysuria, pus discharge	Penicillin or ceftriaxone
Trichomoniasis	Vagina	*Trichomonas vaginalis* (parasite)	Itching, discharge	Metronidazole
Thrush	Vagina	*Candida albicans* (fungus)	Itching, white discharge	Clotrimazole (vaginal pessary)
Herpes	Vulva	Herpes simplex virus	Burning red blisters that may recur	Aciclovir if recurrent
Warts	Vulva	Human papilloma virus	Growths on the vulval skin	Podophyllotoxin cream
HIV and AIDS	Systemic	Human immunodeficiency virus	Chronic, progressive immunodeficiency	Combination antiretroviral therapy
Hepatitis	Systemic (liver)	Hepatitis virus B, C and E	Chronic liver disease	(Interferon-α)
Syphilis	Systemic (vulval lesion)	*Treponema pallidum*	Ulcerated nodules, but may become systemic	Penicillin

Fig. 12.25 Common sexually transmitted infections in women.

Candidiasis
Overgrowth of the yeast (a type of fungi) *Candida albicans* is responsible for thrush. Infection causes the following symptoms:
- Vulval itching and soreness.
- Redness of the vulva and vagina.
- Thick, white vaginal discharge.

It is diagnosed by microscopy of the vaginal discharge that reveals dark purple yeast spores or filaments when stained with potassium hydroxide. A vaginal tablet (pessary) of clotrimazole is used to treat the infection; the partner(s) may also need to be treated to prevent recurrence.

Trichomoniasis
This is an infection with the sexually transmitted flagellated parasite *Trichomonas vaginalis*. It is often asymptomatic, but an accompanying rise in vaginal pH may cause symptoms including:
- Thin, watery (may be green or foamy) vaginal discharge.
- Some itching and redness.

It is diagnosed by viewing the vaginal discharge mounted on saline under a microscope; the organism can be identified by the movement seen. Oral metronidazole cures 90% but a second vaginal swab should be performed 2 months later. The partner(s) should also be treated.

Infections of the vulva
The vulva is susceptible to sexually transmitted viral infections similar to those that affect skin on other areas of the body:

Herpes simplex
This is caused by the herpes simplex virus (usually HSV type II though HSV type I can also cause vulval disease). About 25% of patients experience acute symptoms including:
- Localized itching and burning.
- Multiple painful red vesicles.
- Ulceration of vesicles causing more pain.
- Dysuria if the area round the urethra is involved.
- Fever and malaise.

The vesicles appear about 3 days after infection and take up to 2 weeks to heal, during which the virus is shed and can be transmitted. The virus cannot be transmitted if vesicles are not present, but some vesicles may be hidden. The virus enters the dorsal root ganglion of sensory nerves supplying the infected area. It usually lies dormant, but it may cause recurrent attacks in 5% of patients.

Herpes simplex can be diagnosed from viral culture of fluid in the vesicles and antibody screening. It is treated symptomatically (e.g. anaesthetic cream), although the antiviral agent aciclovir can reduce the frequency and duration of recurrent attacks. The virus can be transmitted to other body parts via the hands; the eyes are particularly susceptible. The presence of herpes vesicles in late pregnancy is an indication for caesarean section.

Genital warts

These are benign growths of the epithelium caused by human papilloma viruses (HPV), of which there are many types. Some HPV strains infect the vulval skin causing small cauliflower-shaped warts that may cause itching and burning. The infection can spread to the vagina and cervix, and it is readily passed onto sexual contacts. Certain strains of HPV can predispose to dysplastic changes in the cervix, vulva and anus that may lead to carcinoma; these strains rarely cause obvious growths.

Vulval warts are treated with podophyllotoxin cream, cryotherapy, or minor surgery if they cause the patient distress. Smear tests should be repeated annually.

Bartholin's cyst

This disease is caused by bacterial infection of Bartholin's gland and duct found either side of the introitus (vaginal orifice); it is a common vulval disorder. If the duct becomes obstructed, a cyst can form that presents as a vulval swelling. The main causative bacteria are staphylococci, *E. coli*, and gonococci.

Cysts require surgical treatment and antibiotics to prevent abscess formation.

Systemic sexually transmitted infections

Other sexually transmitted diseases cause systemic illness and often have a chronic course. They are briefly described in Fig. 12.25; recent infection with syphilis can result in a genital lesion at the site of infection.

Neoplasia of the vagina

Tumours very rarely develop in the vagina, though they may spread to the vagina from the cervix and endometrium. Of the tumours that do develop in the vagina, the majority are squamous cell carcinomas found in the upper third of the vagina in elderly women. Adenocarcinoma, melanoma, and sarcoma are even rarer but tend to affect younger women.

Vaginal carcinoma presents with abnormal vaginal bleeding, pelvic pain and the detection of a lump. It is investigated by smear test and ultrasound scan to determine the origins of the carcinoma. The treatment of cervical and endometrial carcinoma are described on p. 140 and p. 134 respectively. Squamous cell carcinoma is mainly treated with radiotherapy, though surgery and chemotherapy may also be required.

Vulval dystrophies

The vulval dystrophies are non-neoplastic, chronic disorders of the vulval skin; they mainly affect menopausal or postmenopausal women, but they may affect prepubescents. They are distinct from the physiological vulval atrophy caused by oestrogen deficiency following the menopause. There are two patterns of vulvar dystrophy:

- Lichen sclerosus.
- Squamous cell hyperplasia (also called hypertrophic dysplasia or leucoplakia).

These diseases are compared in Fig. 12.26. Both conditions present with vulval itching and pain on contact leading to superficial dyspareunia; they can undergo malignant change. They are investigated using a colposcope (p. 139) and biopsy of suspicious areas under local anaesthetic. Treatment is shown in Fig. 12.26.

Neoplasia of the vulva
Malignant tumours
Vulval intra-epithelial neoplasia (VIN)

Pre-cancerous changes can be detected in the vulva (VIN) in a similar manner to the cervix (CIN; see p. 138). These changes are associated with HPV (wart virus) infection, smoking and the vulval dystrophies described above. The affected vulva may feel itchy and sore, and there may be a lump or ulcer. It is investigated using a colposcope and biopsy to determine the extent of cellular dysplasia; it is graded in a similar manner to CIN to determine treatment (Fig. 12.27). Appropriate treatment

Disorders of the Vagina and Vulva

Fig. 12.26 Comparison of lichen sclerosus and squamous cell hyperplasia.

Comparison of lichen sclerosus and squamous cell hyperplasia

Feature	Lichen sclerosus	Squamous cell hyperplasia
Epidermis	Thin	Thick
Dermis	Hyalinized (degeneration and replacement with collagen) and oedematous	Oedematous
Inflammatory cells present	Lymphocytes	Plasma cells
Appearance	Shiny, white and crinkly plaques	Thick, white/grey areas with deep skin-folds
Treatment	Vaseline or topical steroids	Topical steroids

Fig. 12.27 Classification and treatment of vulval intra-epithelial neoplasia (VIN).

Classification and treatment of vulval intra-epithelial neoplasia

Grade	VIN I	VIN II	VIN III
Classification	Mild dysplasia	Moderate dysplasia	Severe dysplasia/carcinoma *in situ*
Treatment	Topical steroids and regular follow up	Topical steroids and regular follow up	Surgical excision or destruction

ensures a 5-year survival of 100%; untreated VIN may progress to vulval cancer. The risk is significantly higher for VIN III.

Vulval carcinoma

Vulval carcinoma is an uncommon malignancy that mainly affects elderly women. It is usually a squamous cell carcinoma predisposed by HPV infection and smoking; 30% occur as a progression from VIN. They present with similar symptoms to VIN and are investigated with colposcope examination and biopsy. The staging system and 5-year survival are shown in Fig. 12.28. Vulval carcinoma is treated surgically (radical vulvectomy and lymph node dissection to various extents) and with radiotherapy in stages III and IV.

Other malignancies

Since the vulva is covered with skin, it can develop similar malignant tumours to other areas of skin. The melanocytes can give rise to malignant melanoma, and the vulval ducts can develop cancerous Paget's disease.

Benign tumours

The vulva is prone to the same benign tumours as other areas of skin. One of the most frequent tumours is papillary hidradenoma. This is a benign growth of the sweat (eccrine) glands.

Fig. 12.28 Staging and prognosis of vulval carcinoma.

Staging and prognosis of vulval carcinoma

Stage	Description	Five-year survival (%)
I	Tumour confined to vulva and perineum, <2 cm	>90
II	Tumour confined to vulva and perineum, >2 cm	80–90
III	Tumour involves lower urethra/vagina/anus and/or unilateral lymph nodes	50–80
IV	Distant metastases, involvement of upper urethra/bladder/rectum/pubic bone and/or bilateral metastases	5–30

The Female Reproductive System

Fig. 12.29 Types of breast lump with associated features.

Disease	Most common age group	Frequency	Features of lump
Fat necrosis	Any	Rare	Single, hard and irregular
Mammary duct ectasia	Older women before the menopause	Common	Tender; near the areola
Nodular fibrocystic change	Older women before the menopause	Very common	Single or multiple firm nodules
Cystic fibrocystic change	Just before the menopause	Very common	Rapidly growing smooth, rounded cysts
Fibroadenoma	Younger women (25–35 yrs)	Common	Single firm, highly mobile, non-tender lump
Phyllodes tumour	Older women	Less common	Single, large, rubbery lump
Duct papilloma	Middle-aged women	Common	Lump near the nipple with bloody discharge
Carcinoma	Middle-aged to elderly women	Common	Single hard lump with stromal interference

Disorders of the female breast

The most common presenting complaint involving the breast is a lump. Fig. 12.29 shows the most common causes and their associated features.

Congenital abnormalities
Supernumerary nipples
Failure of the fetal mammary ridge to regress can cause extra nipples (polythelia) to develop. The extra nipple is usually just below a normal breast, however they can be found anywhere along the line of the mammary ridge (axilla to inguinal rings) and very rarely in other locations. It is a common disorder affecting about 1% of people, though the extra nipple is often mistaken for a mole.

Accessory breast tissue
In females, accessory breast tissue (polymastia) can also develop, although it is usually not noticed until puberty. The extra tissue is usually in the axilla, but it can form in the same locations as extra nipples. The accessory breast tissue may be associated with a nipple to form an extra breast.

Congenital nipple inversion
The nipple is usually inverted at birth; development of the areola should cause it to rise. If this process fails the nipple may be permanently inverted, causing difficulty in breastfeeding. It is important to differentiate between congenital inversion and a recent inversion that may suggest a malignancy.

Inflammatory disorders and infections
Acute mastitis and breast abscess
When the woman is lactating, the breast is prone to bacterial infection through cracks in the nipple and areola. The usual organisms responsible are *Staphylococcus* and *Streptococcus*.

The initial infection causes an acute inflammation of the breast called mastitis, in which the breast becomes tender and enlarged. It should be treated with antibiotics to prevent breast abscesses that must be drained surgically. Chronic mastitis can also develop, but it is very rare.

Mammary duct ectasia
Mammary duct ectasia is a chronic inflammatory condition of unknown origin that causes the lactiferous ducts near the nipple to dilate. It is most common just before the menopause in women who have had children. The dilated duct fills with a creamy, protein-rich fluid that causes a green discharge from the nipple and tender lumps near the areola. The possibility of carcinoma must be excluded by biopsy or surgical excision. The dilated

ducts are prone to infection, which requires antibiotic treatment to prevent abscess formation.

Fat necrosis
Relatively minor trauma to the breast can result in necrosis of the adipose tissue (fat cells). This necrosis prompts an inflammatory reaction that can cause fibrous scarring, producing a hard, irregular lump in the breast. These lesions can mimic carcinomas including characteristic interference with the normal breast connective tissue such as skin dimpling. This condition is relatively rare, and it can only be differentiated from a malignancy by excision biopsy.

Fibrocystic change/fibroadenosis
Fibrocystic change is caused by benign growth of the breast tissue resulting in tender lumps. It is a very common disease that affects 50% of women, though only 10% are symptomatic. It is most common in older women before menopause.

The overgrowth occurs in two tissues:
- Epithelial lining of the ducts and lobules.
- Fibrous stroma.

A variation in hormonal response results in a growth imbalance between the fibrous and epithelial tissues causing solid nodules and fluid-filled cysts to develop. The nature of this imbalance can vary, resulting in three patterns of change (described below); these may coexist.

Simple fibrocystic change
Overgrowth of both tissues is seen in simple fibrocystic change producing a mixture of fibrous nodules and epithelial cysts. The imbalance in tissue overgrowth is a local effect, and many microscopic areas display these changes. Larger growth imbalances can cause symptomatic fibrocystic change.

Single or multiple fibrous nodules can develop, and these frequently become tender towards the end of the menstrual cycle (i.e. premenstrually). It is difficult to differentiate the single lumps from a carcinoma except by biopsy or excision.

Palpable epithelial cysts are most frequent close to the start of the menopause. They can be distinguished from lumps by palpation (smooth, rounded and flocculent) and their appearance on X-ray mammograms and ultrasound scans. Cysts contain a watery fluid that can be aspirated and examined for evidence of malignancy; this aspiration also treats the cyst. Simple fibrocystic change is entirely benign.

Epithelial hyperplasia
If the epithelial overgrowth predominates, the condition is called epithelial hyperplasia. It is a benign change, but it has an increased risk of breast carcinoma (two times normal risk). The risk of carcinoma is higher if the epithelial cells show signs of dysplasia, called atypical hyperplasia (five times normal risk).

Sclerosing adenosis
If the fibrous overgrowth predominates, the condition is called sclerosing adenosis. This rare variation is entirely benign. It can compress the surrounding glands and cause a histological appearance of solid cords that can be confused with invasive carcinoma on histological and X-ray (mammogram) examination.

Benign neoplasia
Fibroadenoma
Fibroadenomas are common, solitary, benign lumps that occur mostly in young women (below 35 years of age). Lumps develop from the mammary ducts and stroma in response to hormonal stimuli. They contain glandular epithelial components from the duct and fibrous components from the stroma, so they may be caused by a type of fibrocystic change instead of being true benign tumours. The lumps are usually firm, rubbery, and non-tender; they are highly mobile and can slip away during palpation (hence the name 'breast mice').

Phyllodes tumour
Phyllodes tumours resemble fibroadenomas, but they are often much larger. They can occur at any age after puberty, but usually affect older women. These lesions are usually benign, but a spectrum of dysplasia is seen: 10% are malignant.

Duct papilloma
Duct papillomas are benign growths of the mammary duct epithelia. These solitary tumours usually develop in the lactiferous duct just below the nipple in middle-aged women. In younger women they usually develop in the smaller ducts. They cause a lump and bloody nipple discharge. The tumour is surgically excised and the breast is examined by mammography to reveal any underlying carcinoma.

Multiple duct papillomas are rare and have an increased risk of breast carcinoma.

Malignant neoplasia

Breast carcinoma is the most common cancer in women, affecting about 1 in 12 at some point in their life. The risk of breast cancer increases with age, being very rare before 25 years but moderately common by 40 years of age. There are a number of risk factors for breast cancer:

- Family history—including the recently discovered autosomal dominant *BRAC 1* and *BRAC 2* mutations.
- Geographical—there is a higher risk in developed countries.
- Excess oestrogen exposure—this can be caused by many factors (e.g. early menarche, late menopause, HRT, no pregnancies) but probably not the 'Pill'.
- Epithelial hyperplasia—see the section on fibrocystic change.

Classification of breast carcinoma

The vast majority of breast cancers are adenocarcinomas. These can develop in three patterns according to their location in the breast:

- Invasive ductal carcinoma (50%).
- Invasive lobular carcinoma (30%).
- Mixed ductal and lobular carcinoma (10%).

The other 10% are rarer forms of breast cancer (see below).

All three patterns of adenocarcinoma can be preceded by non-invasive carcinoma *in situ*. With time, roughly 25% of these will develop into invasive carcinoma, but early mastectomy usually prevents this. Two forms are seen, identified by location:

- Intraductal carcinoma.
- Intralobular carcinoma.

Three other types of carcinoma occur more rarely. They have a better prognosis than invasive adenocarcinoma. They are:

- Tubular carcinoma.
- Mucoid carcinoma.
- Medullary carcinoma.

Presentation

The majority of breast cancers present as lumps or from mammogram screening. All lumps should be investigated as if they were malignant. This includes:

- Ultrasound or attempted aspiration to differentiate cysts and lumps.
- Fine-needle aspiration or excision biopsy of lumps.

Excision biopsy is only carried out on lumps or cysts with a history or appearance suggestive of malignancy. There are a number of signs of carcinoma that suggest excision biopsy is needed or that prompt investigation in the absence of a lump. These are mostly caused by interference with the breast stroma or lymphatic drainage:

- Skin dimpling.
- Recent nipple inversion.
- Bloody nipple discharge.
- Peau d'orange (skin with the appearance of orange peel).
- Surface erythema or ulceration.
- Paget's disease of the nipple.

Paget's disease is an eczema-like rash around the nipple caused by local spread of invasive ductal carcinoma.

The features of a lump or cyst that suggest malignancy are:

- Large, hard or growing masses.
- Fixing of mass to underlying structures.
- Enlargement of the affected breast.
- Blood-stained cystic fluid.
- Recurrence.
- Enlarged axillary lymph nodes.

Spread, prognosis and staging

Invasive breast carcinoma can spread by three routes:

- Local spread—within the breast or into surrounding structures, e.g. skin, nipple, pectoral muscles and the chest wall.
- Lymphatic spread—to the axillary, internal thoracic lymph nodes and the other breast.
- Blood-borne spread—especially to the bone (especially vertebral bodies), lungs and ovaries (called a Krukenberg tumour).

The prognosis of breast cancer is predicted by the stage of the tumour. This is determined from the TNM system, which assigns scores according to three measures (Fig. 12.30): **T**umour size, **N**odal involvement and **M**etastases.

In general, low scores have a better prognosis, e.g. 80% 5-year survival if no nodes are involved. The presence of metastases is a particularly poor prognostic sign, reducing the 5-year survival to 10%.

Treatment

The options for treating breast cancer are the same three for most cancers, namely:

- Surgery.

Disorders of the Female Breast

- Radiotherapy.
- Chemotherapy.

Surgical removal of the tumour is necessary if a cure is to be achieved. There are two options depending on the size of the tumour:
- Simple mastectomy (removal of the affected breast).
- Lumpectomy (removal of the lump) along with a large area of surrounding tissue.

Radiotherapy is used to prevent local and lymphatic spread. The breast, chest wall and surrounding lymph nodes are often irradiated.

Chemotherapy is aimed at preventing metastatic spread. The standard cytotoxic drugs are used along with tamoxifen, which is an oestrogen receptor blocker. This drug causes symptoms of an early menopause in younger women.

TNM staging system of breast cancer

Stage	0	1	2	3
Tumour (T)	None	<2 cm	2–5 cm	>5 cm
Nodes (N)	None	Nodes involved but mobile	Nodes involved but immobile	
Metastases (M)	None	Metastases		

Fig. 12.30 TNM staging system of breast cancer.

- Describe the structure and location of the ovaries.
- Describe the development of an ovarian follicle in the first half of the cycle. How is the oocyte transported from the ovary to the uterus?
- Describe the structure and location of the uterus and cervix. How does the cervix differ from the body of the uterus?
- Describe the structure and location of the vagina and vulva.
- Describe the structure and location of the adult female breast.
- Describe the synthesis and regulation of the three types of ovarian sex steroid.
- List the major effects of oestrogens and progestogens.
- Describe the hormonal changes during the menstrual cycle.
- Describe the ovarian changes during the menstrual cycle.
- Describe the endometrial changes during the menstrual cycle.
- Discuss the presentation, diagnosis and treatment of pelvic inflammatory disease. Which organisms are usually responsible?
- Describe the hormonal changes and symptoms of polycystic ovarian syndrome.
- List the common types of ovarian cysts and tumours along with a brief description of each. How are they treated?
- Describe the presentation, diagnosis and treatment of endometriosis. How does it differ from adenomyosis?
- List the conditions that can be caused by excess oestrogen.
- Describe the presentation, diagnosis and treatment of fibroids and endometrial carcinoma.
- List the common causes of amenorrhoea, menorrhagia, and dysmenorrhoea.
- Describe the diagnosis, treatment and natural progression of cervical intra-epithelial neoplasia.
- Discuss common infections of the vagina and vulva.
- List the common causes of lump in the breast. How are they investigated in clinic?

13. The Male Reproductive System

The male reproductive system must perform two main functions:
- Spermatogenesis—production of sperm (male gamete).
- Ejaculation—expulsion of sperm into the vagina.

After puberty, the testes begin to produce sperm and they continue to do so until death. This process is regulated by three main hormones (Fig. 13.1):
- Follicle stimulating hormone (FSH).
- Luteinizing hormone (LH).
- Testosterone.

The ejaculation of sperm is controlled by neural stimuli from the sympathetic system. The sperm are ejected along with seminal fluid that protects them and provides nutrients.

After reading this chapter you should be able to:
- Visualize the anatomy of the male reproductive system.
- Describe the hormones that regulate the male reproductive system.
- Understand the process of sperm production.
- Discuss the common disorders of the male reproductive system.

Important words:
Inguinal canal: a canal allowing passage of the testes and associated vessels from the abdominal cavity to the scrotum
Androgen: male sex steroids, e.g. testosterone
Spermatozoa: the medical term for sperm
Spermatid: final stage in the development of a spermatozoon
Spermatogenesis: the process of sperm production

Fig. 13.1 Hormonal regulation of the male reproductive system. (FSH, follicle stimulating hormone; GnRH, gonadotrophin releasing hormone; LH, luteinizing hormone.)

Organization

The male reproductive system consists of five main components:
- Testes—produce the sperm.
- Epididymis—store and mature the sperm.
- Ductus deferens—transmit the sperm from the epididymis to the penis.
- Prostate and seminal vesicles—secrete seminal fluid to support and protect the ejaculated sperm.
- Penis—becomes erect to penetrate the vagina and deposit the sperm at the cervix.

All of these components lie outside the peritoneal cavity, however there is no opening into this cavity

149

The Male Reproductive System

comparable to the infundibulum of the fallopian tubes in the female. Each of these components is discussed individually in the following sections. Their locations are shown in Fig. 13.2 and their blood supply, lymphatics and innervation are shown in Fig. 13.3.

Testes

The testes are two oval-shaped organs that produce sperm (the male gametes) in response to gonadotrophins (LH + FSH) from the pituitary gland and testosterone from the Leydig cells of the testes. They are suspended in the sac-like scrotum by the spermatic cord; this keeps their temperature 2–3°C lower than body temperature. If their temperature rises then sperm production ceases. Each testis is surrounded by a capsule of three layers (starting nearest the testis):

- Tunica vasculosa—loose connective tissue with blood vessels.
- Tunica albuginea—fibrous connective tissue.
- Tunica vaginalis—derived from peritoneum, it also surrounds the epididymis.

Microstructure

Fibrous septa divide each testis into about 300 lobules, each containing 1–4 seminiferous tubules that produce sperm. The seminiferous tubules are closed loops lined with a specialized epithelium that contains two types of cells:

- Germinal epithelium containing spermatogonia that undergo meiosis to give rise to haploid sperm.
- Sertoli cells that support the developing sperm and secrete testicular fluid into the tubules.

The process of sperm production (spermatogenesis) is discussed on p. 154. In between the seminiferous tubules there are vessels and clusters of testosterone-secreting interstitial (Leydig) cells. The microstructure of the seminiferous tubules is shown in Fig. 13.4.

Inside the testis, the loops of the seminiferous tubules drain into the rete testis. These are convoluted networks of ducts that are lined by simple cuboidal epithelial cells with microvilli and a single flagella. All the vessels that support the testis enter the testis capsule at the location of the rete testis. The sperm pass from the rete testis to the epididymis via about 15 efferent ductules lined by ciliated epithelium.

Epididymis

The epididymis is a firm, comma-shaped structure attached posteriorly to the top of the testis within

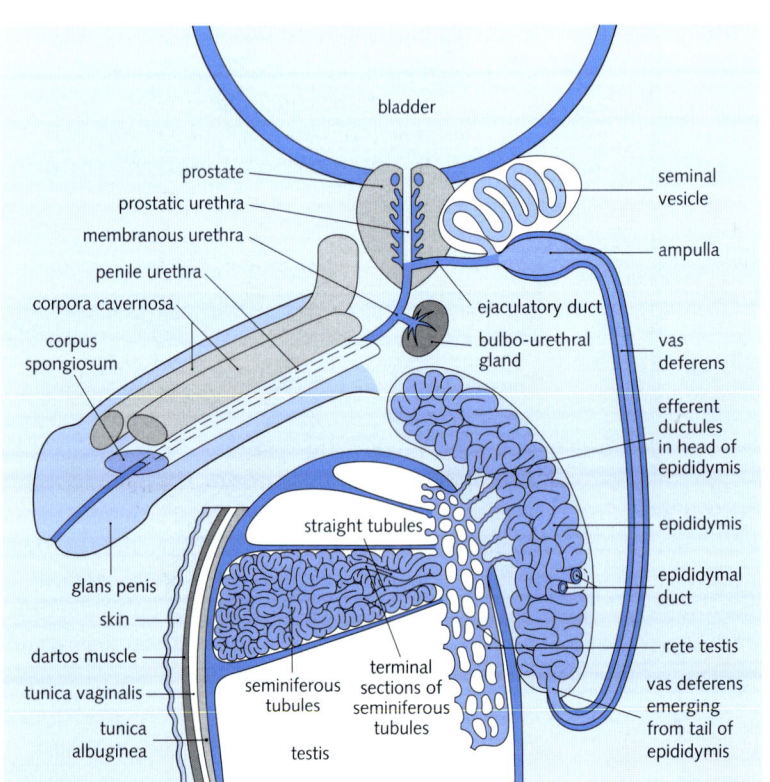

Fig. 13.2 Arrangement of the male reproductive system. This diagram is not to scale.

Organization

Fig. 13.3 The blood supply, lymphatics and innervation of the male reproductive organs.

Blood supply, lymphatics and innervation of the male reproductive organs

Organ	Arterial supply	Venous drainage	Innervation	Lymphatic drainage
Testis	Testicular arteries from the aorta via the spermatic cord	Pampiniform plexus, which forms the testicular veins. Left drains into the left renal vein, right into the inferior vena cava	Sympathetic innervation via the splanchnic nerves	Para-aortic lymph nodes
Scrotum	Pudendal arteries	Scrotal veins	Branches of the genitofemoral, ilioinguinal and pudendal nerves	Superficial inguinal lymph nodes
Prostate	Vesicular and rectal branches of the internal iliac artery	Prostatic venous plexus drains into the internal iliac veins	Parasympathetic via splanchnic nerves; sympathetic from inferior hypogastric plexus	Internal iliac and sacral lymph nodes
Penis	Internal pudendal arteries	Venous plexus, which joins the prostatic venous plexus	Branches of the pudendal nerve	Superficial inguinal lymph nodes

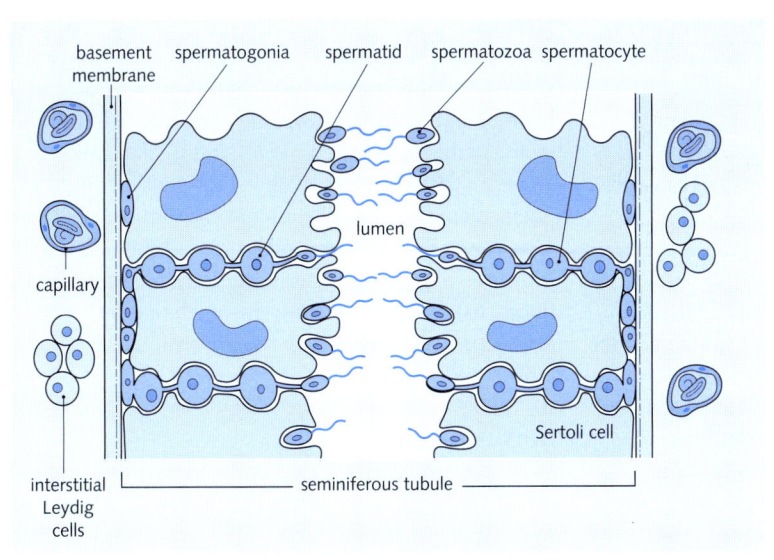

Fig. 13.4 Microstructure of the seminiferous tubule.

> The prefix 'orch-' is used to describe processes relating to the testes and is derived from the Greek word for the testes: 'orchis'. This prefix is also used for a group of plants (orchids), which could be because their roots resemble testes or due to potential aphrodisiac properties!

the scrotum. Normally it can be distinguished from testis by palpation. It is described in three sections:
- Head—superior section where the efferent ductules enter.
- Body—in between the head and tail.
- Tail—inferior section, continuous with the ductus deferens.

The epididymis contains a single, 5 m long, tightly coiled tube formed by the fusion of the efferent ductules from the rete testes. Testicular fluid is

151

partially reabsorbed by the tall columnar epithelium with long microvilli (stereocilia). The sperm pass slowly through this tube to reach the ductus deferens at the tail of the epididymis. The epididymis is surrounded by a fibrous capsule that is separated from the testis by the tunica vasculosa and tunica albuginea, except at the head where the testis and epididymis join. The tunica vaginalis surrounds the testis and the fibrous capsule of the epididymis.

Scrotum

The scrotum is a sac-like structure that contains both testes, epididymides and the start of ductus deferentia. The skin is usually pigmented and wrinkled with a line down the midline called the scrotal raphe.

Microstructure

The wall of the scrotum is formed by five layers; three of these continue as the covering of the spermatic cord. All five are continuous with the abdominal wall from which they are derived. They are described in Fig. 13.5 and their development is shown in Fig. 11.5. Two muscles are found in the scrotum:

- Dartos muscle—contracts the scrotal skin in response to cold.
- Cremasteric muscle—retracts the testis towards the abdomen.

Ductus (vas) deferens

The ductus deferens is the continuation of the epididymis on each side of the scrotum. It transports sperm from the epididymis to the ejaculatory ducts during the emission phase of an ejaculation (see Ch. 14). It has three muscular layers to propel the spermatozoa along its 40 cm length and a lining epithelium similar to that found in the epididymis. The ductus deferens is palpable in the scrotum; this allows male sterilization with only minimal incisions.

Once the ductus deferens enters the abdomen in the spermatic cord it passes along the lateral wall of the pelvis, external to the peritoneum. It crosses the ureter and descends to the base of the bladder. The duct widens into the ampulla before the junction with the duct of the seminal vesicle.

Spermatic cord

The spermatic cord is formed at the deep inguinal ring of the abdominal wall and passes along the inguinal canal and then inferiorly in the scrotum to the testis. It contains five structures:

- Ductus deferens.
- Testicular artery.
- Pampiniform plexus of veins.
- Autonomic nerves (including the sensory nerves that transmit pain from the testis).
- Lymph vessels.

These structures are surrounded by three layers (see Fig. 13.5).

> The contents of the spermatic cord can be remembered as 'Nymphomaniacs Love Trying Daring Positions', i.e. Nerves, Lymphatics, Testicular artery, Ductus deferens and Pampiniform plexus.

Seminal vesicles and ejaculatory ducts

The seminal vesicles are two 6 cm long, pear-shaped structures found above the prostate between the bladder and rectum. They were originally believed to store sperm (hence their name), but in fact they secrete a fructose-rich, alkaline fluid that forms 70% of ejaculated semen. The duct of the seminal vesicles joins the ductus deferens at the ampulla behind the prostate. Together they form the 1 cm long ejaculatory ducts that pass forward into the prostate gland. The ejaculatory ducts open into the prostatic urethra just before it leaves the prostate (see Fig. 13.2).

Microstructure

Each seminal vesicle is formed from a slightly coiled tube that is 15 cm long. It is covered by two layers of smooth muscle and an external fibroelastic layer. The tube is lined by tall, secretory columnar epithelium.

The ejaculatory ducts have no muscular coverings. They are lined by tall, columnar cells and smaller, rounded cells.

Prostate

The prostate is a walnut sized gland that surrounds the prostatic urethra at the base of the bladder. The posterior surface is palpable by rectal examination; it should feel regular and firm with a midline groove. The prostate secretes seminal fluid into the urethra during ejaculation. This fluid is rich in acid phosphatase and citric acid.

Organization

Layers of the abdominal wall, spermatic cord and scrotum		
Abdominal layer	**Spermatic cord layer**	**Scrotal layer**
Skin	None	Skin
Superficial fascia	None	Dartos muscle
External oblique aponeurosis	External spermatic fascia	External spermatic fascia
Internal oblique muscle and fascia	Cremasteric muscle and fascia	Cremasteric muscle and fascia
Transversalis fascia (not the muscle)	Internal spermatic fascia	Internal spermatic fascia

Fig. 13.5 Layers of the abdominal wall, spermatic cord and scrotum.

Inferior to the prostate, the two small bulbourethral glands secrete sugar-rich mucus into the membranous urethra. This fluid may lubricate the urethra prior to ejaculation.

Microstructure

The prostate gland is composed of three concentric rings of glands surrounded by smooth muscle and a fibrous capsule. The smooth muscle contracts during ejaculation to squeeze the prostatic secretions into the urethra. The three types of gland are:
- Inner periurethral (mucosal) glands—these secrete directly into the urethra, which they surround; they can undergo benign prostatic hypertrophy.
- Outer periurethral (submucosal) glands—these secrete into the urethra via short ducts; they can undergo benign prostatic hypertrophy.
- Peripheral zone glands—this ring of glands is incomplete anteriorly; they secrete via long ducts and can give rise to prostatic cancer.

The glands are lined by tall columnar epithelium with a few flat basal cells. Their secretory activity is dependent on testosterone.

Penis

The penis is the outlet for urine and semen. The internal structure of the penis is shown in Fig. 13.6. It is described in three sections:
- Root—the section that is attached to the perineum.
- Body—the free portion, also called the shaft; it is suspended from the pubic symphysis.
- Glans—the sensitive, distal end of the body that includes the external opening of the urethra.

The edge of the glans that joins the body of the penis is called the corona. The penis is composed of three cylinders of erectile tissue, each surrounded by a fibrous capsule called the tunica albuginea. Surrounding these structures is a layer of thin, pigmented skin; over the glans this skin is called the foreskin or prepuce. The three cylinders of erectile tissue are:
- Two corpora cavernosa—these large cylinders form the dorsum (upper surface) and sides of the penis; they form the crura that support the erect penis in the root.
- One corpus spongiosum—this smaller ventral (lower surface) cylinder surrounds the spongy urethra; it forms the whole of the glans and is called the bulb at the root.

The majority of vessels and nerves supplying the penis are found on the dorsal (upper surface) side. The arteries within the corpora cavernosa are called the deep arteries.

Microstructure and mechanism of erection

The erectile tissues consist of interconnecting vascular spaces that fill with blood during an erection. Erection is normally prevented by shunts between the helicine arteries (branches of the deep arteries) and the deep veins. Stimulation by the parasympathetic nervous system constricts these shunts so that blood fills sinuses of the cavernosa. The increase in pressure collapses the thin veins preventing blood from leaving and erection occurs. Ejaculation requires sympathetic stimulation, which inhibits the parasympathetic system causing a loss of erection.

The Male Reproductive System

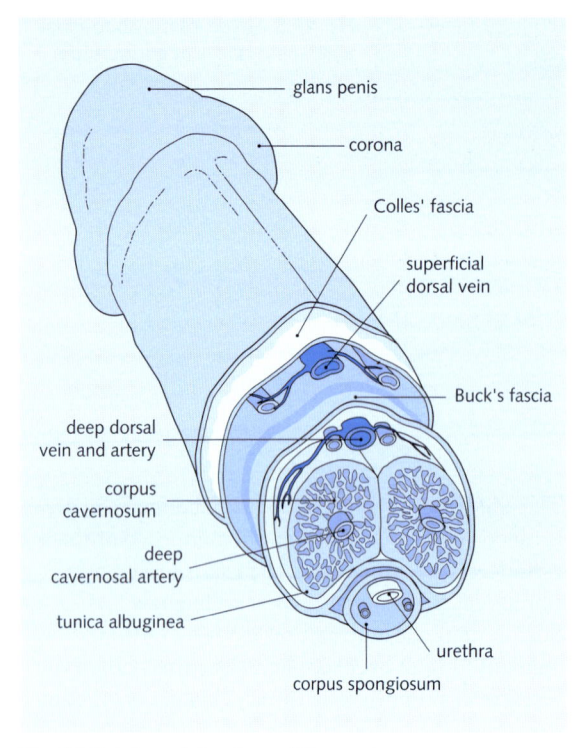

Fig. 13.6 Internal structure of the penis.

Labels: glans penis; corona; Colles' fascia; superficial dorsal vein; Buck's fascia; deep dorsal vein and artery; corpus cavernosum; deep cavernosal artery; tunica albuginea; urethra; corpus spongiosum.

Remember 'Point and Shoot' to recall the function of the Parasympathetic and Sympathetic innervation of the penis.

Spermatogenesis

Early spermatogenesis
Spermatogenesis is the process by which haploid (23 chromosome) spermatozoa are formed from diploid (46 chromosome) stem cells called spermatogonia. Spermatogonia are found close to the basement membrane of the seminiferous tubules. There are three types:
- Dark A cells (Ad)—these are the true stem cells; they divide by mitosis to produce more Ad cells and a few Ap cells.
- Pale A cells (Ap)—these represent the first step towards differentiation into spermatocytes; they divide by mitosis to produce more Ap cells and a few B cells.
- B cells—these divide by mitosis to produce more B cells and a few primary spermatocytes.

The mitotic cell divisions of spermatogonia are incomplete, so that the daughter cells remain connected by thin cytoplasmic bridges. These bridges remain throughout the development of the spermatocytes and spermatids until the spermatozoa are released into the lumen of the seminiferous tubule.

Meiosis
The primary spermatocytes undergo meiosis, which is the special form of cell division that produces haploid gametes. The first division forms two haploid secondary spermatocytes. The second meiotic division occurs soon afterwards to form four haploid spermatids.

Meiosis in the male gives rise to four functional, haploid sperm. Meiosis in the female only produces one functional haploid oocyte; the other haploid cells are non-functional polar bodies.

As meiosis progresses, the germinal cells migrate from the basal membrane to the apex of the Sertoli cells (towards the lumen of the seminiferous tubule). The Sertoli cells perform a number of functions that are essential for spermatogenesis:
- They provide nutrients and remove waste.
- They phagocytose excess cytoplasm or poorly developing spermatids.
- Tight junctions between cells prevent antibodies reaching the haploid cells; they form a blood–testes barrier.
- They produce androgen binding protein to raise the local androgen concentration.

Spermiogenesis
Having completed the meiotic division, the haploid spermatids are small, spherical cells that must still develop the structure of a mature spermatozoon. This process is called spermiogenesis and takes place on the surface of Sertoli cells in four stages:
- Golgi phase—the golgi apparatus begins to form an acrosomal vesicle over the nucleus while the

centrioles migrate to the other end of the cell. One of the centrioles begins to form the tail.
- Cap phase—the acrosomal vesicle surrounds the front of the nucleus while the nucleus condenses.
- Acrosome phase—the nucleus becomes smaller and lengthens squeezing the acrosome and cell membrane together. Mitochondria migrate towards the tail to form the 'middle piece' while microtubules condense behind the nucleus to form the neck or 'manchette'.
- Maturation phase—excess cytoplasm is 'pinched off' and phagocytosed by the Sertoli cell. The completed spermatozoon is then released into the lumen of the seminiferous tubule by breaking off the cytoplasmic bridges.

Final maturation

The maturation from Ad cells to released spermatozoa in the seminiferous tubule lumen takes about 64 days. Further maturation occurs as they pass through the epididymis to make the spermatozoa motile and capable of fertilization. The spermatozoa are pushed through the rete testis and efferent ductules by the movement of testicular fluid, caused by the action of the cilia, to reach the epididymis.

The epithelium of the epididymis secretes glycoproteins that bind to the surface of the spermatozoa causing the phospholipid membrane to be remodelled. The epididymis is also capable of removing degenerate or poorly formed spermatozoa by phagocytosis. Spermatozoa pass through the epididymis by the movement of testicular fluid and peristalsis.

The fully mature spermatozoa are stored in the epididymis until they are ejaculated or broken down. The epididymis contracts during orgasm to transport the spermatozoa into the ductus deferens. The ductus deferens has a thick muscular layer that propels the sperm along the duct at ejaculation.

Continuous fertility

After puberty males are always fertile, though their fertility may decline with age. Continuous fertility is achieved because there is a population of stem cells in the testes (cf. ovaries):
- A new cycle of spermatogenesis starts every 16 days (and lasts 64 days) at each point in the tubule.
- Sertoli cells are arranged in bands in which cycles of spermatogenesis begin at different times.
- Cycles are at different stages in different segments of the seminiferous tubules.

Mature spermatozoon

Mature spermatozoa have a distinct head and tail (Fig. 13.7). The head is composed largely of condensed chromatin in the pointed nucleus; the front is surrounded by a giant lysosome called the acrosome that allows the spermatozoa to penetrate the oocyte.

The tail is a long and specialized flagellum derived from one of the centrioles. It has the usual pattern of microtubules with nine outer pairs around a central

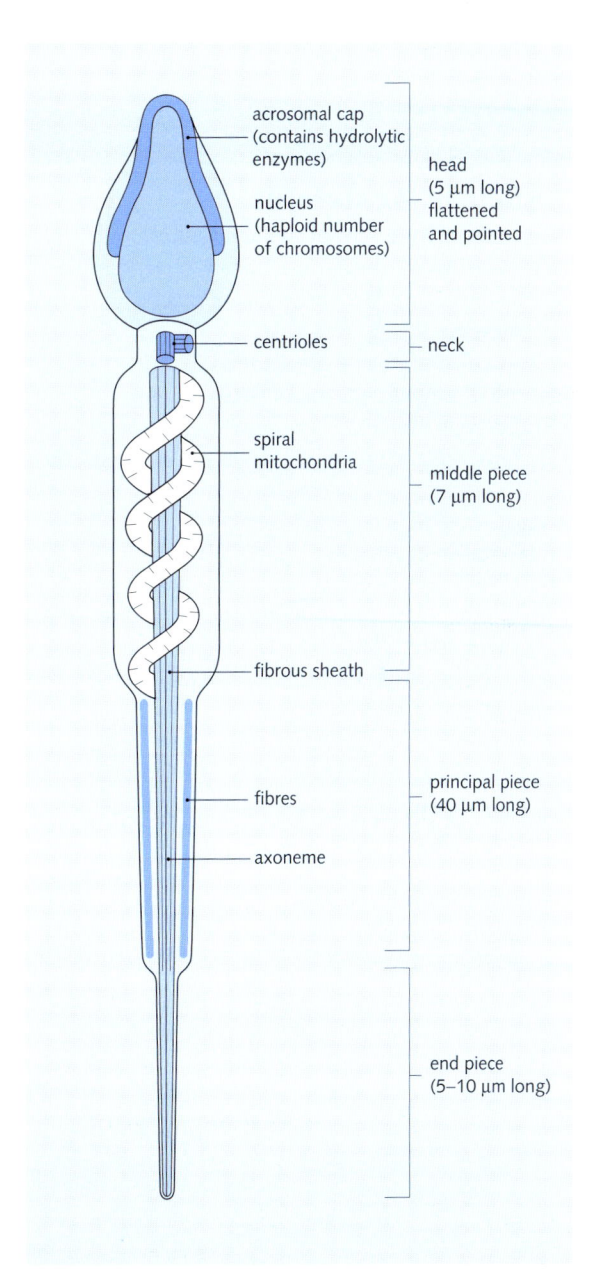

Fig. 13.7 Microstructure of a mature spermatozoon. This diagram is not to scale.

pair; this structure is called the axoneme. There are four sections of the tail:
- Neck—this narrowing contains the centrioles connected to the axoneme.
- Middle piece—the axoneme is surrounded by elongated spiral mitochondria. These release energy to drive the axoneme by the anaerobic respiration of fructose.
- Principal piece—this forms the majority of the tail. It contains fibres, which are not present in the smaller end piece.
- End piece.

Hormones

Testicular sex steroids

The testes secrete 95% of the male sex steroids called androgens; the adrenal cortex is responsible for the remaining 5%. The main androgen is testosterone.

Testicular androgens are secreted by the interstitial Leydig cells found between the seminiferous tubules. They convert cholesterol into the steroid testosterone by a series of reactions. The Leydig cells also secrete small quantities of oestrogens and progestogens as by-products of testosterone synthesis.

Testosterone is a strong androgen. However, some target tissues can convert it to the more potent form called dihydrotestosterone (DHT). The conversion requires the 5-reductase enzyme and occurs in the:
- Seminiferous tubules.
- Prostate gland.
- Skin.

Testosterone is transported in the plasma by sex-hormone binding globulin (SHBG) or albumin. It acts via intracellular receptors to regulate protein synthesis producing the actions shown in Fig. 13.8. The main actions are:
- Growth and development of male reproductive tract.
- Development of male secondary sexual characteristics (e.g. male hair pattern, muscle growth).
- Stimulation of spermatogenesis.
- Stimulation of growth and the fusion of the growth plates of the long bones (see Ch. 9).

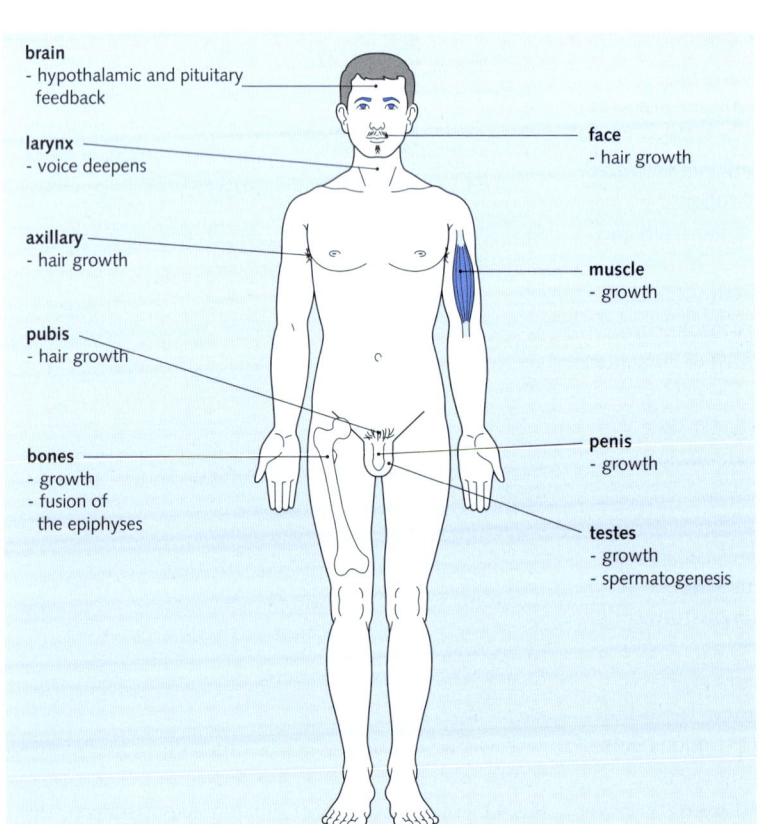

Fig. 13.8 Actions of testosterone.

Control of testicular steroid production

Testosterone synthesis and release is controlled by the same hormones in the male as oestrogen synthesis in the female. Gonadotrophin releasing hormone (GnRH) from the hypothalamus is transported to the anterior pituitary gland by the portal veins. It stimulates the gonadotroph cells to secrete gonadotrophins (LH and FSH). This process is described in more detail in Chapter 2.

LH acts on the Leydig cells to stimulate the first step in testosterone production. Testosterone feeds back to the hypothalamus and pituitary gland to inhibit LH release, but it has little effect on FSH.

FSH acts on the Sertoli cells; it increases the number of testosterone receptors to stimulate spermatogenesis. It also causes inhibin release from the Sertoli cells, which feeds back to the hypothalamus and pituitary gland to inhibit further FSH release. It has little effect on LH.

Other testicular hormones

The fetal Sertoli cells produce the Müllerian inhibiting substance (MIS; described in Ch. 11). This hormone prevents development of the female internal genitalia by causing the Müllerian ducts to regress.

Disorders of the testes and epididymis

Congenital abnormalities and regression
Cryptorchidism

Cryptorchidism is the medical term for testes that have failed to descend into the scrotum. It is a common finding in newborns (3–4%), especially premature babies, and it can affect one or both testes. An undescended testis must be distinguished from a testis retracted due to cold (by the cremasteric muscle). It is only undescended if it cannot be massaged into the scrotum or cannot be felt at all. An impalpable testis usually lies in the inguinal canal or at its abdominal entrance (deep inguinal ring). The descent of the testis and formation of the scrotum are described in Chapter 11.

In the majority of babies, this condition resolves with no treatment. However, if the testes still have not descended after a year their development can be affected. Failure to treat the condition at this stage causes a high risk of infertility because the spermatogonia (stem cells) need cooler temperatures to survive. Testosterone production is not affected because the Leydig cells are not as sensitive. Cryptorchidism also carries a higher risk of germ-cell testicular cancer in later life. It is corrected by surgically fixing the testes in the scrotum—an operation called orchidopexy.

The cause of cryptorchidism is unknown, but both testosterone deficiency and oestrogen excess (from the environment) have been suggested.

Abnormalities of the tunica vaginalis
Congenital hernia

The processus vaginalis is a tube of peritoneum that connects the tunica vaginalis to the abdominal cavity through the inguinal canal in the fetus. This structure is shown in Fig. 11.5; a persistent processus vaginalis is shown in Fig. 13.9. If it fails to close then intestines may pass into the scrotum causing an indirect inguinal hernia (through the inguinal canal). This condition is more common in the presence of undescended testes. It must be corrected surgically to prevent the risk of bowel obstruction and ischaemia due to the narrow inguinal ring. Femoral and direct inguinal hernias are very rare in neonates.

Congenital hydrocoele

If the processus vaginalis persists but is too small for a hernia to form then peritoneal fluid may enter the tunica vaginalis causing a hydrocoele. This presents as a transluminable swelling within the scrotum that cannot be distinguished from the testis. It often regresses without treatment, but if not it should be drained. Recurring or large hydrocoeles require surgical correction. Throughout life hydrocoeles are the most common cause of intra-scrotal swelling; the fluid may be a sign of inflammation or neoplasia of the testis.

Trauma and vascular disturbances
Testicular haemorrhage
Haematoma

Trauma to the groin can cause bleeding within the testes. If the bleeding remains within the intact tunica albuginea it is called a testicular haematoma. It is extremely painful.

Haematocoele

If the tunica albuginea splits, blood can collect in the tunica vaginalis forming a haematocoele. This can also be caused by a testicular tumour in the absence of trauma. If the haemorrhage is not drained the

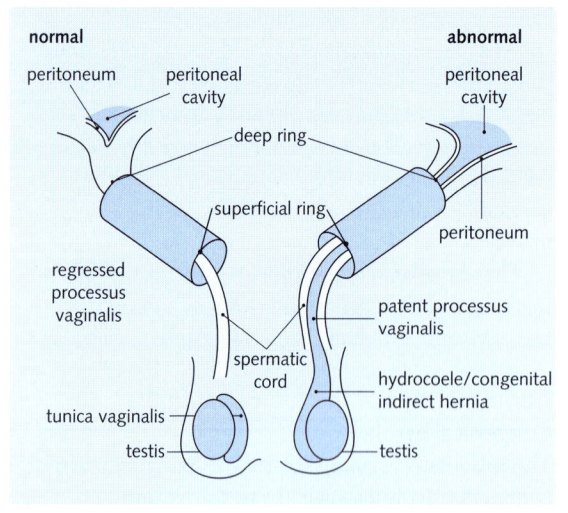

Fig. 13.9 Diagram of a normal testis and one with a persistent processus vaginalis causing a hydrocoele or indirect hernia.

blood can coagulate and constrict around the testis potentially causing ischaemia and necrosis of the testis.

Torsion of the testis

Torsion of the testis occurs when the spermatic cord becomes twisted, blocking the blood vessels from a testis. It can only occur if the testes are free to rotate within the scrotum due to a congenital abnormality. Normally the tunica vaginalis attaches to the spermatic cord close to its origin from the epididymis to prevent rotation. In the abnormal state the tunica vaginalis attaches further along the spermatic cord. If the testis rotates excessively, the pampiniform venous plexus within the spermatic cord can become twisted and occluded. Arterial blood continues to enter the affected testis but cannot leave, so the testis swells and undergoes venous infarction leading to haemorrhagic necrosis.

Testicular torsion is most common in children and adolescents following mild trauma, especially sporting injuries. It presents with the sudden onset of pain and tenderness in one testicle. On palpation the affected testis is high in the scrotum and the spermatic cord may be thickened. With time, the pain becomes severe with vomiting and dull abdominal pain as the affected testicle swells.

It can be difficult to distinguish the later stages of torsion from acute orchitis. Since testicular torsion is an acute surgical emergency an unclear diagnosis needs surgical exploration. Without treatment the testis will undergo necrosis and need to be removed.

Early surgery (before about 6 hours) can untwist and save the testicle; both testes should be secured to the scrotum to prevent recurrence.

Varicocoele

When the pampiniform plexus in the spermatic cord becomes varicose (dilated and tortuous) it is called a varicocoele. It is found in about 10% of young men and in 90% of these the spermatic cord on the left is affected. This may be due to the higher pressure in the renal vein compared with the inferior vena cava (see Fig. 13.3). The underlying cause is not known, although rapidly developing varicocoeles may very rarely be caused by carcinoma of the left kidney obstructing the left testicular vein.

The testes should be examined with the patient standing up because the blood often drains out of the plexus when they lie down. This response is inhibited if there is an underlying obstruction (e.g. renal carcinoma). The full varicocoele feels like a 'bag of worms' above the testes.

Occasionally varicocoeles present with discomfort, but the vast majority are asymptomatic. The excess blood can warm the testes reducing the fertility of the patient. Varicocoeles can be treated surgically, but this is only considered if discomfort or infertility are present.

Inflammation and infection
Acute orchitis and epididymitis

Inflammation of the testis is called orchitis while inflammation of the epididymis is called epididymitis. These two conditions often occur together and they are managed in a similar manner. The following acute bacterial infections are the most common causes of epididymo-orchitis:
- *Escherichia coli* and other coliforms (from a urinary tract infection; UTI).
- *Chlamydia trachomatis* (sexually transmitted disease; STD).
- *Neisseria gonorrhoeae* (STD).

Patients complain of either one or two painful, tender and enlarged testes, which may be accompanied by a secondary hydrocoele, general malaise, fever or headache. It is diagnosed from the history and examination of the urine, but surgical exploration is often performed to exclude testicular torsion. Antibiotics are used to treat the infection otherwise scarring and infertility can develop.

Other infections
Mumps
The mumps virus often causes viral orchitis if it infects post-pubertal males. It tends to cause a single tender and enlarged testis. Rarely it can cause infertility if the disease is bilateral.

Tuberculosis
Tuberculosis (TB) can infect the epididymis and testis from the blood or urinary tract. It causes chronic granuloma with caseous necrosis that can persist long after infections at other sites have been successfully treated.

Syphilis
The rare tertiary stage of *Treponema pallidum* infection (i.e. syphilis) can produce the characteristic chronic syphilitic granuloma (called gumma) in the testes.

Autoimmune granulomatous orchitis
Granulomatous orchitis is a rare autoimmune disease that causes chronic inflammation of the testis and destruction of the seminiferous tubules. The tight junctions between the Sertoli cells normally prevent an immune reaction against the developing haploid sperm, but this barrier appears to break down.

It presents in a similar manner to a testicular tumour; the affected testis becomes firm and enlarged, sometimes with a secondary hydrocoele.

Epididymal cysts
Epididymal cysts often develop in adult life and are found in the head of the epididymis. The cysts contain either a watery fluid or a milky fluid containing sperm (this can be called a spermatocoele). They usually present as painless swellings that can be distinguished from the testis and transluminated. If they cause symptoms they can be removed surgically.

Neoplastic disorders
Tumours of the testes are important because they are the most common type of cancer in young adult males. They are often highly malignant but frequently curable with early detection. If undetected they can spread to the para-aortic lymph nodes or, by blood, to the lungs and liver. Patients often present with:
- A painless, enlarged, hard testis or testicular lump.
- Secondary hydrocoele.
- Gynaecomastia (due to human chorionic gonadotrophin (hCG) or oestrogen secretion).
- Metastases to the lungs or liver.

Testicular ultrasound is used to distinguish cystic swellings from testicular tumours. Two tumour markers can be detected in the blood; these are used in diagnosis and monitoring treatment:
- α-Fetoprotein (AFP).
- Human chorionic gonadotrophin (hCG).

Other tests are aimed at detecting any tumour spread:
- Chest X-ray to check the lungs.
- Abdominal computed tomography (CT) for liver and lymph nodes.

Testicular carcinoma is treated by orchidectomy (removal of testis) using an inguinal incision. The spermatic cord should be clamped before the testis is removed to help prevent venous spread. If the testis appears normal and on-the-spot biopsy analysis does not reveal malignancy then the testis can be returned to the scrotum. Confirmed carcinoma is followed up with radiotherapy or chemotherapy depending on the type of testicular tumour. The prognosis also depends on the type of tumour, but it is generally very good. If the lymph nodes are not involved there is almost a 100% 5-year survival rate.

There are two groups of testicular tumour:
- Germ-cell tumours (97%).
- Sex-cord stromal tumours (3%).

The different types within these groups are shown in Fig. 13.10.

Germ-cell tumours
Germ-cell tumours develop from the gamete-producing spermatogonia in the seminiferous tubules. They are predisposed by undescended testes. There are three types of germ-cell tumours, described below.

Seminoma
This tumour accounts for 50% of germ-cell tumours, and it is most common in 40–50 year-olds. It presents as a painless enlargement of one testis and histology reveals a creamy-white tumour. Ten per cent of tumours secrete hCG as an ectopic hormone. They usually spread via the lymphatics. They are very sensitive to radiotherapy, so the prognosis is good.

Types of testicular tumour and the age group they commonly affect			
Tumour type	Tumour	Main age group (years)	Testicular malignancy (%)
Germ-cell tumours	Seminoma	40–50	50
	Teratoma	20–30	35
	Mixed tumour	20–40	12
Sex-cord stromal tumours	Leydig-cell tumour	Any	<1
	Sertoli-cell tumour	40–50	<1
	Primary lymphoma	65+	2

Fig. 13.10 Frequency of types of testicular tumours and the age group they commonly affect.

Teratoma

These tumours are most common in 20–30 year-olds. Teratomas are composed of a number of tissue types derived from endoderm, mesoderm and ectoderm. Well-differentiated tumours are usually benign; more commonly, the tissues are undifferentiated and immature causing a highly malignant tumour (cf. ovarian teratomas). There are a number of categories of teratoma according to their histological appearance. They often secrete α-fetoprotein and hCG because they develop from yolk sac and trophoblast tissues. They metastasize early to the lungs and have a poorer prognosis than seminomas, but they are very sensitive to chemotherapy.

Mixed tumours

Some tumours contain different types of teratomas or a combination of seminoma and teratoma tissues. Their behaviour and prognosis depends largely on the types of teratoma involved.

Sex-cord stromal tumours

Tumours that develop from the other tissues in the testes are called sex-cord stromal tumours. They account for only 3% of testicular tumours. There are three types, described below.

Leydig-cell tumours

These are also called interstitial or stromal-cell tumours. They can occur at any age and are usually benign. They frequently secrete testosterone or oestrogens which can affect the reproductive system.

Sertoli-cell tumours

These are also called androblastomas or sex-cord tumours. They can occur at any age, but they are most common between 40 and 50 years of age. They are often benign, but they may secrete ectopic oestrogens.

Primary lymphoma

In the elderly, lymphomas can develop in the testes; aggressively malignant non-Hodgkin's lymphomas are the most common type of testicular tumour in those aged 65 years or over.

Disorders of the prostate

Inflammation and infection
Bacterial prostatitis

The prostate can become inflamed due to bacterial infection; this can follow an acute or chronic course.

Acute

Bacteria reach the prostate from the urethra so the common causative organisms are those which cause urinary tract infections or are sexually transmitted, namely:
- *E. coli* and other coliforms (UTI).
- *Neisseria gonorrhoeae* (also called gonococcus; an STD).
- *Chlamydia trachomatis* (STD).

Patients present with increased urinary frequency, penile and testicular pain and systemic illness (e.g. fever). The prostate gland becomes enlarged and acutely tender on rectal examination. Severe inflammation can obstruct the urethra causing urinary retention and abscesses may develop leading to a urethral discharge of pus.

It is diagnosed from urine culture and treated with appropriate antibiotics (e.g. trimethoprim or erythromycin).

Chronic
Failure to cure acute prostatitis can lead to chronic prostatitis. It can also be caused by tuberculosis infection, often from the kidney or epididymis. The symptoms are similar to acute prostatitis but ill-defined and less severe. It is managed in the same manner as an acute infection or with treatment for TB.

Abacterial prostatitis
Chronic prostatitis can also be caused by non-infective disease in which anti-inflammatory drugs may ease symptoms. The exact aetiology is unknown but theories include:
- Allergy, since it is associated with asthma.
- Autoimmune disease.
- Urine reflux.

Benign prostatic hypertrophy
The prostate gland begins to enlarge from the age of 45 years. By the age of 70 years the vast majority of men have some benign enlargement, making it the most common disease of the prostate gland. It is caused by nodular hyperplasia (increased cell division forming nodules) of the inner and outer periurethral glands of the prostate. The underlying fault is believed to be an imbalance between oestrogen and androgen secretion. It does not develop into prostate carcinoma.

The enlarged prostate gland can compress the urethra causing difficulty urinating called 'prostatism'. This only occurs in the more severely affected men and it is characterized by the following symptoms:
- Increased frequency of micturition (urinating).
- Increased urgency.
- Reduced size and force of urinary stream.
- Hesitancy and interruption.
- Dribbling at the end of micturition.

If the hypertrophy is very severe, urinary obstruction may result due to compression of the internal urethral sphincter. Chronic urine retention follows, leading to recurrent urinary tract infections or renal impairment.

Benign prostatic hypertrophy is diagnosed by the history and rectal examination in the absence of evidence for renal failure or carcinoma of the prostate. Mild disease can be treated medically with α-blockers to lower prostate tone. More progressive disease may be treated by transurethral resection of the prostate (TURP), however this procedure can often cause retrograde ejaculation and occasionally impotence. The histology of the resected prostate tissue should be examined under a microscope to exclude malignancy.

Neoplastic disorders
Prostatic carcinoma is the second most common male carcinoma after lung cancer. It usually affects the elderly, but the incidence is currently increasing in younger men. It is an adenocarcinoma that develops in the glands of the peripheral zone of the prostate. It is not preceded by the benign hyperplasia discussed above, since the two conditions affect different zones of the gland.

The risk factors for prostate cancer are not known. They are often sensitive to testosterone, but an endocrine imbalance does not appear to be the underlying cause.

Adenocarcinomas of the prostate show a wide variation in behaviour. Some can remain confined to the prostate for many years, whereas others metastasize early and are highly invasive.

Presentation
Since prostate carcinoma develops in the external glands of the prostate, away from the urethra, it is initially asymptomatic. As the tumour grows it begins to compress the urethra causing similar symptoms to those of benign prostatic hyperplasia, although they tend to progress more rapidly. This is a relatively late feature. Systemic and metastatic symptoms can be the first signs in advanced disease, including:
- Weight loss.
- General malaise.
- Anaemia.
- Back pain.

Rectal examination can reveal a hard nodule on the posterior surface of the prostate gland. As the tumour grows the median groove of the gland is obliterated. The staging system of prostate tumours is shown in Fig. 13.11.

Diagnosis and investigation
The possibility of prostate cancer should be excluded in all patients with symptoms of prostatism. Raised levels of the blood-borne tumour marker called 'prostate-specific antigen' (PSA) suggest carcinoma. This marker can also be used to monitor treatment. Blood specimens for PSA measurement should be taken prior to rectal examination, since this can raise the levels. Further investigation includes:

The Male Reproductive System

Fig. 13.11 Staging and prognosis of prostate carcinoma.

Staging and prognosis of prostate carcinoma

Stage	Description	Symptoms	Five-year survival (%)
A1	Microscopic, focal and well-differentiated tumours	None	95
A2	Microscopic, diffuse and/or poorly differentiated tumours	None	80
B	Larger tumours that are palpable rectally	None	75
C	Tumours that involve the entire prostate gland +/− local invasion	Prostatism	50
D	Metastatic spread to lymph nodes or organs including bones	Prostatism and systemic/metastatic symptoms	35

- Transrectal ultrasound.
- Rectal or transurethral needle biopsy.
- Bone X-rays to locate metastases.

Treatment

The prostate can be removed by TURP, though this procedure carries a high risk of retrograde ejaculation (into the bladder causing infertility) or, less commonly, impotence. Resection is performed in patients with urinary obstruction or late-stage tumours. Early-stage, asymptomatic tumours are often simply observed or treated with radiotherapy, though resection is recommended by some clinicians.

If the patient has incurable metastatic disease then the following medical treatments may help:
- Cyproterone acetate blocks androgen receptors, which can reduce tumour size and invasion.
- Continuous GnRH analogues can inhibit LH secretion from the pituitary gland reducing testosterone secretion.

> Because prostate cancer develops in the outer part of the prostate gland, it can be missed following transurethral biopsy.

Invasion and metastasis

Prostate carcinoma can spread in the following manners:
- Local spread within the gland and to the bladder and seminal vesicles.
- Lymphatic spread to the pelvic and para-aortic lymph nodes.
- Blood-borne spread to the bones, especially the pelvis, spine and skull. It causes characteristic osteosclerotic lesions, i.e. bone is formed not destroyed.

Disorders of the penis

Structural abnormalities
Congenital abnormalities
Hypospadias

The body of the penis is formed by the fusion of the urogenital folds round the urethra as described in Chapter 11. This fusion can fail to varying degrees, resulting in hypospadias in which a meatus (opening) of the urethra forms along the ventral surface (underneath) of the penis, usually at the base of the glans (Fig. 13.12); it is the most common structural abnormality of the penis.

Hypospadias is caused by a deficiency of fetal testosterone production. It is associated with undescended testes, a 'hooded' foreskin and a downward curvature of the penis called congenital chordee. It can be corrected surgically using the foreskin.

Epispadias

This is a much rarer condition than hypospadias. It is a similar abnormality except that the urethral meatus is on the dorsal (top) surface of the penis, usually at the base of the body. It can cause urinary incontinence and recurrent urinary infections; it is corrected surgically. It is often associated with

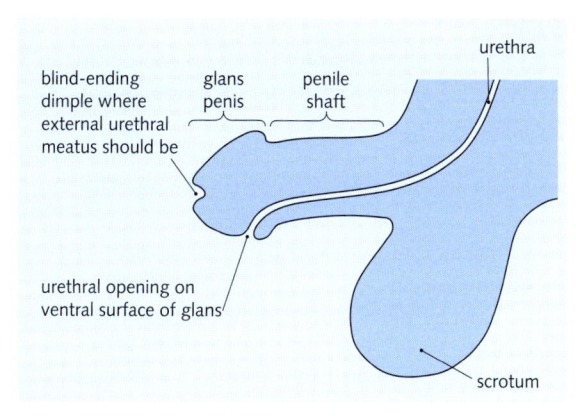

Fig. 13.12 Congenital hypospadias of the penis.

other abnormalities of the genitalia or urinary tract.

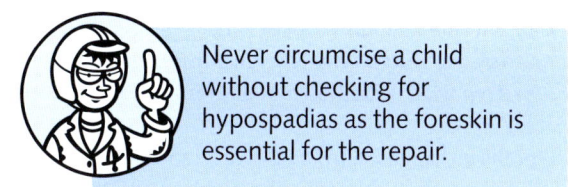

> Never circumcise a child without checking for hypospadias as the foreskin is essential for the repair.

Phimosis
When the foreskin is too narrow to retract over the glans of the penis the condition is called phimosis. It is normal after birth, but the foreskin should become retractable within a few years as the adhesions between the glans and foreskin breakdown. Phimosis can be congenital, but this is rare. More commonly, inflammation and fibrosis can cause narrowing following chronic or recurrent *Candida* infection. It is treated by topical steroids and exercising (gently stretching) the foreskin. Fibrosis can also be caused by balanitis xerotica obliterans (BXO), a disease of unknown aetiology that causes inflammation of the glans and is treated by circumcision.

Phimosis causes dyspareunia (pain on intercourse) and, in severe cases, it can obstruct urinary flow. Attempts to retract the fibrosed foreskin during erection can cause paraphimosis in which the foreskin gets stuck behind the glans. The venous drainage becomes obstructed restricting blood flow and causing oedema. This is treated by squeezing the glans hard to allow the foreskin to return.

Inflammation and infection
The penis, especially the glans and urethra, is very susceptible to the same sexually transmitted infections that affect the female reproductive tract. A summary of these infections is shown in Fig. 13.13.

Viral infection
The penis is prone to the same viral infections as the female vulva.

Herpes simplex
This is a sexually transmitted disease caused by herpes simplex virus (HSV, usually type II). It causes a recurrent, acute and itchy skin infection. This

Fig. 13.13 Summary of infections of the male genital tract.

Summary of infections of the male reproductive tract

Infection	Organism	Symptoms	Treatment
Herpes	Herpes simplex virus	Burning red blisters that may recur	Aciclovir if recurrent
Warts	Human papilloma virus	Growths on the penile skin	Podophyllotoxin cream
Gonorrhoea	*Neisseria gonorrhoeae*	Dysuria, white urethral discharge	Penicillin or ceftriaxone
Non-gonococcal urethritis	*Chlamydia trachomatis*	Milder dysuria, white urethral discharge	Erythromycin
Syphilis	*Treponema pallidum*	Ulcerated nodules, but may become systemic	Penicillin
Balanitis	*Candida albicans* (fungus)	Itching, white urethral discharge	Nystatin cream

progresses to form vesicles and painful skin erosions on the glans. The inguinal lymph nodes are often enlarged and painful.

The primary infection is often subclinical, but the virus then remains latent in the dorsal root ganglion that supplies that area of skin. The virus can be reactivated causing recurrent attacks.

HSV can be diagnosed from viral culture of the fluid in the blisters. It is treated symptomatically (e.g. anaesthetic cream) but aciclovir is often used for asymptomatic contacts, the initial infection or in the immuno-suppressed. The virus can be transmitted to other body parts on the hands: the eyes are particularly susceptible.

Genital warts
This is a sexually transmitted disease caused by human papilloma virus (HPV), of which there are many types. They cause benign, warty skin lesions called condyloma acuminatum, usually on the glans or inner surface of the foreskin.

The warts are treated with cryotherapy, in which a liquid nitrogen spray freezes and kills infected cells, or with podophyllin cream, which burns the infected cells chemically. The warts often recur despite treatment. Some strains of HPV predispose to malignant change of the penile skin (see below).

Bacterial infection
Gonorrhoea
This is a sexually transmitted disease caused by the bacterium *Neisseria gonorrhoeae* (often called gonococcus) that infects the distal urethra. Symptoms include dysuria (pain on urinating) and a discharge of a whitish pus. It is diagnosed by microscopy and culture of the pus to identify the organism and its antibiotic sensitivity; many gonococcus bacteria are now resistant to penicillin, in which case ceftriaxone is used.

Untreated gonorrhoea can lead to the following complications:
- Epididymo-orchitis.
- Proctitis (infection of the rectum).
- Infective arthritis.
- Septicaemia.

Non-gonococcal urethritis
More commonly the urethra can be infected by other bacteria, particularly *Chlamydia trachomatis*. It causes similar but milder symptoms. Epididymo-orchitis can also develop, but non-gonococcal urethritis is usually cured by a single dose of erythromycin.

Syphilis
This is a sexually transmitted disease caused by the bacterium *Treponema pallidum* that initially infects the glans or the inner surface of the foreskin. Today syphilis is a rare disease; the complications are almost never seen because of effective treatment. The untreated disease follows four stages:
- Primary syphilis—a solitary, painless ulcerated nodule appears at the infection site. The inguinal lymph nodes become enlarged but not painful.
- Secondary syphilis—two months later, a systemic disease may develop including scaly rashes and mucosal ulcers over much of the body surface. It normally resolves after several months.
- Tertiary syphilis—rarely the disease may enter a chronic stage in which characteristic rubbery granulomas (called gummata) develop throughout the body.
- Quaternary syphilis—with time syphilis may affect the cardiovascular and central nervous system with potentially fatal effects.

Syphilis is diagnosed by serological (antibody) tests, and it is treated with penicillin, which is also given to recent sexual contacts.

Fungal infection
The penis can be infected by *Candida albicans* causing thrush. This fungal infection is often an endogenous infection (from another body part), but it can be sexually transmitted.

Thrush causes inflammation of the foreskin and glans called balanitis producing red, itchy patches and a white urethral discharge. It is diagnosed by microscopy and culture of the discharge. Treatment is with topical antifungals (e.g. nystatin). Chronic balanitis can cause phimosis.

Neoplastic disorders
Benign tumours
Benign tumours of the penis take the form of warts caused by HPV. These are called condyloma acuminatum; they are discussed with the other viral infections of the penis.

Malignant tumours
Carcinoma of the penis is rare, but squamous cell carcinoma can develop in the penile skin. A series of dysplastic changes from carcinoma *in situ* to invasive carcinoma are grouped together as 'penile intra-epithelial neoplasia' or PIN. Dysplastic change

affects elderly uncircumcised men. It is predisposed by poor hygiene and HPV infection.

The initial neoplasia is called Queyrat's erythroplasia. This is a raised, red plaque that is non-invasive but has the histological appearance of squamous cell carcinoma. It is usually found on the inner surface of the foreskin or the base of the glans.

With time this carcinoma *in situ* can develop into invasive squamous carcinoma of the penis. This is a slow growing tumour with a warty appearance. It can spread to the inguinal lymph nodes.

PIN is treated with radiotherapy, partial or complete penile amputation and lymph node dissection, depending on the extent of the tumour and its spread.

Disorders of the male breast

Gynaecomastia

The male breast contains the same tissue components as the female breast in the undeveloped, prepubescent state. If the male breast enlarges, the condition is called gynaecomastia. The development of the male breast in gynaecomastia resembles the changes that the female breast undergoes during puberty; there is growth and development of the mammary ducts and connective tissues. It is caused by any condition or process that raises oestrogen or lowers testosterone levels. It often occurs during puberty, but will usually resolve with time. The underlying cause should be treated. Surgical removal of breast tissue may be performed in severe cases for cosmetic reasons.

Carcinoma

Breast carcinoma in men is very rare compared with that in females; it accounts for <1% of all breast cancer. The carcinoma is usually of the intraductal or infiltrating duct types (described in Ch. 12). It presents in elderly men as a breast lump or nipple discharge. The tumours have often metastasized to the axillary lymph nodes, lungs, brain, bone or liver by the time of presentation, so the prognosis is poor.

The Male Reproductive System

- Describe the structure and location of the testes.
- Describe the structure and layers of the epididymis and scrotum.
- Describe the course and layers of the spermatic cord; how do these relate to the abdomen and scrotum?
- Draw a diagram to illustrate the structure and location of the prostate.
- Describe the structure of the penis and the mechanism of erection.
- Name three hormones involved in the male reproductive system, along with their origins and actions.
- How is the male reproductive system regulated hormonally?
- Describe the process of spermatogenesis. Explain the roles of the Sertoli and Leydig cells.
- How is continuous fertility achieved?
- Draw a diagram to illustrate the structure of a mature sperm. What are the functions of the different components? Where are mature sperm stored?
- Name and briefly describe three congenital abnormalities of the testes.
- Compare testicular torsion and epididymo-orchitis. How are they treated?
- Describe the differential diagnosis of a scrotal lump.
- Name three types of testicular carcinoma, the cells they are derived from and the age groups they affect.
- Describe the aetiology, presentation and treatment of benign prostatic hypertrophy. Is it pre-malignant?
- Describe the location, presentation and treatment of prostatic carcinoma.
- Describe the common congenital malformation of the penis, along with three other conditions it is associated with.
- Discuss four sexually transmitted diseases, including their presentation and treatment.
- Describe the series of changes seen in penile intra-epithelial neoplasia.
- Compare gynaecomastia and male breast carcinoma.

14. The Process of Reproduction

The previous two chapters have described the organization of the male and female reproductive systems along with how the gametes are produced. This chapter describes how the gametes meet and the processes that occur from fertilization until the baby is born and being breastfed.

During pregnancy many processes and changes take place under hormonal control. The main hormones are:
- Progesterone.
- Oestrogens.
- Human chorionic gonadotrophin (hCG).
- Human placental lactogen (hPL).
- Prolactin.
- Oxytocin.

After reading this chapter you should be able to:
- Explain the process of sexual intercourse and the common associated disorders.
- Understand the processes involved in fertilization and implantation.
- List the most common methods of contraception and describe their modes of action.
- Give an account of the development and functions of the placenta.
- Appreciate the adaptations required for a successful pregnancy.
- Picture the stages of childbirth.
- Explain the physiology and process of lactation.

Important words:
Amenorrhoea—when menstruation fails to occur
Dyspareunia—pain on sexual intercourse
Menorrhagia—large volumes of blood lost during menstruation
Contraception—a method of preventing pregnancy
Trimester—the development of the fetus in the uterus is divided into three periods of equal length (about 13 weeks) each called a trimester

Sexual intercourse and dysfunction

Sexual arousal and sensation
In both men and women sexual arousal is derived from two components: physical stimulation and psychological stimulation.

Sexual arousal activates the parasympathetic nerves to the genitalia to cause erection of the penis or clitoris; the mechanism of erection is described in Chapter 13.

Physical stimulation
Physical stimulation of the glans of the penis in the male or the clitoris in the female results in sexual arousal, however other parts of the body can also have this effect (erogenous zones). These structures contain an abundance of sensory receptors that are stimulated by the massaging action of sexual intercourse. Stimulation of the external and internal genitalia can add to these sensations (e.g. vagina, urethra and prostate). These signals reach the spinal cord via the pudendal nerves and are transmitted to the central nervous system (CNS). They also initiate reflexes in the lumbar and sacral spinal cord. These reflexes produce the physical signs of sexual arousal to the extent that erection and ejaculation is possible in male patients with severed spinal cords, though there is no accompanying sensation.

Psychological stimulation
Psychological stimuli greatly enhance the response to physical stimuli. Erotic thoughts can activate the limbic system in a similar manner to physical stimuli, causing sexual arousal including lubrication and erection.

Physiology of sexual intercourse
Stages of sexual intercourse
Sexual intercourse is also called coitus. The changes in sexual arousal during coitus are often described in the four stages of the EPOR model (Fig. 14.1):
- Excitement phase—initial rapid rise in the level of sexual arousal.
- Plateau phase—sexual arousal is maintained at a high level.

The Process of Reproduction

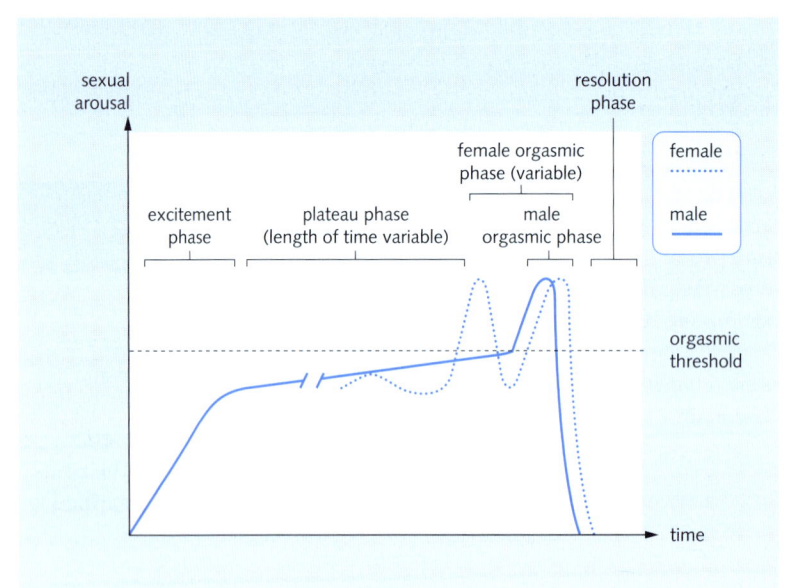

Fig. 14.1 The EPOR model of sexual arousal.

- Orgasmic phase—sexual arousal crosses the threshold resulting in orgasm.
- Resolution phase—the fall in sexual arousal following the cessation of coitus; physical and behavioural changes revert to normal.

Excitement and plateau phases

Sexual arousal causes stimulation of the parasympathetic nerve plexuses supplying the genitalia; this results in the following physical changes in the male:

- Vasocongestion of the penis (leading to erection) and scrotum.
- Secretion of small amounts of pre-ejaculatory fluid that may contain sperm.

Similar changes occur in the female:

- Vasocongestion of the clitoris (leading to erection), vagina, labia minora, and nipples.
- Secretion of lubricating fluid by Bartholin's glands on either side of the vestibule and a fluid exudate from blood vessels in the vaginal wall.

These physical changes increase the level of stimulation, and the increase in stimulation causes further sexual arousal, resulting in a positive feedback system.

Systemic changes occur in both sexes; these changes reach a peak at the point of orgasm:
- Rise in heart rate and blood pressure.
- Rise in respiratory rate.
- Rise in temperature and blood supply to the skin.
- Increase in muscle tone.

The male orgasm and resolution

When sexual arousal reaches the 'threshold' a reflex is initiated in the lumbar sympathetic nerves called an orgasm; it is accompanied by intense physical sensations. In the male the orgasm is accompanied by the expulsion of roughly 200 million spermatozoa in 2–4 mL of seminal fluid. The sperm are projected deep into the vagina, close to the external os of the cervix. This expulsion occurs in two phases:

- Emission—the epididymis, ductus deferens, seminal vesicles and prostate begin to contract, pushing sperm and seminal fluid into the urethra. This urethral sensation heightens stimulation and contraction.
- Ejaculation—the urethra contracts rhythmically expelling the semen in a series of pulses.

The sympathetic stimulation of male ejaculation inhibits the erection maintained by the parasympathetic input. The male enters a short 'absolute refractory' period during which further arousal is not possible; resolution will follow unless there is further stimulation.

The female orgasm and resolution

The female orgasm causes intense physical sensations and muscle contraction throughout the body in a similar manner to the male orgasm. A female orgasm is not necessary for pregnancy to occur, though it may raise fertility. Following the female orgasm sexual arousal returns to the plateau phase without an absolute refractory period. Continued stimulation can result in further orgasms, otherwise resolution will occur.

Sexual Intercourse and Dysfunction

> A single ejaculation contains 200 million spermatozoa, whereas the female only produces 2 million oocytes throughout her entire life of which only 400 are ovulated. Theoretically, a man could fertilize the entire fertile female population of the world with just 10 ejaculations.

Sexual dysfunction

The term 'sexual dysfunction' includes any process that interferes with a normal sex life. This can be caused by stress or physical or mental illness, but the majority of conditions result from ignorance, embarrassment or poor communication between sexual partners. Any therapy should involve both partners and encourage discussion and intimacy.

Male sexual dysfunction
Reduced libido
This is a lack of sexual desire; it is less common in men than women. It can be caused by psychological reactions to a deteriorating relationship or depression. It can also be caused by physical conditions including systemic illness, medications and a decrease in testosterone levels with age.

Impotence
Impotence is the inability to maintain an erection suitable for vaginal penetration despite normal sexual desire. It is the most common presentation of sexual dysfunction in men and it can be caused by many conditions (Fig. 14.2). Impotence can be treated by the following methods according to the cause:
- Sexual counselling.
- Vacuum aids.
- Oral medications (including sildenafil citrate—Viagra®).
- Medications injected directly into the penis.
- Implants.

Premature ejaculation
The definition of premature ejaculation depends on the expectations and desires of both partners. In severe cases ejaculation may occur prior to penetration. It is made worse by anxiety and can be treated by relaxation techniques.

Female sexual dysfunction
Reduced libido
A lack of sexual desire can be present from puberty or it can develop with time, in which case it is often a reaction to a deteriorating relationship or depression. It can be caused by physical illness that prevents the enjoyment of sexual intercourse (e.g. vaginal infection, general illness or medications).

Anorgasmia
The majority of women require more time and greater stimulation to achieve orgasm than men. Anorgasmia is the complete failure to achieve orgasm; it is not infrequent orgasms. The best treatment involves teaching both partners about the

Fig. 14.2 Causes of male sexual dysfunction.

Causes of male sexual dysfunction

Cause	Examples	Mechanism
Psychological	Performance anxiety, stress	Stress inhibits the parasympathetic nervous system that maintains erection
Alcohol	Brewer's droop	Acutely inhibits sensory nerves, chronically damages liver raising oestrogen levels
Medications	Antihypertensives	Reduce blood flow to the penis
	Antidepressants and antipsychotics	Antagonize sexual arousal in the CNS
Endocrine	Diabetes	Long-term complications can damage the nerves and blood vessels of the penis
Vascular disease	Atherosclerosis	Prevents sufficient blood reaching the penis to maintain erection
Neurological	Multiple sclerosis	Inhibits sexual arousal in the CNS

female body; encouraging masturbation may also benefit some women.

Dyspareunia

Dyspareunia is pain on intercourse. There are two kinds:
- Superficial—felt on the external genitalia.
- Deep—felt internally.

Apart from the common vaginismus (see below), dyspareunia tends to have a physical cause (Fig. 14.3).

Vaginismus

This is involuntary contraction of vaginal muscle upon attempted penetration, causing apareunia or dyspareunia. It is a common psychological condition caused by fear of penetration sometimes resulting from previous dyspareunia. It can be treated by teaching relaxation techniques together with insertion of vaginal dilators of increasing size until penile insertion is possible. The woman should be in control of all insertion until she can relax enough to allow her partner to do so.

Apareunia

This is the inability of the vagina to accept penile penetration. It can be caused by congenital malformations of the vagina or severe vaginal infections, but most commonly it is caused by vaginismus.

Fertilization and contraception

Fertilization

Oocyte

At ovulation the oocyte is released from the mature ovarian follicle onto the surface of the ovary in the peritoneal cavity. At this stage, it has completed the first division of meiosis and has arrested at metaphase of the second meiotic division. It is surrounded by two layers:
- Zona pellucida—a glycoprotein layer.
- Corona radiata—granulosa cells from the follicle (also called cumulus oophorus).

The oocyte is wafted into the fallopian tube by the action of cilia on the fimbriae. Further ciliary action and peristalsis move the oocyte along the fallopian tube to the ampulla. It is capable of fertilization for less than 24 hours.

Sperm

About 200 million sperm are ejaculated deep into the vagina close to the external cervical os. They are suspended in the fructose-rich, alkaline seminal fluid; this provides the energy required by the sperm and acts as a buffer to the acidic vaginal environment. The motile sperm must cover a significant distance to reach the oocyte in the fallopian tubes. Their passage is obstructed by two narrow openings that act as filters against damaged sperm:

Causes of female dyspareunia

Type of dyspareunia	Disease process	Symptoms and signs
Superficial	Vulval/vaginal infections	Lesions and discharge
	Previous surgery or trauma	Scars
	Cystitis and UTIs	Urinary frequency, urgency, and pain
	Oestrogen deficiency	Lack of lubrication, post-menopausal
Deep	Endometriosis	Recurrent pelvic pain in time with the menstrual cycle
	Pelvic inflammatory disease	Abdominal pain
	Fibroids	Menorrhagia and pelvic pain
	Ovarian cysts/tumours	Pelvic pain and menorrhagia

Fig. 14.3 Causes of female dyspareunia. (UTIs, urinary tract infections.)

- External os of the cervix.
- Uterine entrance of the fallopian tubes.

Their passage may be assisted by physiological changes brought about by the female orgasm and chemical signals directing them towards the oocyte.

Within the female genital tract, sperm are viable for less than 48 hours. The majority degenerate and are absorbed by the female, with only about 200 sperm reaching the oocyte in the ampulla of the fallopian tube. The quickest sperm can arrive in just 5 minutes.

Capacitation
The ejaculated sperm cannot penetrate the oocyte until they undergo capacitation. The following changes occur in the uterus or fallopian tubes:
- Removal of glycoproteins covering the acrosome.
- Reorganization of membrane phospholipids to alter the membrane potential and charge.
- Influx of calcium that increases flagellar activity and motility.
- Activation of the acrosome allowing the release of enzymes.

Capacitation must be mimicked during in-vitro fertilization (IVF) by incubation in a suitable medium.

Fertilization
Fertilization normally occurs in the ampulla of the fallopian tube. The sperm must penetrate the layers surrounding the oocyte and, in the process, prevent other sperm from fertilizing the oocyte a second time. This process requires several steps:

1. *Penetration of the corona radiata*   The acrosome membrane begins to perforate, releasing the enzyme hyaluronidase, which disrupts the cell matrix allowing sperm to push through the remaining granulosa cells.
2. *Penetration of the zona pellucida*   Receptors on the acrosome bind to ZP3 molecules causing the release of acrosin, an enzyme that digests the glycoprotein chains of the zona pellucida. The sperm can then push through the weakened structure.
3. *Fusion of the plasma membranes*   The membranes of the sperm and oocyte bind via integrin receptors and fuse. The sperm nucleus enters the oocyte cytoplasm leaving its tail and membrane behind. (NB No sperm mitochondria enter the oocyte.)
4. *Fast block*   The membrane fusion causes the oocyte to depolarize preventing other sperm from binding for about a minute.
5. *Slow block*   The fusion also causes calcium to enter the oocyte resulting in the release of granules containing hydrolytic enzymes. These enzymes break down the ZP3 molecules of the zona pellucida, permanently preventing other sperm from binding.
6. *Second meiotic division*   The calcium influx also causes the oocyte to complete the second meiotic division. This produces two haploid cells:
   - Female pronucleus with the majority of the cytoplasm.
   - Second polar body with almost no cytoplasm.
7. *Formation of the pronuclei*   The nucleus of the sperm enlarges to form a pronucleus that is indistinguishable from the female pronucleus. The DNA within each pronucleus is replicated.
8. *Mixing of the chromosomes*   The nuclear membranes surrounding the pronuclei break down and the maternal and paternal chromosomes mix producing a diploid (46 chromosomes) zygote.
9. *First mitotic division*   The zygote immediately begins to divide by mitosis and a copy of the replicated DNA enters each cell. This division takes about 30 hours to complete.

> The fast and slow blocks do not always prevent multiple fertilizations. Triploid (three sets of chromosomes) zygotes are formed, but they usually die within the uterus though the mechanism for this abortion is unknown.

Development and implantation
Early development
The zygote divides into two cells called blastomeres as the embryo is transported along the fallopian tube. The blastomeres continue to divide mitotically, becoming smaller with each division. The cells then reorganize to form a tighter ball; a process called compaction. When the embryo reaches the 12–15 cell stage it is called a morula.

The morula enters the uterus about 3 days after fertilization. Glands in the endometrial lining secrete a glycogen-rich fluid under the influence of progesterone from the corpus luteum. Glycogen can cross the zona pellucida to nourish the morula.

Soon after the morula enters the uterus, a cavity develops forming an inner and outer layer of cells:
- Trophoblast—the outer cell layer that will form the placenta.
- Embryoblast—the inner cell mass that will form the embryo.

The morula is now called a blastocyst. Over the next 2 days the zona pellucida breaks down allowing the blastocyst to increase in size. It is now ready for implantation. These changes are shown in Fig. 14.4.

Implantation

About 6 days after fertilization the blastocyst binds to the endometrial lining. This usually occurs in the body of the uterus with the embryoblast nearer to the endometrium and the cavity closer to the uterus lumen. The trophoblast layer grows rapidly and differentiates to form two layers:
- Cytotrophoblast—the layer nearest the embryoblast, composed of dividing cells that join the syncytiotrophoblast.
- Syncytiotrophoblast—the layer nearest to the endometrium; it is composed of cytoplasm containing many nuclei without cell boundaries.

The syncytiotrophoblast develops finger-like projections that invade the surrounding endometrium to hold the blastocyst in place. It releases enzymes that break down the glycogen-rich endometrial stroma to release nourishment for the developing embryoblast. This entire process is called implantation (Fig. 14.5).

Contraception

Humans are most unusual animals since neither the female nor the male is aware of the time of ovulation. The majority of female animals display ovulation with dramatic changes in colour, smell and behaviour, to which the male responds appropriately. On the good side this allows humans to engage in sex at any time throughout the year. However it makes pregnancy very difficult to avoid.

A wide range of reliable contraceptive methods are now available allowing people to choose a method that suits their lifestyle and beliefs. No method of contraception can prevent pregnancy 100% of the time in 100% of women. The failure rate is measured as the percentage of women who become pregnant

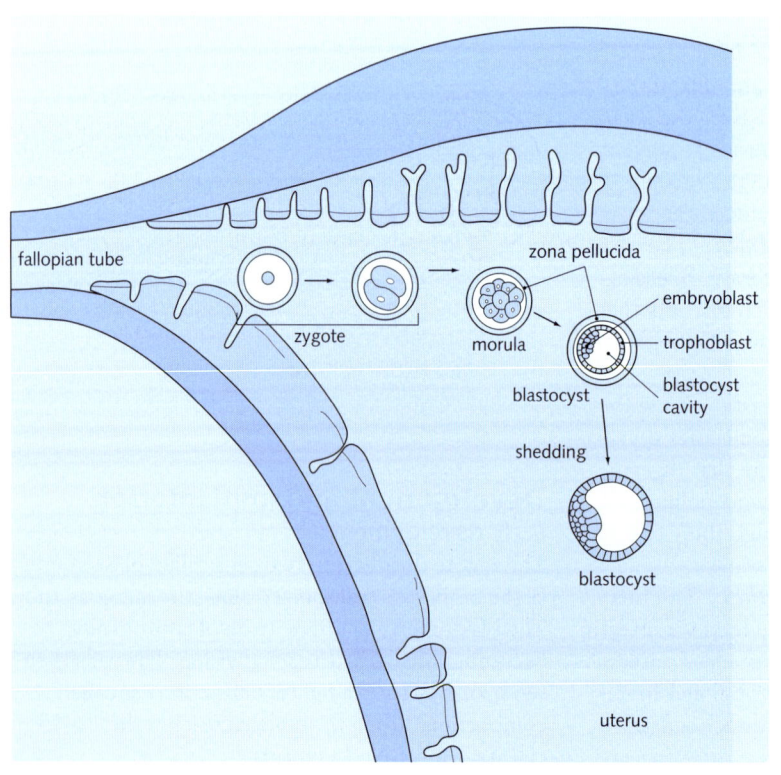

Fig. 14.4 Early development of the zygote, morula and blastocyst.

Fertilization and Contraception

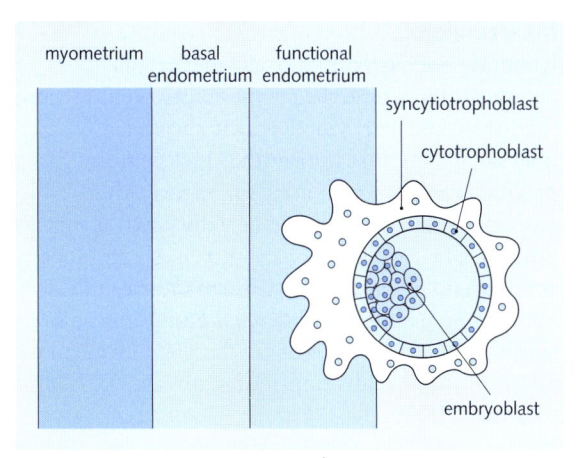

Fig. 14.5 Implantation of the blastocyst.

during 1 year of use. This percentage varies widely between contraceptive methods and the reliability of the user; the lower the figure the more effective the contraception. Fig. 14.6 shows examples of common contraceptive methods and Fig. 14.7 shows where their contraceptive actions take place.

Natural contraception
None
The average age of first contraceptive use is 8 months later than the average age of first sexual intercourse. Many adolescents have disproved the myth that you cannot get pregnant the first time you have sex.

Without any form of contraception 90% of women become pregnant during 1 year of regular sex.

Postcoital douche
Since the sperm are deposited in the vagina, washing the vagina out after sex reduces fertility. This method is very ineffective: it has a failure rate of 45% because the sperm are rapidly transported through the cervix.

Coitus interruptus
This is also called the withdrawal method; the man withdraws his penis from the vagina just before ejaculation. While the theory is good, in practice the sympathetic system hijacks the body to propagate the species. Even with successful withdrawal, pregnancy can still result since sperm is also present in the male pre-ejaculatory secretions. The failure rate of 35% speaks for itself.

Rhythm method
The oocyte can only survive for 24 hours without fertilization whereas the sperm can survive for 48 hours. In theory, fertilization can only occur following sexual intercourse in the 3 days around ovulation. By determining the exact date of ovulation and avoiding sex before and after this date pregnancy can be avoided. In practice, a wider period of sexual abstinence is used, usually about 9 days per cycle. There are a number of methods for detecting the date of ovulation:

Examples of methods of contraception

Method	Rhythm method	Condoms	COCP	IUD	Mirena® (IUS)	Vasectomy
Action	Natural	Barrier	Hormonal	Prevents implantation	Hormonal	Surgical sterilization
Failure rate (%)	2.5–30	1.5–7	1	2	0.1	0.05 per lifetime
Advantages	Free, approved by Catholic church	Cheap, easy, portable and protects from STDs	Cheap, effective, light periods	Lasts up to 5 years	Very effective, light periods, minimal effort	Very effective, simple operation
Disadvantages	Requires high motivation and control	Reduces spontaneity and may affect sensitivity	Several side effects including risk of cardiovascular disease and breast cancer	Must be fitted by doctor, risk of PID and heavy menstrual bleeding	Must be fitted by doctor, risk of PID	Irreversible

Fig. 14.6 Examples of methods of contraception. (COCP, combined oral contraceptive pill; IUD, intrauterine device; IUS, intrauterine system; PID, pelvic inflammatory disease; STDs, sexually transmitted diseases.)

The Process of Reproduction

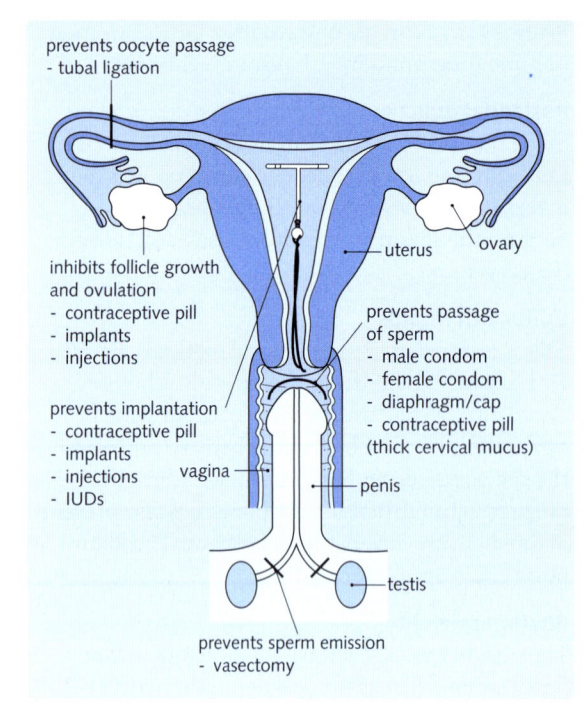

Fig. 14.7 Locations of contraceptive actions. (IUDs, intra-uterine devices.)

Labels in figure:
- prevents oocyte passage – tubal ligation
- inhibits follicle growth and ovulation – contraceptive pill, implants, injections
- prevents implantation – contraceptive pill, implants, injections, IUDs
- uterus
- ovary
- prevents passage of sperm – male condom, female condom, diaphragm/cap, contraceptive pill (thick cervical mucus)
- vagina
- penis
- testis
- prevents sperm emission – vasectomy

- Keeping a calendar of the menstrual cycle and changes.
- Changes in cervical mucus; it becomes clear, sticky, and stretchy at ovulation.
- Changes in temperature; there is a rise of about 0.3°C after ovulation.
- Changes in the types of oestrogens and luteinizing hormone (LH) levels in the urine using over-the-counter kits.
- Mittelschmerz; rarely, women experience 'ovulation pain'.

The failure rate ranges from 2.5–30% depending on the motivation of both partners and the regularity of the menstrual cycle.

Barrier contraception

The aim of barrier protection is to prevent the sperm from reaching the oocyte, thus preventing fertilization. All barrier methods offer some protection against sexually transmitted diseases including human immunodeficiency virus (HIV) infection. The failure rate ranges from 1.5–7% depending on the motivation of both partners.

Male condom

Condoms are currently the only method of contraception where the responsibility is entirely on the part of the man. Condoms are sheaths of lubricated latex that fit over the erect penis to prevent sperm from entering the vagina. After intercourse the penis should be withdrawn as soon as possible whilst holding the condom in place. Modern condoms do not interfere with sensation significantly, but they may reduce spontaneity. They are readily available, cheap, portable and offer good protection against many sexually transmitted diseases (STDs) including HIV.

Female condom

These are larger versions of male condoms that fit into the vagina. There is a risk of the penis 'missing' the condom, but a large ring at the open end aims to prevent this by holding it against the vulva. They do not need to be fitted by a specialist. The main disadvantage is the rustling noise, similar to a plastic bag, that accompanies intercourse.

Diaphragm

Diaphragms are reusable circular latex devices that fit in the vagina (or caps that fit over the cervix); they are used with a covering of spermicidal cream to kill the sperm. They can be inserted a couple of hours before sex, but they must not be removed for at least 6 hours afterwards. They must initially be fitted by a medical professional and require teaching and practise for reliable use.

Hormonal contraception

Hormonal contraception is the artificial administration of oestrogens and/or progestogens to reduce fertility. They mimic the hormonal changes of pregnancy to prevent further ovulation. Oestrogens act in the following ways:

- Strongly suppress follicle stimulating hormone (FSH) to prevent follicle development.
- Suppress LH release to prevent ovulation.
- Weakly inhibit implantation.

Progestogens have slightly different effects:

- Thicken the cervical mucus so that sperm cannot pass.
- Inhibit development of the endometrium to prevent implantation.
- Weakly suppress FSH to prevent follicle development.
- Weakly suppress LH release to prevent ovulation.

The use of oestrogen-containing contraception has a number of beneficial effects:
- Lighter, shorter periods.
- Regular, controllable periods.
- Reduces symptoms of premenstrual syndrome (PMS).
- Reduces the risk of ovarian and endometrial cancer.
- Controls the symptoms of endometriosis.

Unfortunately oestrogens also increase the risk of two serious diseases:
- Thromboembolism—the hormones increase the levels of many clotting factors so the blood is prone to coagulate causing deep vein thrombosis (DVT) and a risk of pulmonary embolus (PE). There is also a risk of hypertension.
- Cancer—there may be a small and temporary increase in the incidence of breast cancer in young women. This is offset by the reduction in risk of ovarian and endometrial cancer.

Since both these diseases are rare in the younger age groups that use hormonal contraception the overall risk remains very small. In fact, it is less than the risk of childbirth or abortion. Smoking greatly increases these risks.

A number of less serious side effects are occasionally experienced:
- Break-through bleeding—irregular menstrual bleeding is common in the first few months of use.
- Slight weight gain—this is usually temporary.
- Headaches and migraine.
- Acne—caused by oestrogen.
- Dry eyes—a concern for contact lens users.
- Loss of libido due to androgen inhibition.

The side effects do result in a number of contraindications for oestrogen-containing contraception; in most cases, progestogen-only contraception can be used instead. These are similar to the contraindications for hormone replacement therapy (HRT) and, likewise, are frequently asked in exams:
- Oestrogen-dependent cancer (including breast cancer).
- Cardiovascular and thromboembolic disorders.
- Liver disease with abnormal liver function tests (LFTs).
- Undiagnosed vaginal bleeding.
- Pregnancy or breastfeeding.
- Smokers aged over 35 years.
- Severe migraines.

Combined oral contraceptive pill (COCP)

The 'Pill' is the most common method of contraception used by young women. It contains low doses of both oestrogen and progestogen giving a failure rate of 1%. It is taken for 21 days followed by a 7 day break, during which withdrawal bleeding occurs; this is not real menstruation. Originally the pill was designed to be taken continuously and this can still be done without risk; the break has been added for the reassurance of the user. The side effects are described above.

Progestogen-only pill (POP)

As the name suggests, this oral contraceptive only contains progestogen. It has a higher failure rate (2%), especially in younger women, but it can be used in women in whom COCPs are contraindicated or not tolerated. It must be taken continuously at the same time (within 3 hours) every day. The main side effect is the menstrual irregularity that occurs in 25% of users.

Depot progestogen

Long acting progestogens can be injected intramuscularly to give 2–3 months of contraception. It has similar side effects and failure rate to POPs, but it carries a risk of amenorrhoea and temporary infertility for several months after discontinuation.

Subdermal implants

A progestogen containing tube (Implanon®) can be implanted in the upper arm to give 3 years of contraception. The side effects and failure rate are similar to POPs. It must be removed after 3 years, but it can be removed earlier to rapidly restore fertility. In the past, six rods were implanted, but this method was discontinued because they were difficult to remove.

Intrauterine system (IUS or Mirena®)

This is a type of plastic intrauterine device (IUD) that releases progestogen; it does not contain copper. The progestogen acts locally to give excellent contraception; it reduces menstrual bleeding and prevents side effects. They last 5 years and the failure rate is about 0.1%. There are some cases emerging where bleeding recommences after 2 years even though the implant is still releasing progestogen, but this is currently under investigation.

Prevention of implantation
Intrauterine devices
These are small plastic devices surrounded by copper wire that are inserted into the uterus lumen. The copper inactivates sperm to prevent fertilization. The device also inhibits implantation. Threads are attached to the device to allow insertion and removal, which must be performed by a trained medical professional. Insertion can be difficult in young or nulliparous women (i.e. women who have never given birth). The woman can check the IUD remains in place by feeling for these threads at the cervix os. The IUD can be left in place for 5 years, throughout which the failure rate is about 1%. There are several side effects and risks:
- Expulsion—especially in the first 2 months.
- Cramping and bleeding—usually diminishes with time.
- Pelvic inflammatory disease (PID)—caused by bacterial infection at insertion.
- Increased risk of ectopic pregnancy.

An IUD can prevent implantation and, therefore, pregnancy if it is inserted within 5 days of sexual intercourse.

Postcoital medication
There are two types of 'morning-after' pill, equivalent to the COCP (called PC4®) or POP (called Levonelle®); the pills simply contain higher doses of the steroids that are used for contraception. The high levels of steroids increase the motility of the fallopian tubes so that the embryo reaches the uterus before the endometrium is prepared for implantation. The emergency hormonal contraception is effective if it is taken within 72 hours (3 days) of unprotected sexual intercourse and the dose must be repeated 12 hours later. PC4 often induces nausea and vomiting so an antiemetic is sometimes prescribed. The side effects of Levonelle® are milder, allowing it to be available over the counter. The failure rate of both pills is about 3%.

Irreversible contraception
Sterilization is the most popular form of contraception in older men and women. It involves surgery to prevent the sperm or oocyte from reaching the site of fertilization. Since the male operation is simpler and more effective it should be considered the preferred technique in stable couples who have completed their family. Both techniques should be considered a permanent form of contraception, although microsurgery can successfully reverse sterilization in about 50% of cases.

Vasectomy
This is a quick operation performed under local anaesthetic in which a section of each ductus deferens is removed. The cut ends are tied or burnt closed by diathermy to prevent sperm from reaching the urethra. It does not interfere with testosterone production or sensation. The man is potentially fertile for 3 months after the operation due to sperm within the proximal ductus deferens; during this time other contraception should be employed. There is a risk of postoperative bruising and bleeding. It is the most effective form of contraception with a failure rate of about 0.05% per lifetime.

Tubal ligation
This is the equivalent operation in the female. It is usually performed under general anaesthetic using a laparoscope to cut, burn or clip both fallopian tubes. The woman becomes infertile after her next period and the failure rate is low at 0.5% per lifetime. The major complications are due to the general anaesthetic.

Infertility

Infertility is defined as the inability to conceive after 1 year of regular unprotected sex. It is a common problem, affecting about 1 in 10 couples. Infertility is caused by:
- Abnormalities in the male (30%).
- Abnormalities in the female (45%).
- Unexplained (25%).

Male infertility
Male infertility is usually caused by abnormalities in sperm production, these include:
- Azoospermia—no sperm.
- Oligospermia—also called a low sperm count (less than 20 million/mL; the average is about 60 million/mL).
- Asthenozoospermia—this is decreased sperm quality due to reduced motility or abnormal morphology (shape).

Some of the conditions that can cause these abnormalities are shown in Fig. 14.8. Less commonly, male sexual dysfunction can prevent normal sexual intercourse.

Fig. 14.8 Causes of male infertility.

Causes of male infertility	
Abnormality	Cause
Azoospermia	Blockage of genital tract
Oligospermia	Testosterone deficiency
	Hyperprolactinaemia
Asthenozoospermia	Raised scrotal temperature, e.g. varicocoele
	Antisperm antibodies
Oligospermia or asthenozoospermia	Genetic disorders, e.g. Klinefelter's
	Genital tract infection (current or previous with scarring)

Female infertility
The causes of female infertility are more varied:
- Oligomenorrhoea or amenorrhoea (45%).
- Structural abnormalities of the fallopian tubes (45%).
- Structural abnormalities of the uterus and cervix (5%).
- Disorders of the cervical mucus (5%).

The main conditions that cause these abnormalities are shown in Fig. 14.9. Some clinicians believe that an inadequate luteal phase can also cause infertility because the blastocyst does not have time to implant successfully before the corpus luteum regresses. Infertility can also be caused by sexual dysfunction.

Investigation
Counselling and reassurance are extremely important in the management of infertility. Many of the questions and procedures can be very embarrassing so the couple must feel they can trust the practitioner. The investigations should be aimed at diagnosing and treating the cause of infertility not finding which partner is 'at fault'.
Infertility is investigated by:
- Examination of sperm under a microscope.
- Hormone investigations—FSH, LH, prolactin, progesterone, oestrogen, testosterone and thyroid hormone.
- Pelvic ultrasound—looks for gross abnormality and can detect the presence of correctly developing ovarian follicles.
- Hysterosalpingography or laparoscopy—contrast media/dye is injected into the uterus to image the fallopian tubes by X-ray or visually (see Fig. 17.9).
- Postcoital test—checks cervical mucus and sexual technique.

Causes of female infertility	
Abnormality	Cause
Oligomenorrhoea or amenorrhoea	Weight loss
	Post COCP pituitary insensitivity
	Polycystic ovaries
	Hyperprolactinaemia
Abnormal fallopian tube	Pelvic inflammatory disease (Chlamydia)
	Endometriosis
	Pelvic surgery
Abnormal cervix	Cervical stenosis
Abnormal cervical mucus	Immunological reaction against sperm

Fig. 14.9 Causes of female infertility. (COCP, combined oral contraceptive pill.)

Treatment
Infertility is rarely absolute; it is usually caused by the reduced fertility (subfertility) of one or both partners. Where possible, treatment is aimed at the underlying cause. Medical treatment includes:
- Bromocriptine—which treats hyperprolactinaemia.
- Clomiphene—an anti-oestrogen, which stimulates FSH release and follicle development.

- hCG—which can trigger ovulation by acting like LH.
- LH and FSH injections; follicle stimulants, if clomiphene fails.
- GnRH analogues; follicle stimulants, if clomiphene fails.

Medications that stimulate follicle development may induce multiple pregnancies.

If these simpler treatments fail, assisted fertilization techniques can be used. These are both expensive and emotionally draining, though they have a 20–30% success rate (live births per cycle); this is compared with a rate of about 15% in fertile couples without assisted fertilization. Assisted fertilization techniques require hyperstimulation of the ovaries using the medications described above. The ovary responds by producing several eggs that are harvested using a transvaginal needle under ultrasound control. The techniques then differ:

- In-vitro fertilization (IVF)—oocytes are fertilized *in vitro* and re-introduced into the uterus. This is the only treatment for blocked fallopian tubes.
- Intracytoplasmic sperm injection (ICSI)—this is used alongside IVF if the sperm are abnormal and are incapable of fertilization.
- Gamete intrafallopian transfer (GIFT)—oocytes are re-introduced into the fallopian tubes along with sperm for in-vivo fertilization.

Therapeutic abortion

In the UK induced abortion is legal until 24 weeks gestation, though it is rarely performed beyond 20 weeks. It is sometimes recommended to prevent physical illness, though 95% are performed for social or psychiatric reasons. It is a relatively safe procedure with no risk of subsequent reduced fertility if performed without complications before 13 weeks gestation. An appropriate method of contraception is usually included as part of the treatment.

Surgical termination of pregnancy (STOP) can be performed until the 13th week of gestation. It is a day case procedure performed under general anaesthetic in which the uterus is evacuated using suction.

Medical termination can be performed until the 24th week of gestation. Two drugs are used:

- Oral mifepristone—this drug is administered first; it raises the sensitivity of the uterus to prostaglandins.
- Prostaglandins—these are given 48 hours after mifepristone; they can be taken orally or applied vaginally and they induce uterine contractions.

The fetus is usually expelled vaginally within 12 hours of the prostaglandin treatment. Medical termination often causes bleeding and pain, but pain relief is available.

The placenta

Development

The placenta develops from the trophoblast surrounding the developing fetus. The development of the trophoblast is described in the section 'Implantation' (p. 172). The placenta develops mainly from fetal tissues and partially from the maternal endometrium; the development is shown in Fig. 14.10.

Lacunar phase

Eight days after fertilization, the syncytiotrophoblast begins to erode endometrial capillaries. Cavities called lacunae develop within the syncytiotrophoblast; they are continuous with the maternal capillaries and fill with maternal blood. This supplies oxygen and nutrition to the blastocyst through diffusion and it represents the early maternal circulation in the developing placenta. The lacunae link together to form a network, whilst the capillaries enlarge to form sinusoids.

Chorionic villi

The chorionic villi are branching projections from the blastocyst into the syncytiotrophoblast and functional endometrium. They increase the surface area in contact with maternal blood. Development progresses in three stages:

- Primary villi—two weeks after fertilization, swellings in the cytotrophoblast extend into the syncytiotrophoblast and begin to branch.
- Secondary villi—at the start of the third week, mesenchyme from the embryo invades the cytotrophoblast of the primary villi.
- Tertiary villi—the mesenchyme forms blood vessels that link up with the newly formed fetal circulation by the end of the third week.

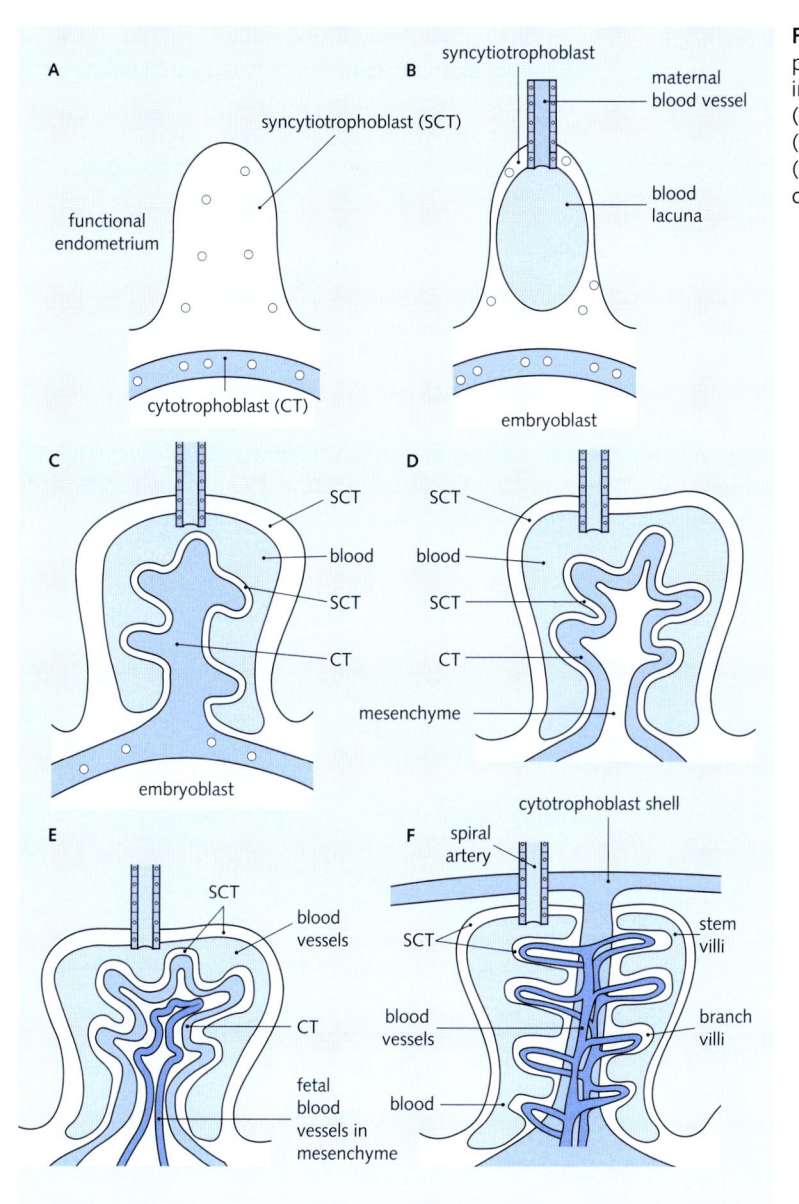

Fig. 14.10 Development of the placenta: (A) Syncytiotrophoblast invasion; (B) Lacunar phase; (C) Primary chorionic villi; (D) Secondary chorionic villi; (E) Tertiary chorionic villi; (F) Mature chorionic villi. See text for details.

At this stage, the placenta has a blood supply from the mother and fetus allowing more efficient exchange of nutrients, gases and waste products. The placenta continues to develop throughout pregnancy, in advance of the growing needs of the fetus.

Further development of the villi

As the tertiary villi form, the cytotrophoblast extends through the syncytiotrophoblast at the tip of each villus. These cells are now in direct contact with endometrial cells, and they form a cytotrophoblast shell that holds the embryo in place. This shell remains directly connected at the top of chorionic villi that are now called stem villi.

Branching villi at the sides of the stem villi are called branch villi. These are surrounded by maternal blood in the lacunae, and they are the site of exchange between fetal and maternal blood. The cytotrophoblast lining of the branch villi breaks down to reduce the distance molecules must diffuse between the two circulations. As the placenta continues to grow, the lacunae become supplied by the spiral arteries and endometrial veins. Deoxygenated fetal blood reaches the placenta via the two umbilical arteries and oxygenated blood returns to the fetal circulation via the umbilical vein.

Further growth of the placenta occurs at the embryonic pole (the side to which the umbilicus is attached) by widening and lengthening; it does not penetrate further into the endometrium. This forms the familiar plate-like shape on one side of the uterus. The endometrium on the same side also changes to form the decidua basalis; this grows into the placenta forming incomplete septa that divide it into sections called cotyledons.

Structure

It is important to remember that maternal and fetal blood do not mix within the placenta. Maternal blood enters large lacunae (cavities) from spiral arteries and is drained by endometrial veins. Fetal circulation in the branch villi is separated from the maternal circulation by two layers:
- Thin syncytiotrophoblast layer.
- Single cell layer of endometrium in the fetal capillary.

This allows for rapid diffusion between the two circulations. A diagram of the fully developed placenta is shown in Fig. 14.11.

Functions

The placenta supplies all the requirements of the developing fetus whilst maintaining an environment in which the fetus can grow. The metabolic rate of the placenta is very high due to protein synthesis,

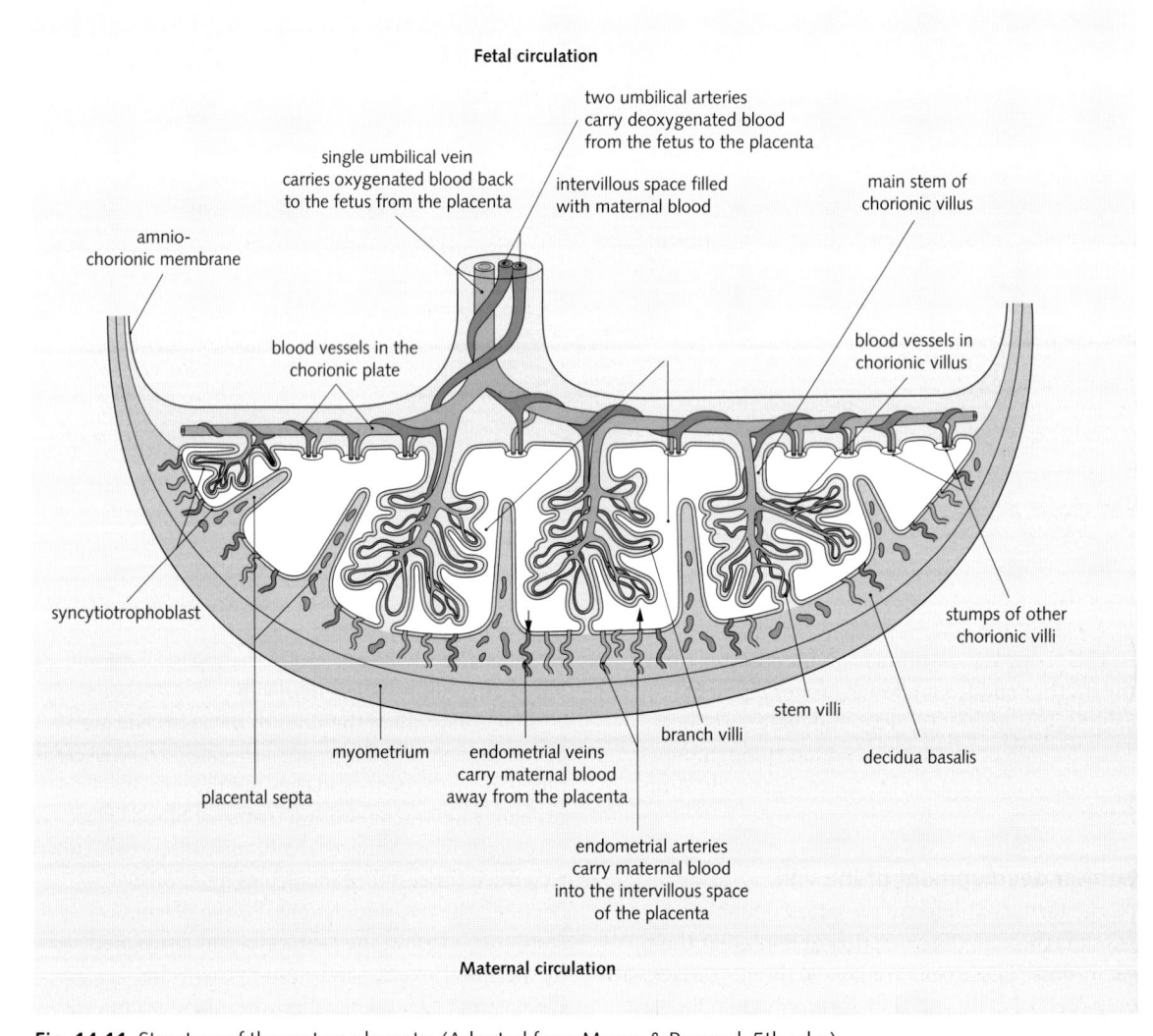

Fig. 14.11 Structure of the mature placenta. (Adapted from Moore & Persaud, 5th edn.)

active transport, and growth. The many actions involved in this task fall into four categories and are described below.

Gaseous transport

Oxygen and carbon dioxide cross the placenta by passive diffusion. The rapid metabolism of the fetus uses up oxygen, so blood in the umbilical arteries has less oxygen than the maternal blood. This forms a concentration gradient that allows oxygen to diffuse across the placenta into the fetal blood. This process is also aided by fetal haemoglobin, which binds oxygen more strongly than adult haemoglobin. High levels of CO_2 are generated, crossing the placenta in the opposite direction.

Nutrient transport

Nutrients cross the placenta by two processes:
- Passive facilitated diffusion.
- Active transport.

Towards the end of pregnancy an excess of nutrients is transported so the fetus can develop energy stores such as glycogen and fat. This includes brown adipose tissue that is broken down within the first few days after birth to create heat.

Immune protection

The process of meiosis and the presence of paternal DNA cause the maternal immune system to recognize the fetus as foreign. The placenta must act as a barrier to prevent immunological rejection.

> A graft from a child to its mother would normally be rejected. However, as a fetus, the same child is protected by the placenta from rejection.

Secretion of hormones

The placenta secretes high levels of steroid and protein hormones that regulate and maintain pregnancy. It also allows maternal and fetal hormones to cross. These hormones are described in the next section.

Reproductive hormones in pregnancy

Sources of reproductive hormones

Endocrine signals are essential for implantation and the maintenance of pregnancy. As the pregnancy progresses, the hormone levels change as shown in Fig. 14.12. There are two phases of hormonal secretion during pregnancy:
- Corpus luteum phase—the corpus luteum secretes hormones to maintain the endometrium and the developing placenta.
- Placental phase—the placenta takes over hormonal secretion to allow maternal adaptation to pregnancy, birth and lactation.

Corpus luteum phase

In the normal menstrual cycle, the corpus luteum secretes progesterone and oestrogen for about 10 days following ovulation, after which it regresses. The dramatic fall in progesterone levels causes degeneration and shedding of the endometrium resulting in menstruation. The blastocyst must prevent the next menstruation by maintaining the corpus luteum and its steroid secretion.

Soon after implantation, around day 6, the syncytiotrophoblast (outer layer of cells) secretes the hormone human chorionic gonadotrophin (hCG). This hormone is equivalent to LH and it acts on the corpus luteum to prevent regression. Progesterone levels continue to rise and the functional endometrium is maintained. This demonstrates the

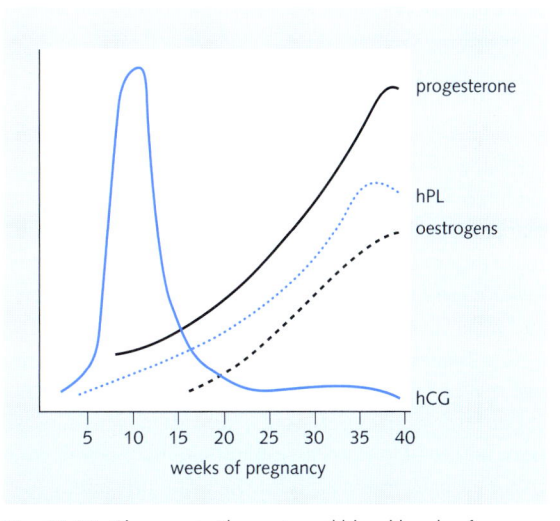

Fig. 14.12 Changes in the maternal blood levels of hormones during pregnancy. (Adapted from D Llewellyn-Jones, 6th edn.)

power of endocrine signals, since the tiny blastocyst can alter the physiology of the mother who is many millions of times larger.

The corpus luteum continues to be the main source of progesterone and oestrogen for the first 6 weeks of development.

Placental phase
The placenta secretes the following hormones:
- Human chorionic gonadotrophin (hCG).
- Progesterone.
- Oestrogens.
- Human placental lactogen (hPL).
- Relaxin.

By the sixth week the placenta is the main source of progesterone and oestrogens, which help the mother's body adapt to pregnancy. hPL helps regulate nutrient levels and metabolism; it also causes the glandular tissue of the breast to develop. Relaxin is secreted towards the end of pregnancy to prepare the body for birth.

Reproductive hormones
Human chorionic gonadotrophin
hCG is a peptide hormone secreted by the syncytiotrophoblast of the blastocyst from implantation. It has a similar structure and actions to LH. These actions include:
- Maintenance of the corpus luteum.
- Regulation of placental oestrogen secretion.
- May be a factor in blocking the maternal immune response to fetus.
- Stimulation of testosterone secretion in the male fetus.

The corpus luteum is initially formed and maintained by the ovulatory LH surge and hCG simply replaces the falling levels of LH to prevent regression. hCG is secreted for the first 10 weeks of pregnancy until the placenta is capable of secreting sufficient sex steroids to maintain the pregnancy. After 10 weeks the corpus luteum is no longer needed for hormone production and hCG levels begin to fall.

The β-subunit of hCG can be detected in the urine just before the first day of a missed period, usually about 10 days after potential fertilization. This allows time for the blastocyst to implant and hCG levels to rise. This principle is used for the urine pregnancy testing available in hospitals and over the counter in pharmacies. Since hCG levels fall after the corpus luteal phase these tests no longer work after 20 weeks. False positives may rarely indicate underlying disease (see the section on placental disorders).

Progesterone
The structure and synthesis of progesterone is described in Chapter 12. Plasma levels of progesterone rise throughout pregnancy, secreted initially by the corpus luteum then by the placenta. It is the single most important hormone in the maintenance of pregnancy. Actions include:
- Maintenance and development of the functional endometrium.
- Inhibition of smooth muscle in the uterus to prevent premature expulsion.
- Metabolic changes, including fat storage.
- Physiological adaptation to pregnancy (this is described in the next section).
- Relaxation of smooth muscle throughout the body, which may cause some side effects (e.g. constipation and oesophageal reflux).

Oestrogens
The structure and synthesis of oestrogens are described in Chapter 12. Plasma levels of oestrogens (especially oestriol) rise throughout pregnancy, secreted initially by the corpus luteum then by the placenta. Like the granulosa cells, the placenta lacks several key enzymes for the synthesis of oestrogen from cholesterol. These steps must be performed by the fetal adrenal gland allowing the fetus to regulate placental oestrogen secretion. This does not affect progesterone secretion, which is formed from cholesterol in just two steps.

The actions of oestrogen prepare the body for birth and lactation. They include:
- Growth of the smooth muscle of the uterus (myometrium).
- Increased blood flow to the uterus.
- Softening of the cervix and pelvic ligaments.
- Stimulation of breast growth and development directly.
- Stimulation of pituitary prolactin secretion.
- Inhibition of pituitary LH and FSH secretion.
- Stimulation of synthesis of oxytocin receptors in the myometrium in late pregnancy.
- Water retention.

Human placental lactogen
hPL is a peptide hormone secreted by the placenta; levels rise throughout pregnancy. It is sometimes called human chorionic somatomammotropin

because its actions are similar to growth hormone and prolactin. These actions include:
- Maternal lipolysis (fat break down) and fatty acid metabolism sparing glucose.
- Maternal insulin resistance sparing glucose for the fetus.
- Enhancing active amino acid transfer across the placenta.
- Stimulating the growth and development of the breasts.

Relaxin
Relaxin is a peptide hormone secreted by the placenta late in pregnancy. It relaxes the myometrium, cervix and the pelvic ligaments, allowing the uterus to enlarge and the pelvis to stretch during birth. It acts by stimulating collagenase enzymes which break down collagen in these tissues.

Reproductive hormones from other sources
Inhibin
Inhibin is a peptide hormone secreted by the ovary in the pregnant and non-pregnant state. It may suppress pituitary FSH secretion and stimulate progesterone production during pregnancy.

Prolactin
The structure, synthesis and control of prolactin is described in Chapter 2. Prolactin secretion from the pituitary gland rises throughout pregnancy due to stimulation by oestrogens. It has a similar action to hPL in that it stimulates the growth and development of the breasts and regulates fat metabolism.

During pregnancy the high levels of placental oestrogens prevent the secretion of milk. After birth the fall in oestrogen levels allows prolactin to act on the breast. If the mother breastfeeds the baby, sensory signals from the nipple cause further prolactin secretion after pregnancy. The high prolactin levels have two effects:
- Secretion of milk, though it is oxytocin that causes the milk to be ejected.
- Inhibition of pituitary FSH and LH, which has a contraceptive effect.

Maternal adaptations to pregnancy

Throughout pregnancy the mother's body is undergoing changes that achieve three main purposes:

- Maintenance of a suitable environment for fetal growth.
- Preparation of the mother for childbirth.
- Preparation of the mother for lactation.

A summary of these changes is shown in Fig. 14.13.

Symptoms and signs of pregnancy
Presentation
The diagnosis of pregnancy may be a joyful or disastrous moment, depending on the situation and beliefs of the mother. There is also a wide range of knowledge and experience between women of different backgrounds and age groups. The earliest signs of pregnancy are:
- A missed period.
- Nausea, possibly with vomiting.
- Temporarily increased frequency of urination.
- Fuller breasts and larger nipples, possibly with tenderness.

A positive urine β-hCG test confirms the diagnosis, and many patients will have done this test at home before presenting to a doctor. Further confirmation can be obtained using ultrasound, but this is not normally necessary.

Side effects
In addition to the early signs mentioned above, the maternal adaptations to pregnancy can cause symptomatic side effects; the common side effects are shown in Fig. 14.14.

Genital tract
Placental hormones cause changes mainly in the lower genital tract:
- Uterus—expands dramatically through pregnancy; the muscle layer hypertrophies massively in preparation for birth.
- Cervix—softens and becomes more readily dilated.
- Vagina—hypertrophy of muscle and becomes more readily dilated.

Cardiovascular system
The effectiveness of the cardiovascular system must increase to cope with:
- Increased oxygen demands of the fetus and placenta—about 130% of normal.
- Extra vascular space created by the expanding uterus and placenta.

The Process of Reproduction

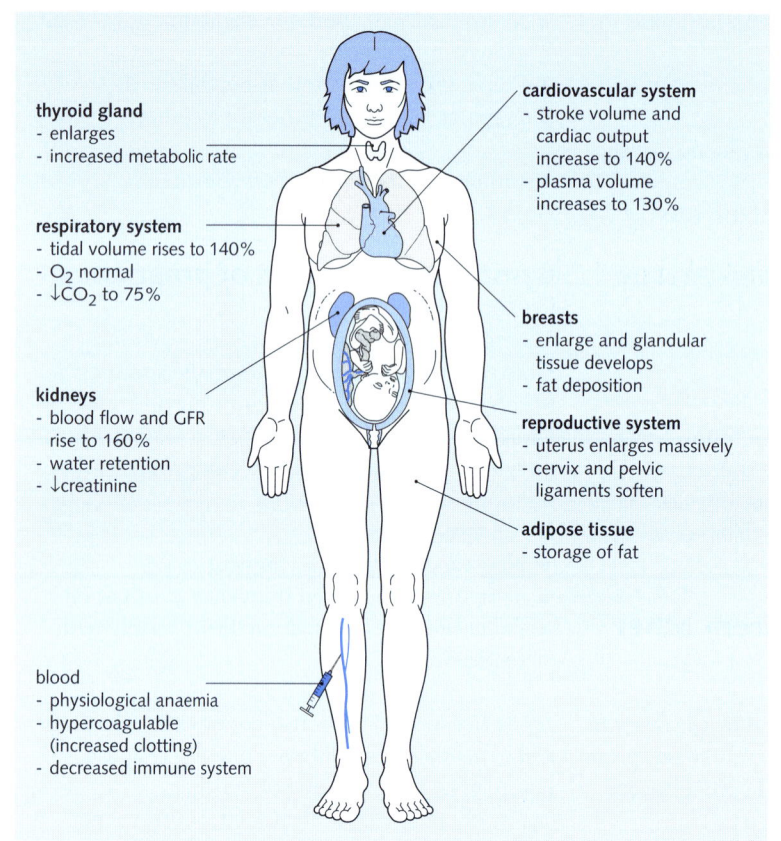

Fig. 14.13 Maternal adaptations to pregnancy.

thyroid gland
- enlarges
- increased metabolic rate

respiratory system
- tidal volume rises to 140%
- O$_2$ normal
- ↓CO$_2$ to 75%

kidneys
- blood flow and GFR rise to 160%
- water retention
- ↓creatinine

blood
- physiological anaemia
- hypercoagulable (increased clotting)
- decreased immune system

cardiovascular system
- stroke volume and cardiac output increase to 140%
- plasma volume increases to 130%

breasts
- enlarge and glandular tissue develops
- fat deposition

reproductive system
- uterus enlarges massively
- cervix and pelvic ligaments soften

adipose tissue
- storage of fat

The early adaptations are mostly caused by an increase in stroke volume with a slight increase in heart rate. Together these factors raise cardiac output to about 140% of normal by the sixth month. Cardiac output then begins to fall because pressure from the enlarged uterus inhibits venous return. Blood pressure does not increase for the first 7 months because of the fall in peripheral resistance caused by the developing placenta. It then begins to rise until birth.

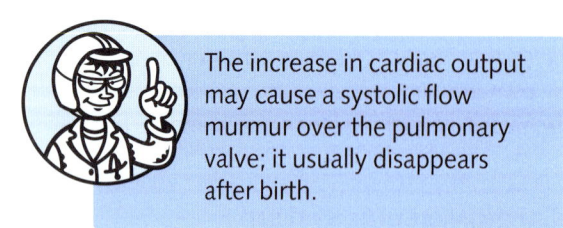

> The increase in cardiac output may cause a systolic flow murmur over the pulmonary valve; it usually disappears after birth.

Blood volume rises in the latter half of pregnancy to about 130% of normal, though there is a slight rise throughout pregnancy. Aldosterone and oestrogen cause renal water retention so the plasma volume increases. There is also an increase in red blood cells, but this is less than the fluid retention so the blood becomes diluted causing a physiological anaemia. The excess blood volume fills the placental vasculature and also protects the mother from haemorrhage during birth.

Respiratory system

The increased oxygen demands of the fetus and placenta also require adaptations in the respiratory system. Progesterone makes the chemoreceptors more sensitive to CO$_2$, causing the tidal volume to increase to 140% of normal while the respiratory rate remains the same. This maintains arterial oxygen saturation at normal levels whilst arterial CO$_2$ levels are about 75% of normal. These adaptations often give a sensation of breathlessness during pregnancy.

Renal system

Blood flow to the kidney increases during early pregnancy and then remains high. This causes the glomerular filtration rate (GFR) to rise to about 160% of normal. This would normally result in

Fig. 14.14 Common symptoms of pregnancy.

Common symptoms of pregnancy		
Symptom	Cause	Stage of pregnancy
Morning sickness (nausea ± vomiting)	Rising oestrogen levels	From 4 weeks but then declines
Increased pigmentation	Raised levels of melanocyte stimulating hormone (MSH) from the pituitary causing pigmentation of the face (chloasma) and abdominal striae	Gets progressively worse through pregnancy
Breathlessness	Changes in the cardio-respiratory system	
Gestational diabetes	Impaired glucose tolerance due to cortisol and hPL	
Constipation	Relaxation of smooth muscle caused by progesterone	
Heartburn and reflux		
Carpel tunnel syndrome	Water retention caused by oestrogen	
Ankle oedema		
Goitre	Raised thyroid stimulating hormone (TSH) acting on the thyroid gland	
Severe abdominal distension	Fetus developing inside the uterus	
Prurigo of pregnancy	Itchy rash over abdomen and limbs	Late
Back ache	Softening of ligaments due to oestrogen	
Urinary frequency	Fetal head presses on the bladder	

sodium loss, but increased secretion of renin, angiotensin II and aldosterone counteract these changes.

Progesterone causes the smooth muscle of the collecting ducts and ureters to become dilated. This slows the excretion of urine making urinary tract infections (UTIs) more common. The urethra is also relaxed and the fetus exerts pressure on the bladder, so urinary incontinence is relatively common in late pregnancy.

Endocrine system
The secretion of the anterior pituitary hormones is altered during pregnancy (Fig. 14.15):
- FSH and LH secretion is almost completely stopped.
- Prolactin secretion rises throughout pregnancy.
- Thyroid stimulating hormone (TSH) secretion initially falls then increases.
- Adrenocorticotrophic hormone (ACTH) secretion increases.
- Melanocyte-stimulating hormone (MSH) secretion increases.

The anterior pituitary gland enlarges as a result of these changes.

The rise in secretion of most hormones is caused by direct actions of placental hormones and an increase in plasma binding proteins (caused by the action of oestrogens on the liver) that reduces negative feedback.

Thyroid glands
hCG inhibits TSH secretion in the 1st trimester but, as the hCG levels fall in the 2nd and 3rd trimesters, TSH then rises above normal. The increase in TSH together with a reduction in iodine from the overactive kidneys causes the thyroid gland to enlarge. Thyroid hormone synthesis also increases but so does the synthesis of thyroid hormone binding proteins stimulated by oestrogen. Overall active/free thyroid hormone levels remain normal.

Adrenal glands
In contrast, free cortisol levels do rise, despite the increase in plasma binding proteins. This raises amino acid and glucose levels in the blood to improve fetal growth.

Secretion of anterior pituitary hormone	Hormone secretion in pregnancy (compared with non-pregnancy)	Effect of altered plasma hormone level in pregnancy
Prolactin ↑↑↑	Enhanced by placental oestrogens	Promotes growth and development of the breasts and regulates fat metabolism
FSH ↓ and LH ↓	FSH secretion is suppressed by inhibin and placental oestrogens LH secretion is suppressed by the combined effect of progesterone and oestrogen	Prevents further follicular development and ovulation during pregnancy
GH ↓	Suppressed by hPL	Unknown (hPL has similar effect to GH)
ACTH ↑	Rises	Stimulates increased cortisol secretion from the adrenal cortex
TSH	Falls in first trimester but then rises in second and third	Changes in thyroid hormone secretion are counteracted by changes in plasma protein synthesis

Changes that occur in pituitary hormone secretion during pregnancy and the effects caused by these changes

Fig. 14.15 Changes in secretion of anterior pituitary hormones during pregnancy. (ACTH, adrenocorticotrophic hormone; FSH, follicle stimulating hormone; GH, growth hormone; hPL, human placental lactogen; LH, luteinizing hormone; TSH, thyroid stimulating hormone.)

Aldosterone secretion from the adrenal cortex also rises slowly in response to the rising ACTH levels. It helps prevent the sodium loss caused by the raised GFR in the kidney.

Changes in metabolism

The mother usually gains 9–15 kg during pregnancy though the majority of this is caused by the fetus, placenta and fluid retention. Six months after birth, maternal weight is usually just 1 kg higher than before the pregnancy. Women have a larger appetite during pregnancy to supply the developing fetus, placenta and breasts. The excess of nutrients is regulated by changes in metabolism.

Carbohydrates

The hormone hPL causes insulin resistance to develop and this effect is enhanced by the raised cortisol. As a result, the maternal metabolism uses a higher proportion of fatty acids and glucose use decreases. This glucose is spared for the growing fetus.

If the mother already has a degree of impaired glucose tolerance (e.g. obesity) then diabetes mellitus can result. The glucose levels should be strictly controlled because hyperglycaemia predisposes to large babies, difficult births and other paediatric complications. After birth this gestational diabetes mellitus usually resolves.

Amino acids

Progesterone inhibits the breakdown of amino acids in the liver to increase their availability for fetus. The raised cortisol also increases the blood levels while hPL aids transport across the placenta.

Fat

Fat stores are initially broken down through the action of hPL to drive maternal metabolism. Towards the end of pregnancy fat is stored in the breasts and subcutaneous tissues. Fat only accounts for a fraction of weight gain through pregnancy.

Other changes

Other minor changes occur during pregnancy:
- Increase of clotting factors so coagulation occurs more easily. In fact, intravascular coagulation and embolism is the most common cause of pregnancy-related mortality in the UK.
- Immune system is repressed to prevent rejection of the fetus and this can predispose to infection.
- Pelvic ligaments soften to allow the fetus to pass during birth.
- Venous congestion in lower limbs due to the pressure of the fetus on venous return; it can cause varicose veins.

Parturition and labour

Position of the fetus

It is important to determine the position of the fetus before the onset of labour so that potential problems can be identified and preparations made. The position is assessed through palpation and ultrasound scans. There are three aspects to the fetal position, described below.

Lie

The lie describes the orientation of the baby's long axis; in simple terms, it is the orientation of the back. It can be:
- Longitudinal—this is the normal position with the back lying along the uterus.
- Oblique—the back is at an angle across the uterus.
- Transverse—the back lies across the uterus.

Palpation can also reveal which side the back is on, in a longitudinal lie.

Presentation

This is simply the part of the fetus that is nearest the cervix and, therefore, most likely to come out first. There are three main presentations:
- Cephalic—head first; this is normal.
- Breech—the bottom or feet first.
- Shoulder—associated with a transverse lie.

Cephalic presentations are further divided according to which part of the head is presenting. The term 'denominator' is used to describe the foremost part of the head. This is important because it affects the widest part of the head that must be born. The different types of cephalic presentation are shown in Fig. 14.16; occipitoanterior is the normal presentation.

Fig. 14.17 shows some important points on the fetal skull along with the widest diameters of each presentation.

Clinically the presentation is described along with the extent to which the presenting part is palpable; this is described in fifths so that a fully palpable head scores 5/5, which decreases as the head descends, e.g. 3/5.

Position

The position of the fetus describes the direction that the denominator is facing compared with the pelvis. This is divided into two sections, which are then subdivided.

Firstly:
- Left (L)—faces the mother's left.
- Right (R)—faces the mother's right.
- Straight—faces the pubic symphysis or sacrum directly.

Secondly:
- Anterior (A)—faces the pubis bone.
- Transverse (T)—across the pelvis.
- Posterior (P)—faces the sacrum.

These two sections are combined with the name of the denominator to describe the position, e.g. left occipitoanterior (LOA) or occipitoposterior (OP). The occipitoposterior position is often associated with a deflexed presentation.

Sequence of labour

Labour is the sequence of actions leading to childbirth (parturition), including the expulsion of the placenta. In normal pregnancies childbirth occurs after 37–42 weeks; the average is 40 weeks (280 days). Labour begins with regular, painful uterine contractions accompanied by cervical

Fig. 14.16 Comparison of the four variations of cephalic presentation.

Comparison of the four variations of cephalic presentation

Presentation	Vertex (occipitoanterior)	Deflexed (occipitoposterior)	Brow	Face
Position of the neck	Flexed	Deflexed	Extended	Very extended
Denominator	Occiput	Vertex	Bregma	Chin
Widest diameter	Suboccipito-bregmatic	Occipito-frontal	Mento-vertical	Submento-bregmatic
Width	9.5 cm	11.5 cm	13.5 cm	9.5 cm

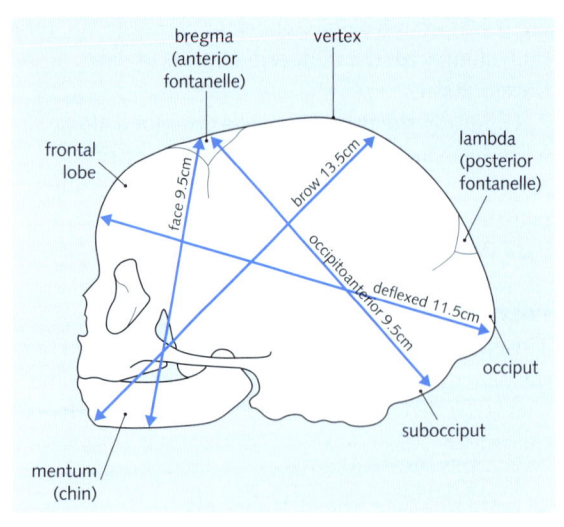

Fig. 14.17 Anatomy of the fetal skull and the widest diameters in the four types of cephalic presentation.

dilatation. It is often preceded by several weeks of false labour with irregular, painful contractions (called Braxton Hicks contractions) and no cervical dilatation. True labour is divided into three stages: first, second and third.

First stage
This is from the onset of true labour until full cervical dilatation (10 cm). The length of the stage varies widely between women, however the cervix should dilate at a rate of 1 cm/h:
- 8–10 hours in first labour (nulliparous women, known as 'primips').
- 2–6 hours in subsequent labours (multiparous women or 'multips').

As labour progresses, the uterine contractions become stronger and more frequent. The contractions push the fetal head into the pelvis towards the cervix. The pain experienced is due to hypoxia of the uterus caused by occlusion of the blood vessels during the muscular contractions. The amniotic membrane often ruptures during this stage, resulting in the loss of amniotic fluid (breaking of the waters). The baby performs two actions before or during the first stage:
1. Engagement of the head into the pelvis.
2. Descent of the head through the pelvis, usually in left occipitoanterior (LOA) position.

The first stage is subdivided into two phases:
- Latent phase—the cervix dilates slowly from 0 to 4 cm.
- Active phase—the cervix dilates more rapidly from 4 to 10 cm.

Second stage
This is from full cervical dilatation until the birth of the baby. It usually lasts 40–60 minutes in primips and 10–15 minutes in multips. The baby must perform eight actions for normal birth to occur:
1. Flexion of the neck so its chin is on its chest and the occiput will be presented.
2. Internal rotation of the head so that it faces the sacrum.
3. Crowning of the occiput (when the baby is visible between contractions).
4. Extension of the neck as the head is born (support the head and check for the umbilical cord at the back of the neck).
5. Restitution as the head rotates back to the normal position outside of the mother.
6. External rotation of the head towards the mother's thigh as the shoulders rotate.
7. Birth of the anterior (top) shoulder (push the baby's head down).
8. Birth of the posterior (bottom) shoulder and body (pull the baby's head up and support the body).

Uterine contractions continue and are assisted by voluntary 'pushing' by the mother (contractions of the diaphragm and abdominal muscles). Once the head has crowned, the mother is asked to stop pushing so that the head can pass the vaginal opening smoothly to prevent tearing.

The pain is most severe during the second stage; it is caused by stretching of the cervix, vagina and perineum. The pain is conducted by normal somatic sensory nerves.

Third stage
This is from birth of the baby until the delivery of the placenta and membranes. Naturally, it lasts between 10 and 45 minutes though current practice is to actively manage this stage. This involves:
- Intramuscular injection of Syntometrine® (5 units oxytocin and 500 μg ergometrine) during the birth of the body.
- Pulling the umbilical cord once there are signs of placental separation (lengthening of the cord, contraction of the uterus or a gush of blood).

The uterus shrinks to the 20-week size and contractions continue. The entire placenta and decidua basalis detach from the uterus and are expelled causing haemorrhage from the ruptured blood vessels. The haemorrhage is stopped by the muscle fibre arrangements within the uterus; this is more effective using active management. The contractions slowly subside once the afterbirth has been expelled. The placenta is inspected carefully to ensure that no sections remain in the uterus.

Initiation of parturition
The myometrium of the uterus becomes more excitable toward the end of gestation causing false labour that blends into the coordinated contractions seen in true labour. The exact mechanism that initiates this increase in excitability and onset of labour are not known, though several factors have been identified.

Oestrogen:progesterone ratio
Progesterone inhibits contractions during pregnancy, but its secretion stabilizes or drops towards the end. Oestrogens stimulate contractions and secretion continues to increase until birth. The balance of these hormones moves in favour of oestrogen, causing increased excitability.

Uterine distension
Stretching the muscle of the uterus increases contractility, so fetal growth and movement may have a stimulatory effect.

Cervical distension
Irritation and stretching of the cervix causes oxytocin release, which stimulates contractions. The fetal head activates this release by pressing against the cervix.

Fetal hypothalamus maturation
At full term the fetal hypothalamus and pituitary secrete more CRH, ACTH and oxytocin. This oxytocin may cross the placenta to act on the uterus.

Fetal adrenal activity
Cortisol secretion from the fetal adrenal glands increases as fetal ACTH rises. This stimulates placental oestrogen secretion and prostaglandin synthesis in the uterine muscle that raises contractility.

Hormonal control of parturition
Oxytocin
Oxytocin is a peptide hormone synthesized in the hypothalamus and secreted by the posterior pituitary gland (it is described in Chapter 2).

During labour oxytocin levels rise due to cervical stimulation by the head. It stimulates uterine contractions that push the fetus against the cervix, stimulating further oxytocin release. A positive feedback mechanism develops called the Ferguson reflex.

Oxytocin receptors in the uterine muscle are increased during late pregnancy by the action of oestrogen. Oxytocin binding stimulates prostaglandin production that causes the increased contractility (especially PGE_2).

Prostaglandins
Prostaglandins are locally acting eicosanoids that regulate many processes throughout the body (see Ch. 1). During labour the prostaglandin PGE_2 is synthesized in the uterine muscle cells in response to oxytocin. It stimulates the release of calcium ions which cause muscle contractions.

PGE_2 is also synthesized in the cervix where it stimulates cervical softening and dilatation.

Relaxin
Relaxin promotes the relaxation of the pelvic ligaments and softens the cervix prior to parturition. This allows both structures to stretch so the fetus can pass through the pelvis.

Induction of labour
Labour can be induced using three methods that can be used in combination or alone:
- Prostaglandins (PGE_2) by vaginal pessary or gel, which acts within a few hours.
- Intravenous oxytocin, which takes about 12 hours to act.
- Rupture of the amniotic membrane (amniotomy), which can only be performed if the cervix is more than 4 cm dilated.

Disorders of labour

Prolonged labour
The progress of labour is plotted on a partogram that records measurements including the frequency and strength of contractions, dilatation of the cervix and descent of the presenting part. The pattern on the partogram can be used to distinguish between two types of prolonged labour (Fig. 14.18).

Primary dysfunctional labour
Primary dysfunctional labour describes slow dilatation of the cervix or descent of the presenting part. It is a common condition especially in primips. It is usually due to inefficient uterine contractions

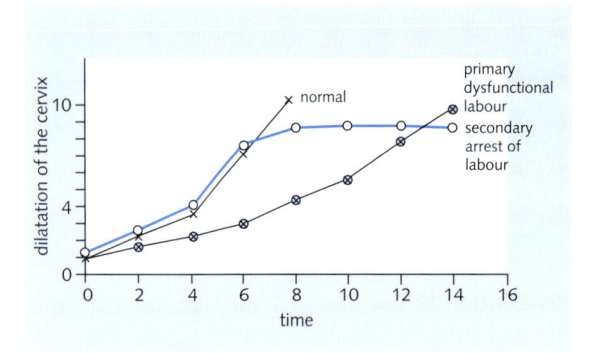

Fig. 14.18 Dilatation of the cervix in normal, primary dysfunctional, and secondary arrested labour.

and it can be treated using intravenous oxytocin to improve the strength of contractions. Alternatively, artificial rupture of the membranes (if they have not already ruptured) using an 'amnihook' can have a similar effect.

Secondary arrest of labour
Secondary arrest describes a labour that 'gets stuck' after progressing normally. The head fails to descend and the cervix remains at the same dilatation. This is less common than primary dysfunctional labour though it is often difficult to distinguish the two patterns. Secondary arrest should be suspected when prolonged labour occurs in a multip; it can also occur in primips. It is caused by:
- Inefficient uterine contractions.
- Cephalopelvic disproportion (the head is too large for the pelvis).
- Malposition of the fetus (e.g. breech).

Since inefficient uterine contractions are the most common cause, intravenous oxytocin is used. If the labour still fails to progress, or if fetal distress is detected, then caesarean section is needed.

Assisted delivery
The second stage of labour is a critical period for the fetus and mother. If the second stage progresses slowly (beyond 1 h in multips or 1.5 h in primips) or fetal compromise is suspected then an assisted delivery may be considered. There are four main types of assisted delivery:
- Kjelland's forceps—these are rotational forceps used to correct a malposition; they are rarely used in developed countries.
- Neville–Barnes forceps—the most common forceps used for mid pelvic-cavity deliveries.
- Wrigley's forceps—short forceps used for low pelvic-cavity deliveries and caesarean sections.
- Ventouse extraction—a suction device that fits onto the fetal vertex.

All assisted deliveries carry the risk of increased trauma to the fetus and mother. There are a number of criteria that must be met before an assisted delivery can be attempted:
- Fetal position known, with cephalic presentation.
- Presenting part descended to the ischial spines or below with <1/5 palpable abdominally.
- Cervix fully dilated and membranes ruptured.
- Maternal bladder empty (to limit trauma).
- Adequate analgesia.
- Consent of the mother.

Lactation

Mammary development
During pregnancy the breast undergoes hormone-induced adaptations in preparation for lactation after birth. This section describes these changes together with the control and process of lactation. The development of the breast is described along with the other organs of reproduction in Chapter 11. The structure and disorders of the adult breast can be found in Chapter 12.

Changes during pregnancy
After puberty the female breast is composed of 15–20 lobes divided into secretory lobules, each with 10–100 acini. These acini are surrounded by fatty connective tissue; they drain into the lactiferous ducts. The breast remains in this state until pregnancy.

Development of the breasts during pregnancy is caused by the rising levels of four hormones:
- Oestrogens—cause the ductal system to grow and branch and fat to be stored in the stroma; inhibit milk production.
- Progesterone—causes growth and an increased number of acini.
- hPL—causes development of the acini cells so they are capable of milk secretion.
- Prolactin—causes development of the acini similar to hPL.

In the last few weeks of pregnancy, oestrogen fails to inhibit breast secretion completely so small quantities of a yellowy fluid called colostrum are secreted. Colostrum contains virtually no fat and it has high concentrations of antibodies.

After birth the oestrogen, progesterone and hPL levels fall because the placenta is expelled. Prolactin secretion continues if the mother breastfeeds the baby; this maintains the breast changes brought about by the other hormones. The lack of oestrogen allows prolactin to stimulate production of milk instead of colostrum, though it takes a few days for the change to occur.

The breast changes caused by oestrogens and progesterone occur to a lesser degree towards the end of each menstrual cycle. The breasts often become swollen and tender.

Hormonal control

Lactation is caused by the effects of two hormones:
- Prolactin—causes milk secretion.
- Oxytocin—causes milk ejection.

Both hormones are secreted by the pituitary gland in response to nipple stimulation. Prolactin is secreted by the anterior pituitary gland and oxytocin is secreted by the hypothalamus and released from the posterior pituitary. The synthesis and secretion of these hormones are described in Chapter 2.

Prolactin

Prolactin secretion increases throughout pregnancy causing the acinar cells of the breast to develop. During pregnancy milk production is inhibited by high oestrogen levels. After childbirth, oestrogen levels fall dramatically so prolactin can stimulate the secretion of milk.

During breastfeeding stimulation of the nipple causes prolactin secretion, resulting in milk secretion and maintenance of the breast. The milk accumulates within the breast causing swelling unless oxytocin triggers the milk to be ejected. This neuroendocrine reflex is shown in Fig. 14.19. Once breastfeeding is stopped, nipple stimulation diminishes so prolactin secretion and milk production cease.

The high levels of prolactin during lactation inhibit LH and FSH secretion giving breastfeeding a contraceptive effect. This is only effective while the baby is suckling regularly. Once lactation ceases, the normal ovarian cycle and fertility return within 4–5 weeks.

Oxytocin

Milk is ejected from the breast by the action of hormonal, rather than neural, signals on the smooth muscle in the breast. Oxytocin induces the smooth

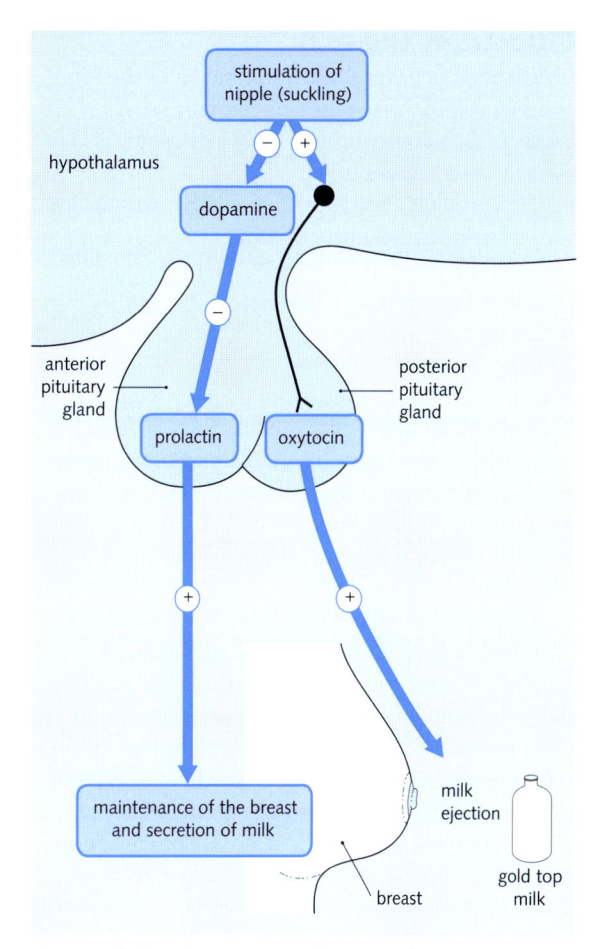

Fig. 14.19 Regulation of lactation by prolactin and oxytocin.

The production of milk is a very good example of biological supply and demand. The more the baby suckles, the more prolactin is secreted and the more milk is produced. Women who are having difficulty producing 'enough milk' should be encouraged to allow the baby to suckle more.

muscle cells surrounding the acini to contract so milk is squeezed out of the nipple. Suckling stimulates this oxytocin release causing milk ejection within about 30 seconds. The reflex is shown in Fig. 14.19. Even the sound of the baby crying can stimulate the release of oxytocin and the ejection of milk. On the other hand, stress can inhibit this reflex and this can be a particular problem if the woman is worried about her ability to breastfeed.

Colostrum and milk

Colostrum is a pale yellow fluid that lacks the fat content of milk. It is richer in antibodies so it protects the neonate against early infection; this is called passive immunity.

Breast milk is a mixture of essential nutrients in water, the main constituents are:
- Lipids (fat).
- Casein (protein).
- Lactose (sugar).

It also contains vitamins, minerals and antibodies. Milk produced by other mammals has a different composition, making it unsuitable for human babies. For example, cow's milk contains less lactose but more casein than human milk. Special formula milks are available for women who choose not to breastfeed.

It is important to consider whether a woman is breastfeeding when prescribing medications. A number of chemicals can enter breast milk and affect the baby, including oestrogen (e.g. combined oral contraceptive pill) and alcohol. HIV positive mothers are advised not to breastfeed, since there is a risk of the virus infecting the baby through the milk.

Lactation is a very energy intensive process, even more so than pregnancy. The woman will require about 120% of her normal energy usage. This extra energy is derived from stored fat and the diet.

Disorders of pregnancy and the placenta

Ectopic pregnancies

It is important to consider the possibility of an ectopic pregnancy in any woman presenting with abdominal or pelvic pain. A pregnancy is described as 'ectopic' if the blastocyst implants in any location other than the endometrium of the uterus. Ninety-nine per cent of the time this means the fallopian tubes, but it can also occur on the ovary or in the abdomen. The incidence is about 1 in 100 pregnancies in developed countries; the risk is increased by factors that slow the transport of the oocyte:
- Pelvic inflammatory disease.
- Previous pelvic surgery.
- Previous ectopic pregnancy.
- Pregnancy despite progesterone-only pill or IUD use.
- Pregnancy from assisted fertilization (e.g. IVF).

Once the blastocyst has implanted, the trophoblast attempts to form a placenta by invading the surrounding structures. Initially this allows the embryo to grow, but the pregnancy usually terminates after 6–10 weeks due to a lack of space and nutrients. This is called tubal abortion.

There is a high risk of complications following ectopic pregnancy, the most serious of which is tubal rupture. The trophoblast erodes through the wall of the fallopian tube causing intra-peritoneal bleeding. In a minority of cases this can be severe and life-threatening.

Symptoms and signs
Subacute
The majority of patients with ectopic pregnancies present following tubal abortion or mild rupture. The most common symptoms are:
- Unilateral abdominal pain and tenderness.
- Recent amenorrhoea.
- Vaginal bleeding.

On vaginal examination there may be a tender mass in a fallopian tube. The abdominal pain may be so mild that it is ignored until vaginal bleeding occurs.

Acute
In severe tubal rupture the patient presents with severe abdominal pain and sudden collapse. They will have hypovolaemic shock and an acute abdomen (tender with guarding).

Investigations
Diagnosis of subacute ectopic pregnancies from the history and examination alone is very difficult due to the non-specific symptoms. Blood tests for hCG will be positive indicating a pregnancy (if measured over a couple of days it may rise slower than expected). An ultrasound scan will show an empty uterus and may reveal the mass in the fallopian tube. If the embryo is not found then a laparoscopy is performed to examine the tubes directly.

Acute tubal rupture is a surgical emergency. It is investigated and treated by laparotomy to remove the entire affected fallopian tube as soon as possible.

Treatment
If ectopic pregnancies are diagnosed before abortion or rupture occur, they are treated surgically using laparoscopy ('keyhole' techniques) or laparotomy (opening the abdomen); there are three treatment options:

- Removal of the entire fallopian tube that contains the embryo.
- Removal of just the embryo through an incision in the tube.
- Injection of methotrexate (cytotoxic drug) into the embryo inducing early abortion.

Since there is a high risk of recurrence or infertility following ectopic pregnancy, any further pregnancies need careful monitoring.

Miscarriage

Miscarriage (spontaneous abortion) is the expulsion of a fetus from the uterus before it is capable of independent survival; clinically, this is before 24 weeks or below 500 g. After 24 weeks, it is termed a premature delivery. Miscarriage is very common, affecting about 10% of pregnancies, usually between the 6th and 10th weeks, though more may occur before the mother realizes she is pregnant. It is caused by:

- Fetal abnormalities, often due to chromosomal disorders (60%).
- Abnormal implantation or placenta.
- Uterine abnormality.
- Maternal illness.
- Idiopathic (unknown).

Symptoms and investigations

Miscarriage is suspected if a pregnant woman experiences vaginal bleeding. Ultrasound is used to visualize the fetus and this normally shows the fetus is alive and well.

If the bleeding is associated with cervical dilatation and uterine contractions (which may be described as pelvic pain) then miscarriage becomes inevitable. The woman must be admitted to hospital to ensure that the entire fetus and placenta are expelled. This is performed by inspecting the expelled material and performing a further ultrasound scan of the uterus. Any retained material must be extracted surgically through the cervix.

Placenta previa

If the placenta is located over the lower uterine segment (sometimes including the cervix) the condition is called placenta previa; the incidence is about 1 in 200 pregnancies. It is important not to perform a vaginal examination if placenta previa is suspected. The severity of placenta previa is graded 1–4 according the distance from the cervix (1 being the furthest). The placenta is prone to bleeding as the uterus grows or when the cervix dilates in labour. The bleeding is usually painless and the uterus remains soft and non-tender; it may also present with the fetus in an abnormal position due to the location of the placenta. The risk of placenta previa is increased by multiple pregnancies and previous caesarean sections. It is diagnosed and monitored by ultrasound. Delivery is performed by caesarean section at 37 weeks (or earlier if bleeding is severe), except grade 1, which may allow normal vaginal delivery.

Placental abruption

Placental abruption is when the placenta separates from the uterine wall before the fetus has been delivered, resulting in bleeding. Vaginal bleeding is usually obvious, however the blood is concealed within the uterus in about 20% of cases. The bleeding ranges from mild to severe and life threatening. It is a common disorder (about 1 in 80 pregnancies) that is predisposed by hypertension, smoking, multiple pregnancies and polyhydramnios (excess amniotic fluid). It presents with:

- Vaginal bleeding.
- Abdominal pain and tenderness.
- Rigid uterus.
- Evidence of fetal compromise.

Placental abruption can be diagnosed using ultrasound to visualize concealed blood clots and distinguish it from placenta previa. The extent of bleeding is determined using signs of shock and blood tests to measure haemoglobin and platelet concentration. Fetal compromise is recorded using fetal heart monitoring. The mother should be resuscitated and stabilized; induced delivery or emergency caesarean section may be indicated.

Pre-eclampsia and eclampsia

Eclampsia is a disease that presents in the second half of pregnancy with convulsions caused by hypertension. It is preceded by an increase in blood pressure called pre-eclampsia.

Pre-eclampsia

Pre-eclampsia (also called pregnancy-induced hypertension or PIH) affects about 5% of women during pregnancy to varying degrees. Treatment is needed if the blood pressure rises above 140/100 or if urine protein is consistently greater than 300 mg/L.

Pre-eclampsia is largely asymptomatic though generalized oedema can occur at any stage. If the following symptoms and signs develop, an eclamptic convulsion is likely:
- Severe headache.
- Irritability.
- Blurred vision.
- Epigastric pain.
- Vomiting.
- Brisk reflexes.

Eclampsia

Careful blood pressure and urine monitoring aims to prevent pre-eclampsia progressing to eclampsia; in developed countries eclampsia affects only 0.0005% of pregnancies, i.e. it is very rare. The woman experiences a brief period of disorientation followed by a tonic–clonic seizure. The initial seizure can be followed by further seizures, coma or haemorrhagic stroke. There is a risk of death for both the mother and fetus, but this is usually prevented by early detection and treatment of pre-eclampsia.

Aetiology

The underlying cause of pre-eclampsia and eclampsia occurs early in pregnancy during implantation. The trophoblast fails to invade the endometrial spiral arteries sufficiently, resulting in poor placental perfusion. This deficient invasion may be due to a maternal immunological response against the embryo.

Both the placenta and fetus become ischaemic, leading to poor development, so fetal growth retardation is often associated with pre-eclampsia. Cells from the ischaemic placenta can be carried (embolize) into the maternal circulation where they trigger the release of thromboplastins. The thromboplastins cause vasoconstriction and poor renal perfusion resulting in:
- Hypertension.
- Proteinuria.
- Oedema.

These signs characterize pre-eclampsia. If the condition continues, cerebral hypoxia and oedema can result, causing the convulsions of eclampsia. There is also a high risk of blood clots forming in the blood vessels (disseminated intravascular coagulation; DIC) that can produce tissue infarctions. The aetiology of pre-eclampsia is shown in Fig. 14.20.

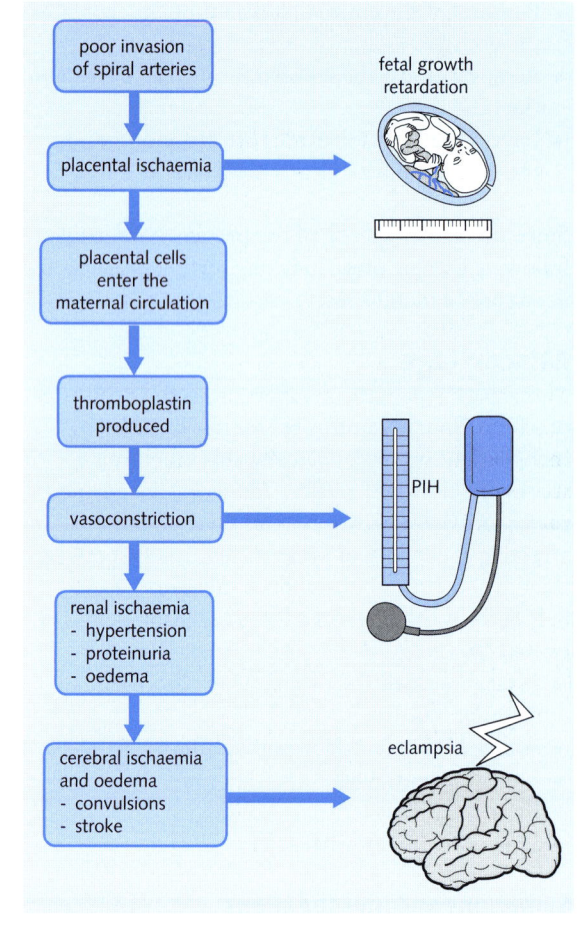

Fig. 14.20 The aetiology of pregnancy induced hypertension (PIH; pre-eclampsia).

Death can occur from:
- Cerebral haemorrhage (stroke) or oedema.
- Cardiac failure.
- Cardiorespiratory arrest.
- Organ failure following DIC.

Treatment

Even mild pre-eclampsia requires frequent blood pressure checks. If the blood pressure exceeds 140/100 or there is significant proteinuria then the woman needs to be admitted and treated.

The only cure for pre-eclampsia and eclampsia is delivery of the baby, often by caesarean section, although eclamptic fits can still occur up to 48 hours later. Several medications may slow the rise in blood pressure so that the baby has more time to mature. Diuretics cannot be used as they lower the blood volume making the placental ischaemia worse.

Magnesium sulphate is used in severe PIH or eclampsia. It causes the arteries to relax, restoring blood flow to the brain and it also inhibits coagulation. Treatment must be monitored as it can depress breathing.

Neoplasia of trophoblastic origin

During implantation the trophoblast normally invades the endometrium. If this process is disrupted there is a high chance of the trophoblast forming an invasive tumour.

Hydatidiform mole

Hydatidiform moles are benign tumours of the chorion that forms the placenta. Chorionic villi enlarge to form grape like vesicles that secrete hCG and progesterone. It is usually caused by major abnormalities of fertilization. There are two types: partial and complete (Fig. 14.21).

> Hydatidiform moles are named from the Greek word *hydratid* meaning water droplet. This is due to the appearance of the vesicles that make up the mole.

Hydatidiform moles occur in about 1 in 2000 of UK pregnancies. The secretion of hCG and progesterone causes exaggerated symptoms of pregnancy:
- Severe morning sickness.
- Early pre-eclampsia.
- Abnormally large and doughy uterus.
- Vaginal bleeding.

Moles produce the following results on investigation:
- Absence of fetal heart sounds.
- 'Snowstorm-like' appearance on ultrasound scans.
- Extremely high hCG levels.

They are treated by suction evacuation of the uterus followed by hCG level monitoring to detect the recurrence that occurs in 10%, often in a malignant form called choriocarcinoma.

Invasive mole

This represents the middle ground between a hydatidiform mole and choriocarcinoma. The tumour invades the myometrium, but it does not spread outside of the uterus. There is a higher risk of recurrence and further invasion.

Choriocarcinoma

This is a highly malignant tumour of the trophoblast without recognizable chorionic villi. It is usually preceded by the recurrence of a hydatidiform mole. It can also occur following pregnancies (1 in 50 000) or miscarriages (1 in 5000). The cancer contains many areas of haemorrhage and necrosis; while villi are not formed, it does secrete high levels of hCG.

It presents in the same manner as a hydatidiform mole or with symptoms and signs of metastasis. Histological examination is used to determine the degree of malignancy.

The prognosis is excellent despite the early blood-borne metastasis to the brain, liver and lungs. It responds to chemotherapy very well and hCG levels can be monitored to guide treatment. Fertility is often not affected.

Fig. 14.21 Comparison of partial and complete hydatidiform mole.

Comparison of partial and complete hydatidiform mole

Feature	Partial mole	Complete mole
Extent of placental involvement	Only a section	Entire placenta
Fetal tissue	Present but fetus is normally non-viable	Not present
Usual cause	Two sperms fertilizing the oocyte (polyspermy)	Sperm entering an oocyte that has lost its nucleus (haploid zygote)
Risk of developing malignancy	Low	High

The Process of Reproduction

- Describe the two components of sexual arousal and the physical changes they cause.
- Describe the physiological changes that accompany sexual intercourse and orgasm.
- Describe the physical changes that occur during the male and female orgasm.
- What are the common causes of male and female sexual dysfunction?
- Describe the mature sperm and mature oocyte and where they meet.
- Describe the process of fertilization. What prevents polyspermy?
- Describe the early development of the zygote and implantation.
- Discuss methods of contraception.
- List the advantages and disadvantages of different types of hormonal contraception.
- Make a table showing the common causes of infertility and their treatments.
- Describe the development of the placenta including the three phases of the chorionic villi.
- Describe the functions of the placenta.
- List the hormonal changes and their source throughout pregnancy.
- Describe the actions of oestrogen, progesterone and hCG during pregnancy.
- Describe four maternal adaptations to pregnancy.
- Describe the three stages of birth and list the factors that stimulate labour.
- Describe the changes of the breast through pregnancy and the stimulation of lactation along with the hormones responsible.
- Describe the presentation, aetiology and treatment of ectopic pregnancies.
- Describe the presentation, aetiology and treatment of pre-eclampsia and eclampsia.
- Describe the presentation, aetiology and treatment of hydatidiform moles and choriocarcinoma.

CLINICAL ASSESSMENT

15. Common Presentations of Endocrine and Reproductive Disease — 199

16. History and Examination — 209

17. Investigations and Imaging — 227

15. Common Presentations of Endocrine and Reproductive Disease

This section gives examples of presenting complaints that are commonly associated with endocrine and reproductive disorders. It does not include the many presentations in which these systems are uncommon causes. Since the disorders of these two systems have a substantial overlap the disorders are presented together in alphabetical order. The important questions in the history are outlined along with a guide to differential diagnosis. The following disorders are included:

- Amenorrhoea.
- Breast lumps.
- Galactorrhoea.
- Gynaecomastia (male).
- Hirsutism (female).
- Loss of consciousness and coma.
- Menorrhagia and intermenstrual bleeding (female).
- Polyuria.
- Scrotal lumps.
- Sexual dysfunction (male and female).
- Thyroid lumps and goitres.
- Weight gain and obesity.
- Weight loss.

Amenorrhoea (Fig. 15.1)

This is a complete failure in menstruation lasting longer than 70 days; there are two types:
- Primary—failure to start menstruating by 16 years of age.
- Secondary—absent menstruation after starting during puberty.

Important questions in history taking are shown in Fig. 15.1. Primary amenorrhoea is usually just late puberty, while secondary amenorrhoea is most commonly caused by low body weight. A number of

Important questions and causes of amenorrhoea

Find out	Findings	Differential diagnosis
Age	>50 years	Menopause, can also be premature
Weight	Low	Low weight is a very common cause
Growth and sexual development	No secondary sexual characteristics, short stature	Turner syndrome
	Minimal pubic hair	Testicular feminization
Sexual history and contraception	Recent intercourse, no contraception	Pregnancy
	Recently started a progestogen-only form of contraception	May cause amenorrhoea
	Recently came off the Pill	Pituitary insensitivity
Systems review	Galactorrhoea, previous sparse periods, weight gain	Hyperprolactinaemia
	Weight gain, hirsutism, acne	Polycystic ovarian syndrome
	Weight loss, irritability, sweating	Hyperthyroidism
Social and medical history	Recent stress or illness	Pituitary insensitivity

Fig. 15.1 Important questions and causes of amenorrhoea.

endocrine disorders can also be responsible, including hyperprolactinaemia.

Breast lumps (Fig. 15.2)

Breast lumps are a common presentation, especially since breast self-examination has recently been encouraged. While breast cancer is very common and can occur at any age, the majority of breast lumps are benign. Despite this every lump needs careful examination and further investigation.

Galactorrhoea (Fig. 15.3)

This is the inappropriate production of milk from the breasts (i.e. without a recent birth). It usually affects women, but rarely can affect men.

Hyperprolactinaemia is the most common cause, but the reason for this excess is often not found.

Gynaecomastia (Fig. 15.4)

This is growth of the breasts in men caused by an abnormal balance between testosterone and oestrogen. It is not the same as simple fat deposition caused by obesity or old age. It is a normal finding during puberty, but otherwise medications or drugs are the most common cause.

Hirsutism (Fig. 15.5)

This is when a woman develops a male pattern of facial and body hair. It should not be confused with excessive hair growth (hypertrichosis) or development of male secondary sexual characteristics (virilism). Polycystic ovary disease is the most common cause, but in many cases a cause is never found (idiopathic hirsutism).

Loss of consciousness and coma
(Fig. 15.6)

Loss of consciousness can be caused by many disorders and endocrine and reproductive disturbance are not the most common. Other causes are excluded from this table.

Menorrhagia and intermenstrual bleeding (Fig. 15.7)

Menorrhagia is excessive bleeding during menstruation (> 80 mL per period). Intermenstrual bleeding is blood discharged from the vagina between periods. The most common cause of both conditions is dysfunctional uterine bleeding, especially at the extremes of reproductive age.

Fig. 15.2 Important questions and causes of breast lumps.

Important questions and causes of breast lumps

Find out	Findings	Differential diagnosis
Age	Young	Fibroadenoma
	Pre-menopause	Fibrocystic change, duct ectasia or duct papilloma
	Elderly	Fibrocystic change, fat necrosis, breast cancer or phyllodes tumour
Obstetric history	No pregnancies	Slightly higher risk of breast cancer
	Recent pregnancy	Breast abscess
Systems review	Bone pain or jaundice	Metastasis from breast cancer
	Creamy nipple discharge and nipple retraction	Duct ectasia
	Bloody nipple discharge and nipple retraction	Breast cancer
	Eczema round the nipple	Paget's disease (breast cancer)
Family history	Strong family history of breast or ovarian cancer	Breast cancer (BRAC genes)

Common Presentations of Endocrine and Reproductive Disease

Important questions and causes of galactorrhoea		
Find out	Findings	Differential diagnosis
Sexual history and contraception	Recent intercourse, no contraception	Pregnancy
Obstetric history	Recent miscarriage or termination	The hyperprolactinaemia of pregnancy takes time to return to normal
Systems review	Female: amenorrhoea, weight gain	Hyperprolactinaemia
	Male: impotence, less facial hair, visual disturbance, gynaecomastia	Hyperprolactinaemia
	Visual disturbance, headache	Pituitary tumour loss of dopamine inhibition
	Weight gain and lethargy	Hypothyroidism is a rare cause
Medical history	Chronic renal failure	Can cause hyperprolactinaemia
Drug history	Methyldopa, oestrogens, tricyclic antidepressants, haloperidol	Drug-induced hyperprolactinaemia

Fig. 15.3 Important questions and causes of abnormal milk secretion called galactorrhoea.

Important questions and causes of gynaecomastia		
Find out	Findings	Differential diagnosis
Age	10–16 years	Puberty
Medical history	Testicular torsion, infection or maldescent	Testosterone deficiency
	Chronic renal failure	Excess oestrogen
Systems review	Small genitalia, tall stature, female fat distribution	Klinefelter syndrome
	Impotence, less facial hair, visual disturbance	Hyperprolactinaemia
Drug history	Spironolactone, tricyclic antidepressants, oestrogens, griseofulvin	Induce gynaecomastia
Family history	Very strong history of breast or ovarian cancer	Male breast cancer (*BRAC* genes)
Social history	Frequent use of amphetamines or cannabis	Induce gynaecomastia
	Chronic alcoholism	Liver disease, excess oestrogen

Fig. 15.4 Important questions and causes of male breast enlargement called gynaecomastia.

201

Common Presentations of Endocrine and Reproductive Disease

Important questions and causes of hirsutism

Find out	Findings	Differential diagnosis
Age of onset	Childhood	Congenital adrenal hyperplasia
	>50	Menopause
Systems review	Associated virilism	Androgen-producing tumours or congenital adrenal hyperplasia
	Amenorrhoea, weight gain, acne	Polycystic ovarian syndrome
	Weight gain, skin bruising, muscle weakness	Cushing's syndrome
Family history	Other relatives affected	Familial hirsutism
Drug history	Use of high dose steroids	Induces hirsutism in the same manner as Cushing's syndrome
	Danazol for endometriosis	Androgenic effects
Social history	Use of androgens to enhance sporting ability	Excess androgens

Fig. 15.5 Important questions and causes of male pattern body hair in females (hirsutism).

Important questions and endocrine/reproductive causes of loss of conciousness and coma

Find out	Findings	Differential diagnosis
Age	Young	IDDM (diabetic ketoacidosis)
Obstetric history	Over 20 weeks pregnant	Eclampsia
Sexual history	Recent intercourse, no contraception	Ruptured ectopic pregnancy
	Current use of combined contraceptive pill	Thromboembolism
Systems review	Weight loss, sweating, palpitations, cardiac arrhythmia	Thyrotoxicosis (hyperthyroidism)
	Weight gain, hypothermia, lethargy, recent illness	Myxoedema coma (hypothyroidism)
	Weight loss, polyuria, thirst for several months	NIDDM (HONK)
	Recent weight loss, polyuria, thirst and sweet-smelling breath	IDDM (diabetic ketoacidosis)
	Pigmentation, weight loss, anorexia, nausea and vomiting	Addison's disease
	Headache, visual disturbances, gynaecomastia	Raised intra-cranial pressure following pituitary adenoma

Fig. 15.6 Important questions and endocrine/reproductive causes of loss of consciousness and coma. (IDDM, insulin dependent diabetes mellitus; NIDDM, non-IDDM.)

Common Presentations of Endocrine and Reproductive Disease

Important questions and causes of menorrhagia and intermenstrual bleeding

Find out	Findings	Differential diagnosis
Age	Young	Dysfunctional uterine bleeding
	Pre-menopause	Dysfunctional uterine bleeding, fibroids, uterine polyps, ovarian cysts, endometriosis
	Post-menopause	Endometrial or cervical carcinoma
Sexual history	Recently started using the Pill	Breakthrough bleeding is common for a few months
	Use of copper IUD	Causes menorrhagia
	Risk of pregnancy	Ectopic pregnancy, miscarriage
	Pregnant	Miscarriage, placenta previa, placental abruption
Systems review	Pelvic pain, dyspareunia	Fibroids, polyps, endometriosis and adenomyosis
	Fever, malaise, pelvic pain	Pelvic inflammatory disease
	Hirsutism, weight gain, acne	Polycystic ovarian syndrome
	Weight loss, irritability, sweating	Hyperthyroidism
	Weight gain and lethargy	Hypothyroidism
Drug history	Oestrogen (e.g. HRT)	Endometrial hyperplasia

Fig. 15.7 Important questions and causes of heavy periods (menorrhagia) and intermenstrual bleeding. (HRT, hormone replacement therapy; IUD, intra-uterine device.)

Polyuria (Fig. 15.8)
A number of endocrine disorders can upset the kidneys to cause polyuria. This is the excretion of an excess volume of dilute urine causing dehydration and thirst (polydipsia). It is not the same as frequency, in which only small amounts of urine are passed frequently so the total volume is not great.

Scrotal lumps (Fig. 15.9)
Lumps in the scrotum are common presenting complaints and self-examination is being encouraged. Scrotal lumps can arise from the testis, other structures in the scrotum or from the abdominal cavity. There is a wide variety of underlying causes including tumours, infections and trauma. Testicular cancer is the most common malignancy in young adult males.

Sexual dysfunction
Female (Fig. 15.10)
In women the presenting complaints are:

- Lack of sexual desire (decreased libido).
- Failure to reach orgasm (anorgasmia).
- Pain on intercourse (dyspareunia).

Decreased libido and anorgasmia often stem from psychological causes or difficulties within the relationship. Dyspareunia tends to have more physical causes.

Male (Fig. 15.11)
In men the presenting complaints are decreased libido and impotence. Medications and alcohol are the most common causes.

Thyroid lumps and goitre
(Fig. 15.12)
A goitre is an enlarged thyroid gland. The enlargement may be caused by the entire gland or by a nodule; any lumps must be investigated to exclude malignancy. Thyroid lumps are often associated with disorders of the thyroid hormones and they are especially common in women.

Common Presentations of Endocrine and Reproductive Disease

Important questions and causes of polyuria

Find out	Findings	Differential diagnosis
Age	Young	IDDM
	Middle-aged or elderly	NIDDM
Medical history	Renal disease	Nephrogenic diabetes insipidus
	Trauma or surgery to the head	Cranial diabetes insipidus
Fluid intake	Excess intravenous fluids	Iatrogenic diabetes
Systems review	Tiredness, thirst, weight loss	Diabetes mellitus
	Bone pain, muscle weakness, headaches, confusion	Hypercalcaemia
Drug history	Any diuretic	Iatrogenic diabetes
	Opiates	Cranial diabetes insipidus
	Lithium, demeclocycline	Nephrogenic diabetes insipidus
	Anticholinergics	Cause a dry mouth and excessive fluid intake
Family history	Diabetes mellitus	Especially NIDDM
	Diabetes insipidus	X-linked inheritance (rarely)
Social history	Psychological problems, abuse	Psychogenic polydipsia

Fig. 15.8 Important questions and causes of polyuria.

Important questions and causes of scrotal lumps

Find out	Findings	Differential diagnosis
Age of onset	Congenital	Indirect hernia, hydrocoele, varicocoele
	Puberty	Testicular torsion, epididymo-orchitis
	Young	Teratomas are common
	Middle-aged	Epididymal cyst or may be a seminoma
	>50 years	Hydrocoele, chronic epididymitis or lymphoma
Medical history	Recent trauma	Haematoma, haematocoele
	Recent vasectomy	Sperm granuloma, haematocoele
	Tuberculosis or syphilis	Infectious granuloma
Systems review	Infertility	Varicocoele
	Fever, malaise, scrotal pain, UTI	Epididymo-orchitis
	Weight loss, scrotum feels heavy	Testicular tumour
Social history	Intensive exercise	Testicular torsion
	Recent lifting, e.g. moved house	Indirect hernia

Fig. 15.9 Important questions and causes of scrotal lumps. (UTI, urinary tract infection.)

Common Presentations of Endocrine and Reproductive Disease

Important questions and causes of sexual dysfunction in females

Find out	Findings	Differential diagnosis
Age	Post-menopause	Oestrogen deficiency causing a lack of lubrication
Sexual history	Muscles of the vagina tense on attempted intercourse	Vaginismus causing dyspareunia
	Never achieved orgasm	Anorgasmia
Medical history	Previous surgery or trauma	Dyspareunia
Systems review	Fever, malaise, dyspareunia	Infections of the urethra, vulva, vagina or pelvic inflammatory disease
	Menorrhagia, pelvic pain, deep dyspareunia	Ovarian cysts and tumours, endometriosis, fibroids
	Amenorrhoea, galactorrhoea and decreased libido	Hyperprolactinaemia
Social history	Lack of communication with partner	Decreased libido and anorgasmia
	Lack of sexual awareness	Anorgasmia

Fig. 15.10 Important questions and causes of sexual dysfunction in females.

Important questions and causes of sexual dysfunction in males

Find out	Findings	Differential diagnosis
Medical history	Diabetes mellitus	Impotence is a chronic complication of diabetes mellitus
	Multiple sclerosis	Inhibits sexual arousal
Systems review	Gynaecomastia, visual disturbance	Hyperprolactinaemia
	Small testes, deficient male pattern hair	Hypogonadism
Drug history	Antihypertensives and diuretics	Iatrogenic impotence
	Antidepressants, antipsychotics, oestrogens	Decreased libido
Social history	Lack of communication with partner	Psychological impotence and decreased libido
	Excessive alcohol intake	Causes acute and chronic impotence

Fig. 15.11 Important questions and causes of sexual dysfunction in males.

Weight gain and obesity (Fig. 15.13)

Obesity is defined as a BMI (see box) greater than 30, while 25–30 is classified as overweight. The cause of obesity is unknown in the vast majority of patients, though current research into the regulation of eating may change this. Currently obesity alone does not warrant investigation, whereas unexplained weight gain does. The causes of weight gain can lead to obesity so the two are considered together.

205

Common Presentations of Endocrine and Reproductive Disease

Fig. 15.12 Important questions and causes of thyroid lumps and goitres. (MEN, multiple endocrine neoplasia.)

Important questions and causes of thyroid lumps and goitres

Find out	Findings	Differential diagnosis
Age	10–16	Temporary physiological goitre
	Young	Papillary carcinoma
	Middle-aged	Autoimmune causes, medullary or follicular carcinoma
	Elderly	Multinodular goitre, medullary, anaplastic carcinoma or lymphoma
Obstetric history	Pregnancy	Temporary physiological goitre
Systems review	Weight loss, sweating, palpitations, irritability, heat intolerance	Graves' disease, multinodular goitre, toxic adenoma
	Weight gain, cold intolerance, tiredness, lethargy, dry skin and hair	Hashimoto's thyroiditis, de Quervain's thyroiditis
	Fever, malaise and painful neck	Infectious goitre (e.g. de Quervain's thyroiditis)
	Bone pain	Metastases from thyroid cancer
Family history	Thyroid disease	Autoimmune thyroid disease
	Medullary carcinoma	MEN IIa and IIb syndromes
Social history	Unusual diet or immigration from inland developing country	Iodine deficiency

BMI is calculated from the weight in 'kg' and height in 'm'. The weight is divided by the height squared. For example, a man weighing 85 kg with a height of 1.65 m has a BMI of $85/(1.65 \times 1.65) = 31.2$.

Weight loss (Fig. 15.14)

Weight loss is often a sign of fairly severe disease, so it must be taken seriously. It can be caused by the failure of the heart, kidneys, or liver, but also by several endocrine disorders.

Important questions and causes of weight gain and obesity

Find out	Findings	Differential diagnosis
Systems review	Abnormal fat distribution, easy bruising, muscle weakness, hirsutism	Cushing's syndrome
	Lethargy, depression, cold intolerance	Hypothyroidism
	Amenorrhoea, acne, hirsutism	Polycystic ovarian syndrome
Drug history	Steroids, antidepressants	Stimulate eating
Family history	Other obese members	Genetic or environmental causes
Social history	Stress or history of binge eating	Psychological cause
	Recently gave up smoking	Often slight weight gain

Fig. 15.13 Important questions and causes of weight gain and obesity.

Important questions and causes of weight loss

Find out	Findings	Differential diagnosis
Age	Elderly	Malignancy or organ failure are most likely
Systems review	Increased appetite, sweating, palpitations, heat intolerance	Hyperthyroidism
	Reduced appetite, malaise, vomiting	Addison's disease, malignancy or organ failure
	Polyuria, thirst, tiredness	Diabetes mellitus
	Light coloured chronic diarrhoea, large appetite	Malabsorption
Social history	Perception of weight	Eating disorders (e.g. anorexia nervosa)
	Unprotected sexual intercourse or intravenous drug use	HIV infection

Fig. 15.14 Important questions and causes of weight loss.

207

16. History and Examination

The process of taking a history and examining a patient form the core of a doctor's job. This chapter includes a description of these two tasks along with the major signs associated with common endocrine and reproductive disorders. Chapter 15 describes common presenting complaints along with useful questions to ask in the history. The diversity of the endocrine and reproductive systems makes both the history and examination difficult to describe in a general manner. Both tasks should be directed towards the presenting complaint.

After reading this chapter you should be able to:
- Take and present a history (or at least try).
- Examine a patient.
- List the common findings of endocrine and reproductive disease on examination.

History

Taking a history seems simple in practice, however the presence of a patient induces amnesia and stuttering in most students. It takes a lot of practice to develop a smooth technique and even longer to learn what questions are important and which answers are relevant. This process is neither quick nor easy, but that is why medical training takes so many years.

A proper history involves many components, each of which are described below. These sections are very important for the correct diagnosis to be made, however they are derived from a doctor-centred approach. To understand the problem from the patient's point of view, the following questions should also be asked:
- Why has the patient sought medical help? Why now?
- What does the patient believe could be causing the problem?
- What does the patient want (e.g. reassurance or treatment)?

In many ways you act as an interpreter for the patient between English and medical jargon; in the same way you should only speak to the patient in plain English.

Preparations

Before taking a history it is important for you, the patient and the surroundings to be suitably prepared. There are a number of factors that can help:
- Check the patient is available and comfortable.
- Find a quiet and private location. This is often difficult in practice, but should always be aimed for, nevertheless.
- Be dressed appropriately.
- Have a visible ID badge.
- Think about a possible differential diagnosis from the presenting complaint.
- Write out the components of a history if you need to.

Structure of a history

Apart from introducing yourself and asking permission, there is no prescribed structure to taking a history. However, the standard structure used is required for presenting the history in the notes and to medical staff. It also serves as a mental framework that will, in the future, help you to remember a patient's history without having to write everything down! That said, it is a good idea to let the patient tell you about the problem in an order they are comfortable with at first and then use specific questions to fill in the gaps. Possible answers to some of these questions are included in Chapter 15.

Introduction and permission

Always introduce yourself including your status (e.g. medical student). Explain to the patient what you would like to do (i.e. take a history) and ask their permission.

Example: My name is Stephan and I'm a medical student. May I have a chat with you about your illness?

Note that patients usually respond well to the use of first-name terms. If you really *must* introduce yourself as 'Dr Huntington-Smythe', all well and good, but remember that 'Call me Stephan' is more likely to put the patient at ease.

1) Patient details

It is good practice to obtain basic details about the patient before starting the history. These are often

written above the history in the notes. It should include:
- Patient's name; check it is the right patient.
- Age.
- Sex.
- Occupation.

Example: Mr D. Vader is a 34-year-old gentleman who works in the military . . .

2) Presenting complaint (PC)
The history should be begun with an open question along the lines of:
- What brought you into hospital?
- Tell me about your problem.
- What led you to go to your doctor?

The presenting complaint is the main symptom(s) that led the patient to seek medical help. It should be written in the patient's words, not medical jargon.

The presenting complaint should be presented as:
- How the patient came to see you.
- The symptom, in the patient's words.
- The duration.

Example: . . . who was referred by his GP with 'strange breathing noises' over the last three days.

> When a patient answers, 'An ambulance.' to the question, 'What brought you into hospital?' the patient will believe this to be an original and funny joke deserving at least a smile in recognition.

3) History of presenting complaint (HPC)
This section forms the bulk of the history and includes all details relevant to the presenting complaint. Ideally, the patient will tell you the complete history with only minor prompting, though in practice many patients will need to be led through it. It should include any previous episodes of a similar nature and the systems review questions relevant to the presenting complaint.

The presenting complaint of pain prompts a series of questions to fully describe it. These can be remembered using 'SOCRATES':

- Site—where the pain is.
- Onset—sudden or gradual.
- Character—sharp/crushing/burning.
- Radiation—is it felt anywhere else?
- Alleviating features—what makes it better?
- Timing—when did it start; does it come and go?
- Exacerbating features—what makes it worse?
- Severity—compare with other types of pain.

When presenting this section of the history it should begin when the patient last felt completely well, and continue in order until the present. If any aspect of other sections of the history is relevant (e.g. family history) then it should be placed appropriately within this section.

Example: Mr Vader felt completely well until 1 week ago when he began to wake at night with a cough. This coincided with his working at the dusty construction site of a new battleship. Three days ago he developed a wheeze on expiration that has become progressively worse.

He remembers going to the doctor with a similar wheeze when he was young but cannot recall any further details. His aunt and uncle on his mother's side both suffer from asthma and his daughter has mild eczema. He has no history of smoking.

4) Past medical history (PMH)
This section includes any previous hospital admissions, chronic illness, acute illness or operations, along with when they occurred. The patient should be asked to describe any previous illness they have suffered and then asked specifically about diabetes, tuberculosis (TB), asthma, jaundice, rheumatic fever, hypertension, heart attacks, strokes and epilepsy.

Extra sections can be added depending on the patient (e.g. obstetric history in women, immunization history in children).

Relevant medical history or negatives should be included in the HPC while the details and any unrelated illnesses are included here. Unrelated negatives are not presented but should be written in the notes, often abbreviated e.g. ^{0}DM/TB.

Example: The patient suffered severe trauma during his late teenage years following a natural disaster that still causes him recurrent back pain. He has no history of surgery or other significant illness.

5) Drug history (DH)
This section is about any medications the patient takes regularly, not the use of recreational drugs.

Accordingly, it is better to ask, 'Do you take any medications (tablets)?' rather than, 'Do you take any drugs?' It includes medications of any form that the patient gets on prescription or over-the-counter (OTC). Two other questions are important:
- In women of reproductive age: Are you on the Pill?
- Everyone: Do you have any allergies? What about penicillin?

The medication should be presented along with the reason for taking it. It is vital to write down the dose and frequency for each drug—don't forget you will be the one having to write out the drug chart!

Example: He takes 200 mg of ibuprofen four times a day for his back pain. He has no known allergies.

6) Family history (FH)

Ask the patient if anyone else in their family has had similar or related problems and ask specifically about asthma, epilepsy, diabetes and heart disease. Record the relationship between the patient and affected relatives along with the age, health/cause of death of parents, siblings and children. It is good practice to sketch the family tree.

Example: Alongside a history of asthma and eczema detailed in the HPC, his grandfather suffered from epilepsy. His father was killed in an accident soon after Mr Vader was born and he has lost touch with his mother. He has no siblings. He has two children (twins) aged 6 years, both of whom are well.

7) Social history (SH)

The social history can be very important to diagnosis and appropriate treatment but must be approached in a sensitive and non-judgemental manner. It is often useful to know about:
- Who they live with and their relationship, e.g. children, wife/partner.
- Leisure activities, exercise, smoking, alcohol, illicit drugs.
- Living status, type of house, financial problems, community help.
- Travel abroad.

Example: Mr Vader lives alone in military accommodation following separation from his partner. He does not visit his children and is unaware of their whereabouts. He does not smoke and consumes about 20 units of alcohol a week. He has travelled extensively with his occupation.

8) Systems review

This is essentially a quick checklist of the other systems in the body to prevent missing important symptoms or other problems. The extent to which each system is investigated depends on the presenting complaint. Learning which questions are necessary takes experience and practice. Examples of these questions are shown in Fig. 16.1.

Only the questions with positive answers should be presented, though all answers can be written in the notes.

Example: He has slight constipation and had a headache 1 day ago.

9) Summary

Having taken the perfect history you should present it back to the patient in simple language covering all the important points. This can be very difficult and again experience and practice are essential. It can be easier to summarize parts of the history as you go along.

When presenting the history, a summary is also required before describing the examination. This allows anyone who was not paying attention to catch up and look knowledgeable. It should include:
- The patient's name and age.
- The presenting complaint along with the duration and means of referral.
- A brief summary of the history of the presenting complaint.
- The differential diagnosis.

Example: Mr D. Vader is a 34-year-old gentleman who was referred by his GP with 'strange breathing noises' over the last three days. He has had a cough for the last week and developed a wheeze three days ago that has progressed since. This is consistent with an episode of asthma or a viral infection.

Communication skills

Many medical schools now place great emphasis on teaching communication skills—partly as a response to public criticism of the medical profession over the last couple of decades. As a student, it is easy to dismiss this teaching as less important than the more factual elements of the course and even as an insult to a group of people who are surely perfectly capable of talking to patients.

However, when it comes to finals, most of which now feature some form of structured clinical

Review of systems

General health
- Weight loss, appetite, night sweats, fevers, any lumps, itch, fatigue, apathy

Cardiovascular system
- Chest pain
- Palpitations (awareness of the heart beating)
- Exertional dyspnoea (quantify exercise tolerance, e.g. number of flights of stairs that can be managed)
- Orthopnoea (breathlessness on lying flat—symptom of left ventricular failure)
- Paroxysmal nocturnal dyspnoea (repeated bouts of breathlessness at night)
- Claudication (calf pain on walking)
- Ankle oedema (sign of right-sided heart failure)
- Skin sores/ulcers

Respiratory system
- Cough
- Sputum (amount, colour)
- Haemoptysis (coughing up blood)
- Shortness of breath
- Wheeze

Gastrointestinal system
- Appetite, weight, diet, taste
- Nausea, vomiting
- Difficulty swallowing
- Haematemesis (vomiting blood)
- Heartburn
- Indigestion
- Abdominal pain (site, severity, character, relationship to eating, previous episodes, etc)
- Bowel habit (frequency, change, difficulties)
- Rectal bleeding

Urinary system ('How are the waterworks?')
- Frequency
- Nocturia (needing to pass urine in the night)
- Urine stream (hesitancy, dribbling)
- Dysuria (pain on passing urine)
- Haematuria (blood in the urine)
- Incontinence

Nervous system
- Headaches, fits, faints, loss of consciousness
- Changes in vision, hearing, speech, memory
- Anxiety, depression, sleep disturbances
- Paraesthesiae (pins and needles)
- Sensory disturbances (numbness)
- Weakness

Reproductive system (female)
- Periods (length, cycle, menarche, menopause)
- First day of last menstrual period (LMP)
- Contraception
- Post-menstrual/intermenstrual bleeding
- Vaginal discharge
- Dyspareunia (pain on intercourse)
- Pregnancies, terminations, births
- Problems in pregnancy
- Breast symptoms

Reproductive system (male)
- Impotence/loss of libido
- Scrotal swelling

Musculoskeletal system
- Aches or pains in muscles, bones, or joints
- Swelling of joints
- Limitation of joint movements
- Weakness of muscles

Skin
- Rashes

Risk factors
- Smoking, alcohol consumption, drug abuse
- Allergies, foreign travel

Fig. 16.1 Symptoms to ask about in the systems review.

examination (OSCEs), you should remember these important points:
- None of us is as good at communicating as we like to think we are.
- A lot of this material seems to be stating the obvious—but the obvious can be easy to forget under pressure and reminding ourselves of the basics is a useful exercise.
- Communication skills account for nearly half the marks in the clinical exams—even in the systems examination stations there are marks for your approach to the patient.
- Practising a few communication scenarios with friends before the exam is an easy way to pick up a lot of extra marks.
- If you get on the right side of the patient or actor in a clinical exam, they are likely to divulge their information much more easily.

Obstacles to communication

There are many factors that can make it difficult to talk with patients and colleagues. It is important to be aware of these and address those you can do something about, while making allowances for those you can't.
- Noisy environment and lack of privacy—try to find a quiet room or cubicle to see your patient if possible.
- Nervousness—both yours and the patient's. You can help yourself with practice. The patient can be

put more at ease by a sensitive approach (more on this later).
- Pain—does the patient need analgesia now, rather than after the history?
- Other medical factors—breathlessness, hearing impairment and confusion (acute or chronic) can all make communication difficult. Patience and persistence are required in these situations.
- Language and cultural barriers—try to take the history with a member of the family who can interpret. If this is not possible, there may be facilities to have an interpreter provided. In an acute situation you will have to make do with smiles and gestures to establish the important points, such as the presence and site of pain.
- Hostility—some people may feel (rightly or wrongly) aggrieved by some aspect of the treatment they have already received. It is vital that you do not take this personally or be drawn into a confrontation. Try to remain calm and civil; empathize with the patient and apologize if appropriate. If all else fails, politely explain that you don't feel anything is being achieved and come back later.

Non-verbal communication skills

A large proportion of our communication 'bandwidth' is non-verbal. This includes body-posture, facial expression, eye-movements and gestures; we are conscious of some of these things, but most are subconscious. Non-verbal cues are very important in a clinical setting, both in achieving a rapport with the patient and in gaining insight into their condition.

The following points may be helpful during a consultation:
- Sit with the patient so that your eyes are on roughly the same level; avoid looking down at the patient. Maintain a comfortable distance between you and try to face them while you are talking. It is also useful to make sure you have a comfortable position to write when you are taking a history—kneeling by the bedside is sometimes the best option!
- Maintain good eye contact, even if the patient doesn't.
- Use non-verbal cues to show you are listening and encourage the patient: nodding, smiling and even appropriate laughter can help to put the patient at ease. Smiling is particularly important!
- It is worth having practice sessions with friends before you go into an exam, as they can point out any nervous habits that you might be unaware of.

Verbal communication skills

The things we say and how we say them. It is important to put the patient at ease during a consultation, although this is often easier said than done as people are often understandably concerned in the clinical situation. The following are important skills:
- Empathize with the patient: this means understanding their point of view and is not the same as sympathy. It is perfectly good practice to use such phrases as 'I understand' or 'That must have been very frightening' when a patient is relating the details of their history.
- Use open questions at first, such as: 'What made you come to see a doctor today?' or 'Have you any other problems that have been worrying you?' It is a good idea to let the patient talk freely for the first minute or so, before you focus the history with closed questions such as: 'Does the pain catch you when you breathe in?'
- Use verbal cues such as 'I see' or 'I understand' to help the flow of conversation.
- Check that you have understood what the patient has told you by repeating a summary back to them.
- Use plain English rather than medical jargon. Having said that, try to assess the educational level of your patient and pitch your vocabulary appropriately.

Objectives in the consultation

It is important to have a mental checklist of objectives when you go into a consultation with a patient—especially when this is part of an exam. Once again, the key to this is practice: preferably on the wards, but you can also run through mock scenarios in a study group if you are short of time. You may also need to produce the following kind of list in a short-answer exam paper or viva:
- Introduce yourself and establish a rapport with the patient.
- Find out why the patient has presented to you.
- Find out what the patient understands about their problem and if they have their own theory as to what has caused it.
- What are the patient's expectations of this consultation—what do they want from you?

History and Examination

- Explore the problem with history and examination and formulate a plan for further management—investigations, treatment, etc.
- Explain your findings and plan to the patient clearly.
- Check the patient understands what you have told them.
- Ask the patient if there is anything they are not happy with.
- Literature—provide leaflets or write things down for the patient to take away.
- Follow-up—make sure the patient knows what the next point of contact will be (e.g. an outpatient appointment).

The last four points can be remembered with the mnemonic 'CALF' and are a good way to use the last few minutes of an OSCE station to your advantage.

Examination

The diversity of the endocrine system makes examination especially difficult. There is no specific 'endocrine examination' comparable to the examination of the cardiovascular system; instead the examination must include aspects of six major examination sequences:

- Cardiovascular.
- Respiratory.
- Abdominal.
- Cranial nerves.
- Peripheral nerves.
- Examination of a lump.

The extent to which these systems are examined depends on the presenting complaint and differential diagnosis. It takes experience to know which systems to examine, though the history should guide this decision.

The situation is slightly better for the reproductive system though there are still three sequences that must be learnt:

- Female reproductive system.
- Breast.
- Male reproductive system.

This section includes a general examination sequence that covers the major systems, followed by a description of the three reproductive examinations. Common findings together with the potential disorders are shown throughout.

General examination

Introduction and permission

Before examining a patient you must introduce yourself, unless you have already done so for the history. You must always explain what you intend to do and ask the patient's permission.

General inspection

When meeting a patient and taking a history, it is important to be aware of signs that give information about the patient's condition. These can include their surroundings (e.g. walking sticks), speech and mental state. Some physical signs can also be seen whilst taking the history, in particular:

- Facial features and obvious eye signs (Fig. 16.2).
- Skin complexion (Fig. 16.3).
- Body physique and posture (Fig. 16.4).

The area to be examined should also be inspected specifically from the end of the bed before any active examination. The area must be exposed before inspection and the patient should be reassured that you are not just staring.

Examination of the face	
Findings	Diagnostic inference
Harsh facial features, large nose, protruding jaw and large hands	Acromegaly
Infant with a broad flat face, widely spaced eyes and a protruding tongue	Cretinism (resulting from hypothyroidism)
Eyes that appear to be bulging out of their sockets, i.e. exophthalmos	Graves' disease (not in other forms of hyperthyroidism)

Fig. 16.2 Common findings on inspection of the face.

Examination of the skin	
Findings	Diagnostic inference
Generalized pigmentation of the skin	Addison's disease, Cushing's disease (not in Cushing's syndrome)
Flushed, red skin with excessive sweating	Thyrotoxicosis, phaeochromocytoma
Boils/skin infections	Undiagnosed or poorly controlled diabetes mellitus

Fig. 16.3 Common findings on inspection of the skin.

Examination

Fig. 16.4 Common findings on inspection of the body.

Common findings on inspection of the body

Findings		Diagnostic inference
Short stature	Failure to grow	Dwarfism
	Infant with flat face	Hypothyroidism (cretinism)
	Bone deformities	Rickets (vitamin D deficiency)
	Female with masculine body shape and webbing of the neck	Turner syndrome (45 chromosomes, XO)
Tall stature	Male with female fat distribution (breasts and hips)	Kleinfelter syndrome (47 chromosomes, XXY)
	Child with excess growth	Gigantism
Overweight	Abnormal fat distribution, wasted arms and legs	Cushing's syndrome
	Purely abdominal	Pregnancy
	Lethargic	Hypothyroidism
Underweight	Young with recent weight loss and wasting	Diabetes mellitus (IDDM)

Examination of the hands, limbs and feet

Examination always begins with the hands. This provides a lot of information and is a non-intrusive start that reassures the patient. The specific features to look for in the nails, hands, limbs and feet are outlined in Figs 16.5 to 16.8.

Examination sequence

1. Inspect the nails.

Fig. 16.5 Common findings on examination of the hands. (ACTH, adrenocorticotrophic hormone.)

Examination of the hands

Findings	Diagnostic inference
Hands are enlarged, greasy, spade-like, with thickened skin	Acromegaly
Palms are warm and moist ± tremor	Thyrotoxicosis
Palms are cold and dry	Hypothyroidism
Palmar creases are pigmented	Addison's disease, Cushing's disease, ectopic ACTH syndrome
Note the extent of the areas where the patient complains of 'pins and needles' (paraesthesia) or numbness (anaesthesia) in the fingers and hands	Diabetes mellitus (complication), hypocalcaemia
Trousseau's sign, showing neuromuscular irritability—test by occluding the blood flow to the hands, using an inflated blood pressure cuff around the upper arm, which causes a typical contraction of the hand (thumb adducts, fingers extend) within 2 minutes	Hypocalcaemia
Tinel's sign, showing carpal tunnel syndrome – diagnosed by tapping over the flexor retinaculum and causing paraesthesia in the medial fingers	Hypothyroidism, acromegaly
Decreased skin turgor, signifying dehydration—present if skin on the back of the hand does not return to normal immediately after being pinched	Uncontrolled diabetes mellitus, diabetes insipidus, hypercalcaemia

History and Examination

Examination of the nails	
Findings	**Diagnostic inference**
Separation of the nail from its bed (onycholysis) and nail tips appear white	Thyrotoxicosis
Clubbing of the fingertips, caused by swelling of the soft tissue at the base of the nail	Graves' disease (not in other forms of thyrotoxicosis)
Nails look broken and weak (fragile nails)	Hypocalcaemia
Deformed nails with inflammation of the surrounding skin is a sign of infection (often caused by *Candida albicans*)	Uncontrolled diabetes mellitus

Fig. 16.6 Common findings on inspection of the nails.

Examination of the limbs		
Feature	**Findings**	**Diagnostic inference**
Pulse rate and rhythm	Rapid pulse rate of >100 beats per minute (tachycardia)	Thyrotoxicosis, phaeochromocytoma
	Slow pulse rate of <60 beats per minute (bradycardia)	Hypothyroidism
	Irregular pulse rhythm (signifying cardiac arrhythmias)	Thyrotoxicosis and hypercalcaemia
	Reduced or absent pulses in the feet and legs (caused by peripheral vascular disease)	Diabetes mellitus (complication)
Skin, muscle and bone structure of the limbs	Infected or ulcerated skin (look especially on the lower leg and ankles)	Diabetes mellitus (complication), Cushing's syndrome (both cause poor wound healing)
	Multiple bruising over the skin, with no history of trauma	Cushing's syndrome
	Thickened skin over the tibia, with elevated dermal nodules and plaques (pretibial myxoedema)	Graves' disease (not in other forms of thyrotoxicosis)
	Proximal muscle wasting (observe and feel the biceps and quadriceps muscles)	Cushing's syndrome, hypothyroidism, thyrotoxicosis
	Bone deformity, e.g. 'bow-legs' or 'knock-knees' (observe when the patient is standing)	Rickets (vitamin D deficiency; rare in UK)
	Pitting oedema at the ankles (caused by salt and water retention)	Cushing's syndrome (SIADH does not cause oedema)
Blood pressure	High blood pressure (hypertension)	Cushing's syndrome, diabetes mellitus (complication), acromegaly
	Low blood pressure whilst moving from lying to standing position (postural hypotension)	Addison's disease, diabetic autonomic neuropathy

Fig. 16.7 Common findings on examination of the limbs.

Examination

Examination of the feet

Findings	Diagnostic inference
Feet are large and wide and patient's shoe size has recently increased	Acromegaly
Skin ulcers and/or gangrene	Diabetes mellitus
Dry, cold, hairless skin of the feet and weak or absent foot pulses may signify ischaemia caused by peripheral vascular disease (check by testing capillary refill)	Diabetes mellitus (complication)
Note the extent of the areas where the patient complains of 'pins and needles' (paraesthesia) or numbness (anaesthesia) in the feet and lower legs	Diabetes mellitus (complication), hypocalcaemia

Fig. 16.8 Common findings on examination of the feet.

2. Test capillary refill.
3. Inspect the hand and skin creases.
4. Feel and count the pulse.
5. Take the blood pressure.
6. Inspect the limbs and feet.
7. Test tone and power of the limbs.
8. Test reflexes of the limbs.
9. Test sensation of the limbs.
10. Test coordination of the limbs.

Examination of the head and neck

Signs found in the head, eyes and neck that suggest endocrine disorders are shown in Figs 16.9 to 16.12.

Examination sequence
1. Inspect the face, eyes and neck.
2. Test the visual acuity and visual fields.

Examination of the head

Feature	Findings	Diagnostic inference
Bone structure, facial features and complexion	Increased head circumference, protruding jaw, coarse facial features, nose and jaw enlarged, malaligned teeth with spaces between them in the lower jaw, thickened facial skin folds	Acromegaly
	Protruding forehead	Rickets (vitamin D deficiency; rare in UK)
	Pale, puffy face with coarse features	Hypothyroidism
	Round 'moon-face'	Cushing's syndrome
	Acne on face, neck and chest	Cushing's syndrome, acromegaly and polycystic ovarian syndrome
	Chvostek's sign, showing neuromuscular irritability—diagnosed by gently tapping the facial nerve where it passes through the parotid gland and causing the facial muscles to twitch briskly on the same side of the face	Hypocalcaemia
Hair distribution	Lack of normal beard growth in postpubescent males	Delayed puberty, hypopituitarism causing gonadotrophin deficiency
	Excessive facial hair (hirsutism) in females	Polycystic ovarian syndrome
Mouth	Hyperpigmented buccal mucosa	Addison's disease, Cushing's disease (not in Cushing's syndrome)
	Malaligned teeth, enlarged tongue (possibly causing dysarthria, i.e. difficulty in pronunciation)	Acromegaly
	Swollen tongue and a hoarse, croaky voice	Hypothyroidism

Fig. 16.9 Common findings on examination of the head.

217

History and Examination

Examination of the eyes

Findings	Diagnostic inference
Lid lag—slow descent of the upper lid, lags behind the descent of the eyeball	Hyperthyroidism
Lid retraction—at rest, the superior limbus of the iris and possibly even some sclera above it (white of the eye) is visible (see Fig. 16.11)	Hyperthyroidism
Exophthalmos—the eye appears to bulge out of its socket and it is possible to see the whole of the iris and sometimes even sclera surrounding its circumference (see Fig. 16.11)	Graves' disease (not in other forms of thyrotoxicosis)
Anaemia—the inner surface of the lower lid looks pale if anaemia is present	Menorrhagia
Retinal disease—look for ischaemic change and neovascularisation using an ophthalmoscope (appearance of 'dots' and 'blots' signify presence of microaneurysms and microhaemorrhages, respectively)	Diabetes mellitus (complication)
Papilloedema (caused by raised intracranial pressure)—both optic discs appear convex and their margins appear blurred	Pituitary tumour
Impaired visual acuity—test the visual acuity in both eyes separately using an eye chart	Diabetes mellitus (complication)
Bitemporal hemianopia visual field deficits—test the visual fields in each eye separately	Pituitary tumour

Fig. 16.10 Common findings on examination of the eyes.

appearance of normal eye
—the upper lid lies over the superior limbus of the iris

appearance of exophthalmos
—both lids are moved away from the centre with sclera visible below or all around iris

mild exophthalmos   severe exophthalmos

appearance of lid retraction
—the upper lid is raised, lower lid normal

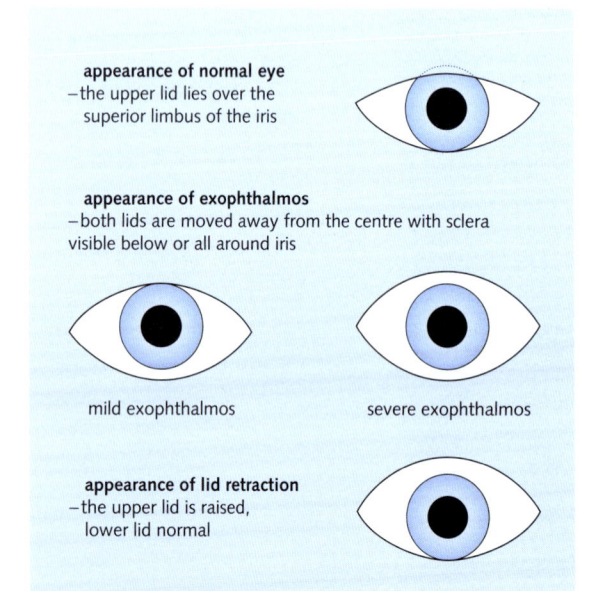

Fig. 16.11 Appearance of the eyes indicating hyperthyroidism.

Examination of the neck

Findings	Diagnostic inference
Anterior neck swelling in the thyroid position which ascends during swallowing	Goitre
Swelling in the neck between the chin and the 2nd tracheal ring which rises when the tongue is stuck out	Congenital thyroglossal cyst

Fig. 16.12 Common findings on examination of the neck.

3. Test the eye movements and look for lid lag.
4. Inspect the fundi.
5. Test the sensation and power of the face.
6. Listen to their speech.
7. Feel for lymph nodes.

8. Examine any lumps present.
9. Look for the height of the jugular venous pressure (JVP).
10. Feel for tracheal deviation or tug.

Examination of the thorax

Signs found in the thorax suggestive of endocrine disorders are shown in Fig. 16.13. The patient should be examined whilst he or she is reclined at an angle of 45°.

Examination sequence

1. Inspect the front, sides and back of the chest.
2. Count the respiratory rate.
3. Locate the apex beat.
4. Feel for heaves or thrills.
5. Palpate the chest movement on the front and back.
6. Percuss the lungs on the front, sides and back.
7. Auscultate the heart (four positions).
8. Listen at the apex with the bell while the patient lies on the left side (mitral stenosis).
9. Listen in the tricuspid area with the patient sitting forward and the breath held on expiration using the diaphragm (aortic regurgitation).
10. Auscultate the lungs on the front, sides and back.

Examination of the abdomen

Signs found in the abdomen which suggest endocrine and reproductive disorders are shown in Figs 16.14

Examination of the thorax	
Findings	Diagnostic inference
A 'pigeon chest' or 'rickety rosary' (outward bowing and thickening of the costochondral junctions)	Rickets (vitamin D deficiency; rare in UK)
Truncal obesity (abnormal fat distribution) and increased chest hair in men and women	Cushing's syndrome
Respiratory distress—deep, rapid hyperventilation ('air-hunger'), called Kussmaul's breathing	Diabetic ketoacidosis

Fig. 16.13 Common findings on examination of the thorax.

Observation of the abdomen	
Findings	Diagnostic inference
Scars	Previous surgery possibly to treat an endocrine or reproductive system disorder
Wide purple striae (linear wrinkled 'stretch' marks) in both sexes and increased abdominal hair (hirsutism) in women	Cushing's syndrome
Abdominal distension (can be caused by fat, fluid, fetus, flatus, faeces or large solid tumours)	Cushing's syndrome (fat), hypothyroidism (fat), pelvic mass (e.g. fibroids, ovarian disease or pregnancy)
Excessive outward curvature of the spine (kyphosis) or excessive inward curvature of the spine (lordosis) can be caused by vertebral collapse	Osteoporosis secondary to menopausal hormone failure, Cushing's syndrome or thyrotoxicosis

Fig. 16.14 Common findings on inspection of the abdomen.

Palpation and percussion of the abdomen	
Findings	Diagnostic inference
Enlarged liver, spleen and kidneys (organomegaly)	Acromegaly
Mass with impalpable lower border	Pelvic mass (e.g fibroids, pregnancy)
Body tenderness	Osteoporosis secondary to menopause, Cushing's syndrome or thyrotoxicosis
	Bony metastases
Lower abdominal tenderness	Pelvic inflammatory disease, ectopic pregnancy
Shifting dullness	Ascites following malignancy

Fig. 16.15 Common findings on palpation and percussion of the abdomen.

and 16.15. The patient should be examined whilst lying flat.

Examination sequence
1. Inspect the abdomen.
2. Lightly palpate the abdomen starting away from tender areas.
3. Palpate the abdomen more deeply.
4. Examine any lump present.
5. Palpate for the liver edge from the right iliac fossa.
6. Palpate for the spleen edge from the right iliac fossa.
7. Ballot the kidneys.
8. Percuss for the liver, spleen and ascites.
9. Auscultate for bowel sounds or fetal heart sounds.
10. Rectal examination.

Fetal heart sounds can be heard over a pregnant uterus using Doppler ultrasound at 10 weeks gestation and with a stethoscope at 25 weeks.

Examination of a lump or mass
If a lump is detected it must be assessed for the features shown in Fig. 16.16. This is a very common presentation and it applies to lumps found in all locations.

Examination of a lump

Feature	Findings
Skin changes	Colour, scarring, ulceration, oedema
Temperature	Hot, cold
Tenderness	Is it painful when touched?
Location	Accurate anatomical description
Size and shape	Estimates of diameter
Surface	Irregular, lobular, smooth
Edge	Sharp, rounded, indistinct
Consistency	Firm, hard, soft
Translucency	Cysts allow light to pass through
Relations	What structures is it attached to?

Fig. 16.16 Examination and description of a lump.

Examination of the female reproductive system
It is essential to carefully explain the examination you wish to perform when it involves intimate body parts. A member of staff of the same sex as the patient should be present at every examination for medico-legal reasons and to reassure the patient.

Examination of the vulva (external genitalia)
The vulva should be examined with the patient lying on her back with the legs apart and knees bent. The common signs of vulval disease are shown in Fig. 16.17.

Examination sequence
1. Inspect the entire vulva.
2. Look for vaginal discharge.
3. Ask the patient to cough or push down.
4. Look for vaginal prolapse and urinary incontinence.

Internal examination with a speculum
This examination should not be performed on a virgin. A warmed vaginal speculum can be used to visualize the cervix and obtain swabs or a cervical smear. Three swabs are routinely taken if there is a risk of infection:
- High vaginal swab.
- Cervical swab.
- Cervical swab for *Chlamydia*.

If a smear is required then the speculum should be inserted without lubricant. A wooden spatula is used to sample the cells of the cervix; the sample is smeared onto a slide and fixed immediately, before being sent for analysis.

Bimanual examination of the vagina and uterus
This examination often follows examination with a speculum and, likewise, it should not be performed on a virgin. The signs of disease of the internal female reproductive system are shown in Fig. 16.18.

Examination sequence
1. Lubricate your index and middle finger.
2. Insert gently into the vagina.
3. Rotate upwards.
4. Palpate the cervix.
5. Press above the pubis with the other hand to

Examination

Fig. 16.17 Common findings on inspection of the vulva.

Examination of the vulva

Findings	Diagnostic inference
Rashes (redness, swelling or white thickened areas called leucoplakia)	Often caused by infections, dermatological conditions (e.g. lichen sclerosus), chemical irritants or allergies
Injury or scars	Can be due to trauma (e.g. childbirth, female circumcision or sexual abuse)
Enlarged clitoris (clitoromegaly)	Congenital adrenal hyperplasia (excessive androgen secretion)
Red painful cystic lump beneath the posterior part of the labium majus	Bartholin's cyst or abscess
Bloody vaginal discharge	Menstruation, miscarriage, cancer, cervical polyp or erosion
Purulent vaginal discharge	Infection, e.g. vaginitis, cervicitis, endometritis
Frothy, watery, pale, yellow-white or purulent discharge and pruritus	Infection caused by *Trichomonas vaginalis*
Thick, white, cottage-cheese-like discharge and inflammation of the skin and mucous membranes	Infection caused by *Candida albicans*

Examination of the female internal genitalia

Findings	Diagnostic inference
Difficult to insert a lubricated finger	Vaginismus
Impalpable uterus	Retroverted uterus
Palpable mass laterally	Mass in the ovary or fallopian tube
Enlarged, nodular uterus	Fibroids
Enlarged, smooth uterus	Pregnancy or cancer

Fig. 16.18 Common findings on examination of the female internal genitalia.

   feel the uterus.
6. Palpate laterally; the ovaries should not be palpable.

Breast examination

Breast examination is usually aimed at finding and describing a lump (see Fig. 16.16). If a lump is found, further investigation is always required to exclude the possibility of malignancy. This cannot be determined from the history and examination alone. The potential findings of breast examination are shown in Figs 16.19 and 16.20. Always examine both breasts. To avoid the patient wondering what you are doing, explain that you always 'start with the good side so you can compare it with the affected side'.

Examination sequence
1. Inspect sitting up.
2. Inspect sitting forwards.
3. Inspect with arms lifted.
4. Inspect with hands pressed on hips.
5. Palpate both breasts and describe any lumps.
6. Examine axillary and supraclavicular lymph nodes.

Examination of the male genitalia
Examination of the external genitalia
The common abnormalities of the male external genitalia found on examination are shown in Figs 16.21 and 16.22.

221

History and Examination

Fig. 16.19 Common findings on inspection of the breast.

Observation of the breast	
Findings	**Diagnostic inference**
Scar due to mastectomy (removal of breast)	Previous breast carcinoma
Breast size decreased bilaterally in women	Hypopituitarism
Enlargement of the female breast (unilateral or bilateral)	Benign hyperplasia of the breast, breast infection/ inflammation, breast neoplasia
Enlargement of the male breast (unilateral or bilateral)	Gynaecomastia (see Fig. 15.4), breast carcinoma
Skin appears pulled in and puckered	Underlying breast carcinoma
Skin has an 'orange peel' appearance (peau d'orange) because of oedema-induced widening of the orifices of sweat glands and hair follicles	Breast carcinoma that is blocking the lymphatic drainage and causing oedema
Skin nodules, abnormal skin texture and colour	Skin infiltrated with tumour cells from a breast carcinoma
Skin is erythematous (reddened) + hot	Infection of the breast or the overlying skin
Skin ulceration (determine its position, size, shape, colour, edge and base)	Advanced breast carcinoma
Nipple and surrounding skin is thickened, red, encrusted and oozy, with an underlying breast lump	Paget's disease of the nipple (breast carcinoma)
Recent nipple pigmentation increased	Addison's disease or pregnancy
Nipple asymmetry and/or retraction	Underlying breast carcinoma
Nipple discharge	Infection, benign or malignant breast tumours, lactation
Nipple duplication—can occur anywhere along the line from the axilla to the groin	Supernumerary nipples
Redness and swelling of the axilla and arm (caused by lymphadenopathy and oedema)	Metastases in the axillary lymph nodes from a breast carcinoma

Examination sequence
1. Inspect skin colour and texture.
2. Inspect the ventral and dorsal sides of the penis.
3. Locate the urethral meatus (hole).

Examination of a scrotal mass
The possible causes of lumps and swellings in the scrotum are illustrated in Fig. 16.23. The exact location of a lump relative to the testes and abdomen is especially important.

Examination sequence
1. Inspect the scrotum.
2. Palpate the testes, epididymis and spermatic cords.
3. Describe any abnormal lumps present.
4. Determine the location of any lumps.
5. Try to transluminate any lumps.

Examination of the prostate gland
The prostate gland and seminal vesicles can be examined by rectal palpation. The common findings are shown in Fig. 16.24.

Examination

Palpation of the breast	
Findings	Diagnostic inference
A solitary, stony-hard, painless lump with an irregular surface and an indistinct edge, which may involve the skin, underlying muscle and regional lymph nodes	Breast carcinoma
A young patient with a solitary, firm, painless lump with a spherical (or knobbly) surface, which tends to be highly mobile and no lymphadenopathy	Fibroadenoma – but further investigations must be performed
In a pregnant woman with a lump or diffuse swelling that is tender, soft/solid and spherical with hot overlying skin and lymphadenopathy	Breast abscess
Palpable regional lymph nodes	Breast infection, breast carcinoma

Fig. 16.20 Common findings on examination of the breast.

Palpation of the male external genitalia	
Findings	Diagnostic inference
Urethral discharge	Infection or inflammation
Empty scrotum (unilateral or bilateral)	Undescended or retractile testis, previous excision
Small firm testes (bilateral)	Hypogonadism, testicular atrophy due to alcohol or drugs
Small firm testis (other testis normal)	Mumps orchitis
Exquisitely tender testis with oedematous swelling of the entire scrotal contents	Torsion of the testis (epididymo-orchitis)

Fig. 16.22 Common findings on examination of the male genitalia.

Observation of the male external genitalia	
Findings	Diagnostic inference
Skin/mucosal rashes or ulceration	Infection, inflammation, connective tissue disease, squamous cell carcinoma
Decreased pubic hair	Hypogonadism, hypopituitarism
Small penis	Hypogonadism
Abnormal position of the external urethral meatus ± hooded foreskin	Hypospadias

Fig. 16.21 Common findings on inspection of the male genitalia.

223

History and Examination

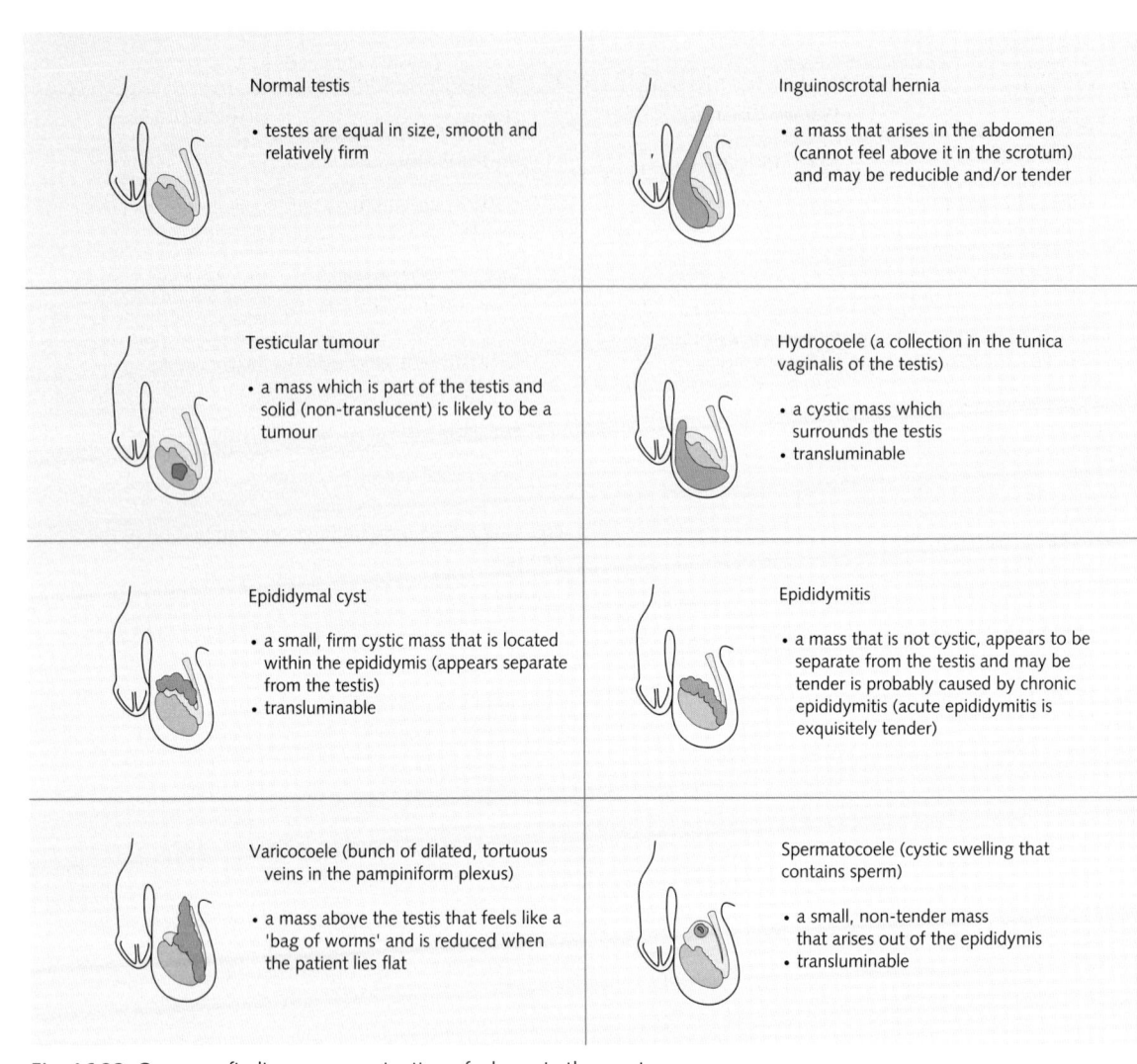

Normal testis
- testes are equal in size, smooth and relatively firm

Inguinoscrotal hernia
- a mass that arises in the abdomen (cannot feel above it in the scrotum) and may be reducible and/or tender

Testicular tumour
- a mass which is part of the testis and solid (non-translucent) is likely to be a tumour

Hydrocoele (a collection in the tunica vaginalis of the testis)
- a cystic mass which surrounds the testis
- transluminable

Epididymal cyst
- a small, firm cystic mass that is located within the epididymis (appears separate from the testis)
- transluminable

Epididymitis
- a mass that is not cystic, appears to be separate from the testis and may be tender is probably caused by chronic epididymitis (acute epididymitis is exquisitely tender)

Varicocoele (bunch of dilated, tortuous veins in the pampiniform plexus)
- a mass above the testis that feels like a 'bag of worms' and is reduced when the patient lies flat

Spermatocoele (cystic swelling that contains sperm)
- a small, non-tender mass that arises out of the epididymis
- transluminable

Fig. 16.23 Common findings on examination of a lump in the scrotum.

Common findings on examination of the prostate

Findings	Diagnostic inference
Smooth, firm gland, 2–3 cm across, with two lobes separated by narrow sulcus	Normal prostate
Enlarged and mobile gland with lobules	Benign hypertrophy
Large, irregular and hard gland fixed to the rectal mucosa with a distorted central sulcus	Prostatic carcinoma

Fig. 16.24 Common findings on examination of the prostate.

- Write out the nine components of a history in the order they are presented.
- Take a history from a friend who pretends to have a suitable condition.
- Present the history.
- Summarize the most important points of the history.
- Perform a cardiovascular examination and present the findings.
- Perform a respiratory examination and present the findings.
- Perform an abdominal examination and present the findings.
- Perform a neurological examination and present the findings.
- Examine a lump and present the findings.
- Describe how the male and female reproductive systems are examined.

17. Investigations and Imaging

Following the history and examination, an endocrine or reproductive disorder may be suspected. Many tests can be used to investigate endocrine function, however the results can be misleading unless the appropriate test is used.

Proper investigation of a suspected endocrine disorder follows the following steps:
- Measure the level of hormone in the blood/urine or its biological effects.
- If a deficiency is suspected use a stimulation test.
- If an excess is suspected use a suppression test.
- If an abnormality is confirmed image the suspected gland.

After reading this chapter you should be able to:
- List the types of tests available.
- Outline the tests relevant to the hormone under investigation.
- Describe other tests that may contribute to diagnosis.
- Discuss the imaging techniques available.

Investigating hormones

Measuring methods
This section outlines the tests used to measure hormone levels and how they change upon stimulation or suppression. These tests are performed if a hormone excess or deficiency is suspected from the history and examination; they aim to confirm the diagnosis and investigate its cause. The four main methods of investigating hormones are described below.

Direct measurement
The levels of hormones in the plasma and urine can be measured using enzyme linked immunosorbent assays (ELISA). The sample is mixed with a known concentration of hormone bound to fluorescent markers. Monoclonal antibodies specific to the hormone are added and the hormone–antibody complexes formed are separated from the solution. The degree of fluorescence is inversely proportional to the original hormone concentration because the native hormone competes with the labelled hormone for antibodies to bind to (Fig. 17.1). In the past radioimmunoassay (RIA) was used, but this is less sensitive. The normal concentrations of commonly measured hormones are shown in Fig. 17.2.

Indirect measurement
Some hormones produce metabolic changes that are easier to measure than the hormone itself. There are two main examples:
- Blood glucose is used to determine insulin levels.
- Urine vs blood osmolality is used to determine antidiuretic hormone (ADH) levels.

Stimulation tests
If a hormone deficiency is suspected then secretion is stimulated to record the response. Blood levels are measured before and after stimulation.

Suppression tests
If a hormone excess is suspected then the hormone secretion is suppressed to record the response. Blood levels are measured before and after suppression.

Hypothalamic function
The quantities of hormones secreted by the hypothalamus are generally too low to be measured clinically. Hypothalamic dysfunction is investigated by measuring the relevant pituitary hormones and their response to stimulation or suppression.

Anterior pituitary function
Hormone assays
Pituitary adenomas can affect the secretion of all the anterior pituitary hormones. For this reason a suspected abnormality in one anterior pituitary hormone prompts investigation of all the others.

Hormones secreted by the anterior pituitary gland are measured along with the hormones that they stimulate. The tests for the following hormones are usually available:
- Thyroid stimulating hormone (TSH), along with tri-iodothyronine (T_3) and thyroxine (T_4).
- Luteinizing hormone (LH) and follicle stimulating hormone (FSH), along with oestrogen or testosterone.

Investigations and Imaging

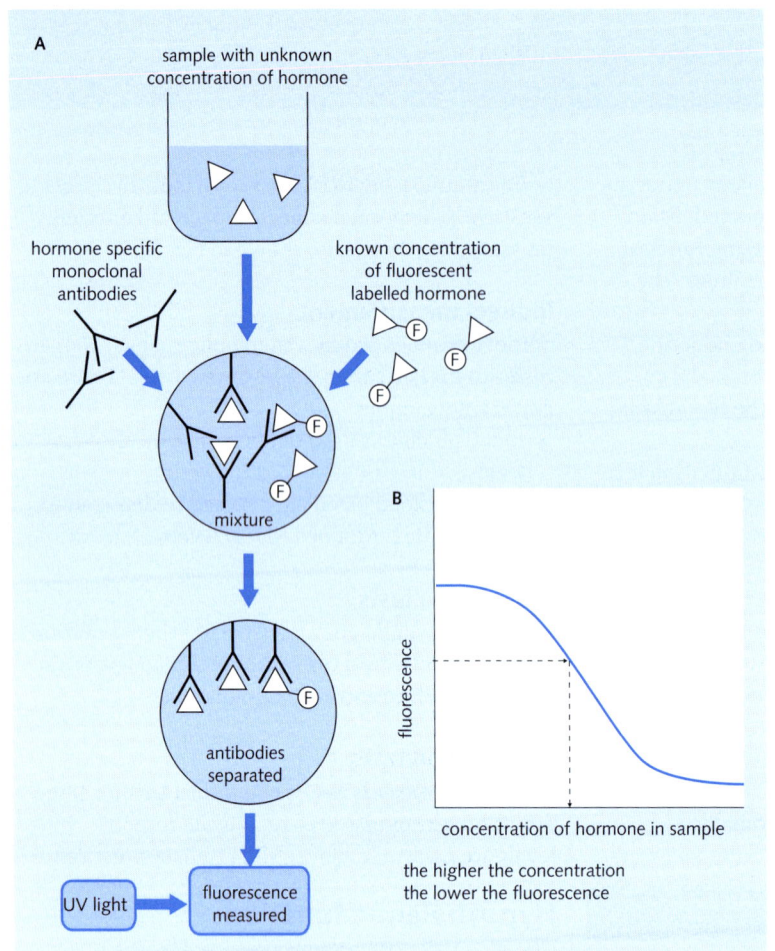

Fig. 17.1 (A) Measuring hormone levels by enzyme-linked immunosorbent assay (ELISA). (B) Interpretation of the results.

- Growth hormone (GH), along with insulin-like growth factor 1 (IGF-1) and glucose; circadian variation must be considered.
- Adrenocorticotrophic hormone (ACTH), along with cortisol; this is not performed commonly and the circadian variation must be considered.
- Prolactin.

Triple stimulation test

The combined pituitary test (CPT) stimulates the anterior pituitary gland to secrete the six major hormones. It is used to investigate hypopituitarism, usually caused by a pituitary adenoma. Three stimulatory substances are injected intravenously:
- Insulin—causes hypoglycaemia that stimulates ACTH, GH and prolactin secretion (Fig. 17.3).
- Thyrotrophin releasing hormone (TRH)—stimulates TSH and prolactin secretion.
- Gonadotrophin releasing hormone (GnRH)—stimulates LH and FSH secretion.

The levels of all six hormones and the hormones they stimulate are measured before and several times after the stimulation.

Individual tests for secretion

Stimulation and suppression tests are especially important for measuring GH and ACTH because of their circadian variation. Hypothalamic function can only be assessed by these tests. The tests for TSH, LH, FSH and ACTH are described under the relevant endocrine organ. There is no further test for prolactin and the tests for GH are described below.

Stimulation test

Used for GH deficiency; GH can be stimulated by:
- Insulin-induced hypoglycaemia (Fig. 17.3).
- Oral clonidine.

Blood samples are measured for glucose and GH before the test and several times over the following 2 hours. To create a suitable stimulus, blood glucose

Normal ranges of commonly measured hormones	
Hormone	Normal levels
Prolactin	Male: <450 µL Female: <600 µL
Adrenocorticotrophic hormone (ACTH)	<80 ng/L
Growth hormone (GH)	<20 mU/L
Thyroid stimulating hormone (TSH)	0.5–5.7 mU/L
Thyroxine (T$_4$)	70–140 nmol/L
Tri-iodothyronine (T$_3$)	1.2–3 nmol/L
Calcitonin	<0.1 µg/L
Cortisol (morning)	450–700 nmol/L
Aldosterone	100–500 pmol/L
Renin (standing)	2.8–4.5 pmol/mL/h
Testosterone	10–35 nmol/L

Fig. 17.2 The normal ranges of commonly measured hormones.

must fall below 2.2 mmol/L causing symptoms of hypoglycaemia. Sugar may need to be given if blood glucose falls too low. GH deficiency is confirmed if secretion does not rise by >20 mU/L.

Suppression test
Used for GH excess; GH is measured every 30 minutes for 2 hours following administration of glucose, i.e. an oral glucose tolerance test (see p. 231). Increased blood glucose normally suppresses GH secretion, but in acromegaly or gigantism GH levels fail to decrease.

Posterior pituitary function
Hormone assays
Antidiuretic hormone
Plasma levels of antidiuretic hormone (ADH) can be measured by ELISA, but they are of little value unless combined with stimulation tests. The comparison between urine and blood osmolality is a more useful clinical measure:
- Dilute urine, concentrated blood: ADH excess (diabetes insipidus).
- Concentrated urine, dilute blood: ADH deficiency (syndrome of inappropriate ADH secretion; SIADH).

Oxytocin
Since abnormal plasma levels do not cause any recognized pathology, secretion is not tested.

Water deprivation test
This is a stimulation test used for suspected ADH deficiency (i.e. diabetes insipidus). ADH release is stimulated by high plasma osmolality caused by water deprivation.

After a night's sleep and fasting the patient is deprived of food and water for 8 hours, during which plasma and urine osmolality is measured repeatedly. The patient is also weighed and the test is abandoned

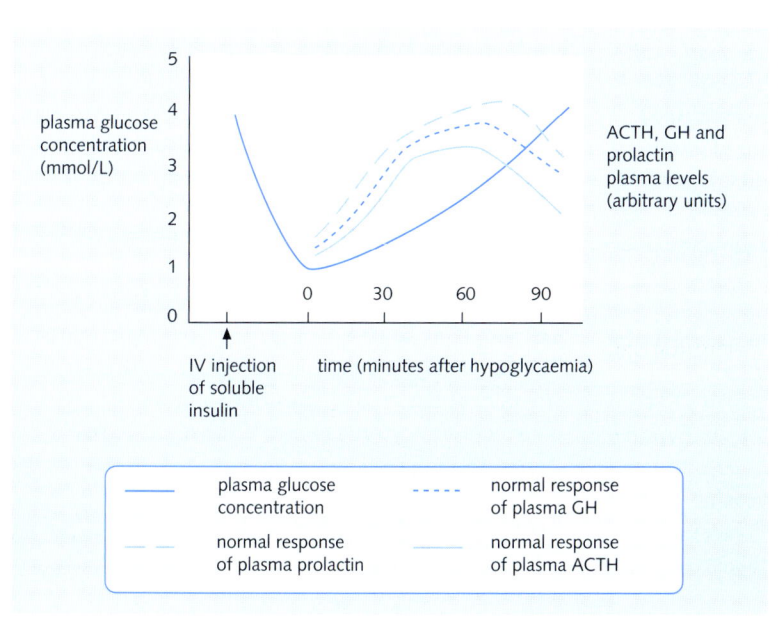

Fig. 17.3 Normal response of adrenocorticotrophic hormone (ACTH), growth hormone (GH), and prolactin to insulin-induced hypoglycaemia. (Adapted from *Lecture Notes on Endocrinology*, 5th edn by WJ Jeffcoate. With permission from Blackwell Science, 1993.)

if the patient loses >3% body weight. At the end of the test desmopressin (an ADH analogue) is given, and urine and plasma osmolality continue to be measured. This section of the test is described below under the desmopressin test.

Diabetes insipidus is diagnosed if urine osmolality is <400 mosmol/kg, the normal range is >600 mosmol/kg. The plasma osmolality should have risen to >295 mosmol/kg to induce this fall.

Desmopressin test
This is a suppression test that identifies the cause of diabetes insipidus (DI) once it has been confirmed by a water deprivation test. It can be caused by:
- Cranial DI—deficient pituitary secretion of ADH.
- Nephrogenic DI—failure of the kidney to respond to ADH.

The ADH analogue desmopressin is injected intramuscularly and the patient is allowed to drink water. Plasma and urine osmolality are measured 1 and 2 hours later. If DI is caused by ADH deficiency (i.e. cranial DI) then this should return urine osmolality to normal:
- Cranial DI—urine osmolality increases to >750 mosmol/kg.
- Nephrogenic DI—urine osmolality remains <400 mosmol/kg.

Thyroid function
Hormone assays
If abnormalities of thyroid function are suspected then plasma levels of TSH, T_4 and T_3 should be measured. Both thyroid hormones are measured because variations in thyroxine-binding globulin (TBG) levels can give inaccurate results. Further investigation is often not needed:
- Low levels of T_4 and T_3—hypothyroidism.
- High levels of T_4 and T_3—hyperthyroidism.

TSH is measured with T_3 and T_4 to locate the lesion. In hypothyroidism a low TSH suggests a pituitary/hypothalamic lesion while a high TSH suggests a thyroid gland problem. In hyperthyroidism TSH will almost always be low.

TSH is also measured to guide thyroxine treatment of thyroid disorders. The correct dose of thyroxine or carbimazole is being administered when TSH levels return to normal.

Thyrotrophin releasing hormone stimulation test
This stimulation test is used if TSH deficiency is suspected. However TSH measurement is now so sensitive that it is rarely used. Hypothalamic TRH is administered intravenously and plasma TSH levels are measured before and after.

Thyroid autoantibody assays
ELISA can also be used to detect thyroid autoantibodies caused by the common autoimmune thyroid diseases:
- Thyroid stimulating antibodies (TsAb) suggest Graves' disease.
- Anti-thyroid-peroxidase (anti-TPO) and anti-thyroglobulin (anti-TgAb) antibodies suggest Hashimoto's thyroiditis.

Adrenal function
Cortisol assays
24-hour urinary free cortisol
Urine is collected over a 24-hour period and the quantity of cortisol is measured; it is normally <280 nmol/24 h. This gives an accurate guide to plasma levels.

Basal plasma cortisol and adrenocorticotrophic hormone
These are measured at 09:00 and 22:00 because of circadian variation. This test is not as reliable as 24-hour urinary free cortisol.

Aldosterone assays
Plasma levels of aldosterone are measured, along with renin and potassium, to assess the appropriateness of aldosterone secretion. Aldosterone is normally secreted in response to high renin or high potassium. Conn's syndrome is suggested by high aldosterone in the presence of low renin and potassium.

Catecholamine assays
Catecholamine levels can be measured directly or more commonly by measuring their metabolites such as vanillylmandelic acid (VMA). Both tests are performed on 24-hour urine samples. Phaeochromocytoma causes VMA levels to rise above 48 mol/24 h.

Dexamethasone suppression test
This test is used if excess cortisol (Cushing's syndrome) is suspected. Dexamethasone is a

synthetic corticosteroid that normally suppresses hypothalamic CRH and pituitary ACTH secretion causing cortisol secretion to drop. High or low doses of dexamethasone may be used, as described below.

Low-dose
This is used to investigate excess cortisol (i.e. Cushing's syndrome). Plasma cortisol is measured in the morning then 0.5 mg/6 h dexamethasone is given orally for 48 hours. Plasma cortisol is measured again in the morning after the last dose. A 24-hour urinary cortisol is also collected on the second day of stimulation. Cushing's syndrome is diagnosed if the test fails to suppress plasma cortisol.

High-dose
This is used if the patient has clear signs of Cushing's syndrome or has had a positive low-dose test. 2 mg/6 h dexamethasone is given orally for 48 hours and the same measurements are collected as in the low-dose test.
- **Slight cortisol depression**—Cushing's disease; ACTH-secreting pituitary tumour.
- **No cortisol depression**—ectopic ACTH-secreting or adrenal tumour.

Synacthen® stimulation test
This test is used to investigate cortisol deficiency (e.g. Addison's disease). Synacthen® (also called tetracosactrin) is a synthetic analogue of ACTH that normally stimulates cortisol secretion. There are two versions of the test (see below).

Short Synacthen® test
This test is used to exclude Addison's disease. 0.25 mg Synacthen® is given intramuscularly and plasma cortisol levels are measured before and 30 minutes later. Addison's is excluded if:
- First plasma cortisol level is >140 nmol/L.
- Second plasma cortisol is >500 nmol/L.
- Second plasma cortisol is >200 nmol/L higher than the first.

Prolonged Synacthen® test
This test is used if the short test fails to exclude Addison's disease. 1.0 mg Synacthen® is given intramuscularly on three successive days and plasma cortisol levels are measured before and 6 hours after each injection. The diagnosis can be made from the plasma cortisol 6 hours after the third injection:
- <690 nmol/L—Addison's disease.
- >690 nmol/L—pituitary ACTH deficiency.

> The name Synacthen® is made by combining SYNthetic with ACTH.

Pancreatic function
Random and fasting blood glucose assays
Endocrine investigation of the pancreas is aimed at diagnosing suspected diabetes mellitus. Insulin deficiency causes abnormally high plasma glucose levels (hyperglycaemia) and sometimes glycosuria. Insulin levels are never measured directly because blood glucose gives a more consistent measure of insulin action, especially if insulin resistance is present.

Diabetes mellitus is diagnosed using blood glucose measurements after an overnight fast on two occasions. Fasting blood glucose levels are normally 3.5–5.5 mmol/L. The results are interpreted as follows:
- >7.8 mmol/L—diabetes mellitus confirmed.
- 6–7.8 mmol/L—impaired glucose tolerance.
- <6 mmol/L—not diabetic.

Oral glucose tolerance test
If the fasting blood glucose measurements show impaired glucose tolerance then an oral glucose tolerance test (OGTT) is indicated. This is a stimulation test to check for insulin deficiency, however glucose is measured not insulin. The patient fasts overnight and then drinks 75 g glucose in water. Plasma glucose is measured before and 2 hours after the drink. The results are shown in Figs 17.4 and 17.5. In patients with normal glucose tolerance the blood glucose should return to the fasting level within 2 hours.

Interpretation of oral glucose tolerance test results

Blood glucose level (mmol/L)	Normal	IGT	Diabetes mellitus
After an overnight fast	<6.0	6.0 – 7.8	>7.8
2 hours after glucose intake	<7.8	7.8 – 11.1	>11.1

Fig. 17.4 Interpretation of blood glucose measurements during an oral glucose tolerance test. (IGT, impaired glucose tolerance.)

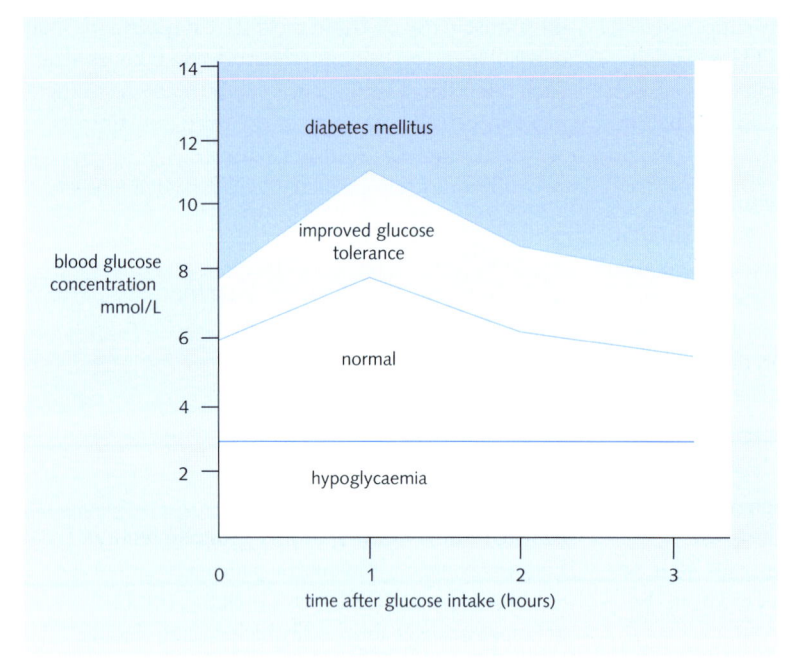

Fig. 17.5 Changes in blood glucose following an oral glucose tolerance test and interpretation of the results.

Gonadal function and pregnancy testing

Hormone assays

Male

The male reproductive hormones are measured to exclude gonadal failure causing infertility following two abnormal sperm counts. Plasma levels of the following hormones are measured:
- FSH and LH.
- Testosterone.
- Prolactin.

Testosterone requires three blood samples to be taken in the morning at 20-minute intervals because it is secreted in a pulsatile manner with a circadian rhythm.

Female

A similar set of hormones are investigated in a woman with amenorrhoea or infertility to exclude gonadal failure. Oestradiol-17β and progesterone are measured instead of testosterone. The stage of the menstrual cycle must be calculated because normal levels of FSH, LH, oestradiol-17β and progesterone vary throughout the cycle.

Primary gonadal failure

Low levels of gonadal steroids along with high levels of LH and FSH indicate primary gonadal failure.

Hypothalamic–pituitary dysfunction

Low levels of gonadal steroids with low or normal levels of LH and FSH indicate hypothalamic–pituitary dysfunction. Raised prolactin levels can cause this pattern.

Gonadotrophin releasing hormone stimulation test

This test is used to investigate gonadal steroid deficiency (i.e. gonadal failure in both men and women). GnRH normally raises secretion of LH and FSH.

100 μg GnRH is injected intravenously and the plasma levels of the reproductive hormones are measured before and four times in the hour after the test. The levels of all the reproductive hormones should be increased following this test. Failure of LH and FSH levels to rise confirms pituitary dysfunction, while an excessive rise suggests hypothalamic dysfunction.

Pregnancy test

Pregnancy can be diagnosed by detecting human chorionic gonadotrophin (hCG) in the urine. The test will be positive approximately 10 days after conception.

Other investigations

Information about endocrine and reproductive disorders can be gathered from tests not specifically designed to investigate hormone levels or organs.

Clinical chemistry

The clinical chemistry laboratory measures the concentrations of ionic and organic components in

Other Investigations

Normal concentration ranges for ionic and organic components of the blood and endocrine causes of abnormally high or low concentrations

Component and its normal range	Endocrine causes of abnormally high concentration	Endocrine causes of abnormally low concentration
Plasma sodium: 135–145 mmol/L	Diabetes insipidus, Cushing's syndrome, Conn's syndrome	Syndrome of inappropriate antidiuretic hormone secretion (SIADH), Addison's disease, diabetes mellitus
Plasma potassium: 3.5–5.0 mmol/L	Addison's disease, diabetes insipidus, diabetes mellitus	Cushing's syndrome, hyperaldosteronism, SIADH, hyperthyroidism
Plasma urea: 2.5–7.5 mmol/L; plasma creatinine: <120 mmol/L	Diabetes insipidus, diabetes mellitus, Addison's disease	SIADH
Plasma osmolality: 270–300 mOsmol/kg	Diabetes insipidus, diabetes mellitus (NB abnormal aldosterone levels do not affect osmolality)	SIADH
pH: 7.35–7.45; bicarbonate: 24–30 mmol/L	Cushing's syndrome, Conn's syndrome	Addison's syndrome, diabetic ketoacidosis, hyperparathyroidism (mild)
Plasma calcium: 2.25–2.55 mmol/L	Hyperparathyroidism, vitamin D toxicity, hyperthyroidism, acromegaly, Addison's disease (NB abnormal calcitonin levels do not affect plasma calcium levels)	Hypoparathyroidism, vitamin D deficiency, Cushing's syndrome
Plasma phosphate: 0.8–1.5 mmol/L	Hypoparathyroidism, hyperthyroidism, acromegaly	Hyperparathyroidism, vitamin D deficiency, diabetes mellitus
Fasting blood glucose: 3.5–6.0 mmol/L	Diabetes mellitus, Cushing's syndrome, acromegaly, hyperthyroidism, phaeochromocytoma	Exogenous insulin overdose, Addison's disease, pituitary insufficiency
Fasting plasma triglycerides: 0.55–1.90 mmol/L	Diabetes mellitus	
Plasma ketone bodies: not normally present in the blood	Diabetes mellitus	

Fig. 17.6 Endocrine causes of abnormal clinical chemistry.

the blood and urine (e.g. U+Es). Fig. 17.6 shows endocrine causes of disordered chemical levels. These results must be viewed alongside the history, examination and other tests, since endocrine diseases are rarely the most common cause.

Haematology
Investigations of the components that make up blood can suggest specific disorders:
- Low haemoglobin—chronic blood loss.
- Raised haematocrit—dehydration.
- Increased percentage of HBA_{1C}—diabetes mellitus.
- Raised erythrocyte sedimentation rate (ESR)—inflammation or infection.

Microbiology and virology
If an infection is suspected then an appropriate specimen can be tested for microorganisms. Specimens can be collected by swabbing, scraping or aspirating the infected region. Infections of endocrine organs are uncommon, but the reproductive system is prone to sexually transmitted diseases.

The following endocrine disorders can predispose to infection elsewhere in the body:
- Diabetes mellitus.
- Cushing's syndrome.
- Hypothyroidism.

Histopathology and cytology
Histopathological examination requires an intact sample of the tissue called a biopsy. Biopsies can be taken from many organs in the endocrine and reproductive systems if an abnormality or tumour is suspected. The specimens are examined for abnormal cells and tissue structure:
- Inflammatory cells—show the presence of inflammation or infection.

- Tissue structure changes (e.g. necrosis, hyperplasia).

Cytology only requires a cell sample (e.g. cervical smear). A full biopsy is not needed, but it can be used. The cells are examined for cancerous changes called dysplasia.

Imaging of the endocrine and reproductive systems

Plain X-ray radiography

Plain X-rays demonstrate bony structures and calcified areas within organs because these tissues are radio-opaque (white), whilst soft tissues are radio-lucent (grey to black). They are used to investigate endocrine and reproductive disorders that cause bone abnormalities:

- Acromegaly—thickened skull, enlarged jaw and hands (Fig. 17.7).
- Hyperparathyroidism and rickets—osteomalacia.
- Cushing's syndrome and thyrotoxicosis—osteoporosis.
- Prostate bony metastases—osteosclerotic lesions.
- Other bony metastases—osteolytic lesions.
- Size of the pituitary fossa—enlarged with large adenomas.
- Organ calcification—following disease in the adrenal glands.

Contrast media

Contrast media and dyes can be used to image soft tissues. Contrast media are used with X-rays, they must be radio-opaque and inert to prevent damage to the organ. The contrast medium or dye can be ingested or injected into a specific body compartment to allow visualization with X-rays.

Angiography

A contrast medium is injected into the arteries (intra-arterially) or veins (intravenously) allowing these blood vessels and the organs they supply to be visualized (Fig. 17.8).

Fluorescein angiography

This is a form of angiography used to visualize the retinal vessels. A fluorescein contrast medium is injected intravenously and the retina is photographed using ultraviolet light that causes the fluorescein to fluoresce. This technique is especially useful in the investigation of diabetic retinopathy.

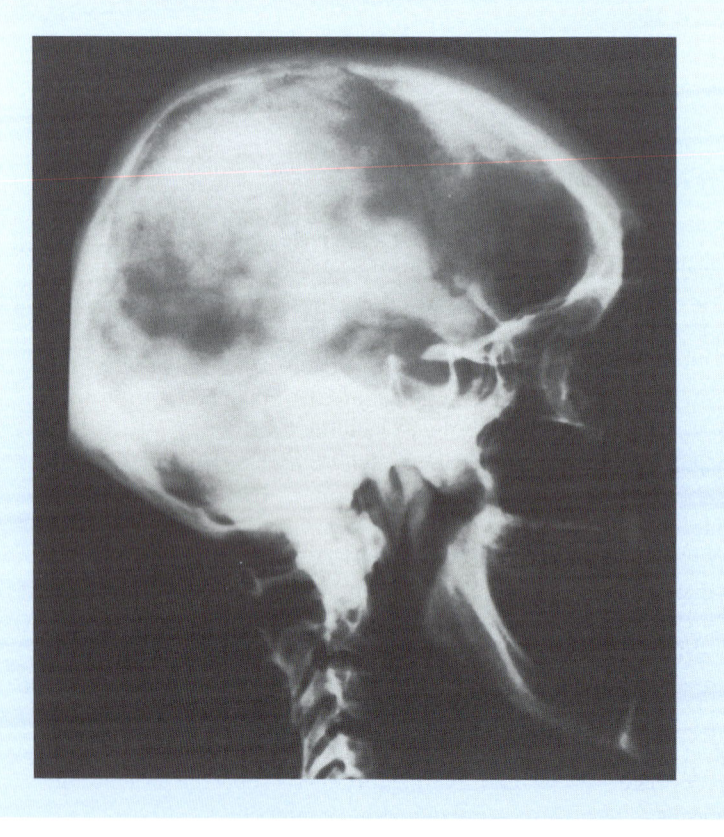

Fig. 17.7 Plain X-ray of the skull of a patient with acromegaly. It shows a large, protruding jaw, a large pituitary fossa and a thick skull. (From Grainger & Allison, 4th edn.)

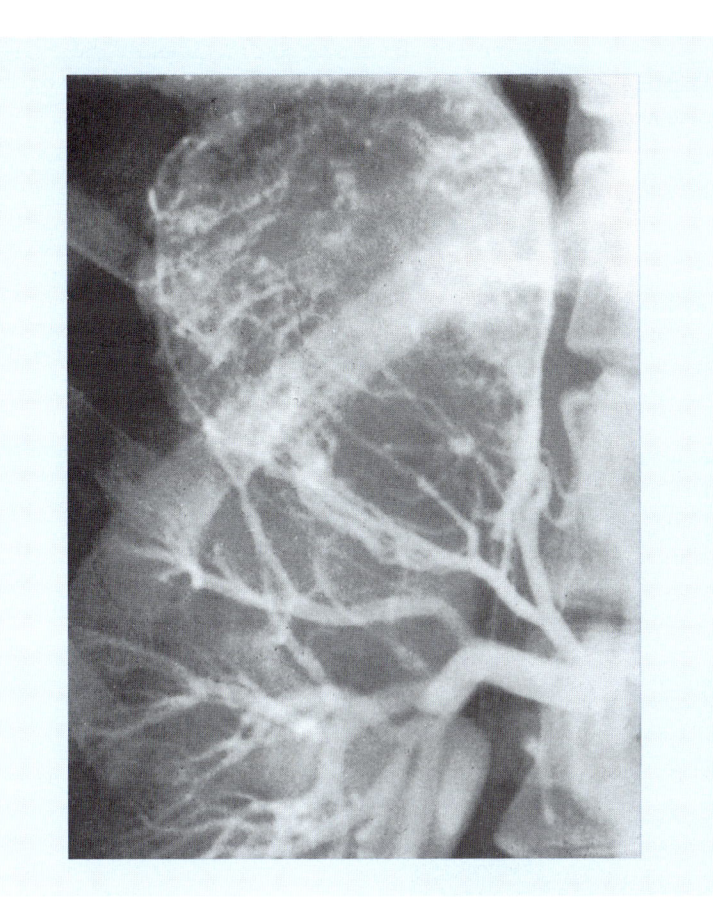

Fig. 17.8 Selective arteriogram of a phaeochromocytoma. (From Sutton, 6th edn.)

Hysterosalpingography

A contrast medium is injected into the uterus and fallopian tubes via the cervix. Its progression along the fallopian tubes is imaged using real time X-ray screening (Fig. 17.9).

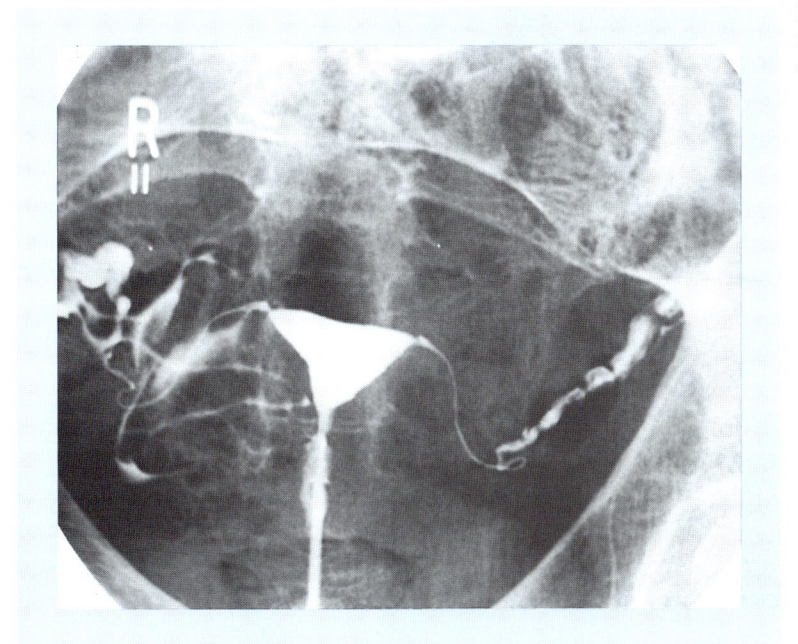

Fig. 17.9 Hysterosalpingogram of a normal uterus and fallopian tubes. (From Grainger & Allison, 4th edn.)

Investigations and Imaging

Mammography

Mammography is one method of investigating and screening for breast lesions. Early detection of breast cancer by mammography may improve the prognosis. An example of a normal breast and one affected by cancer are shown in Fig. 17.10.

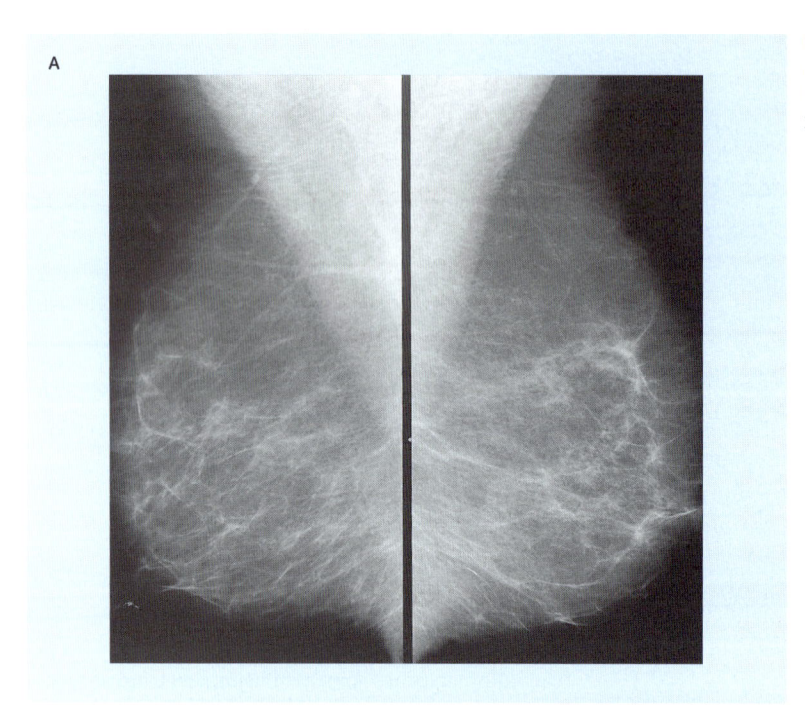

Fig. 17.10 (A) Mammogram of a normal breast. (B) Mammogram of a breast containing a carcinoma. (From Sutton, 6th edn.)

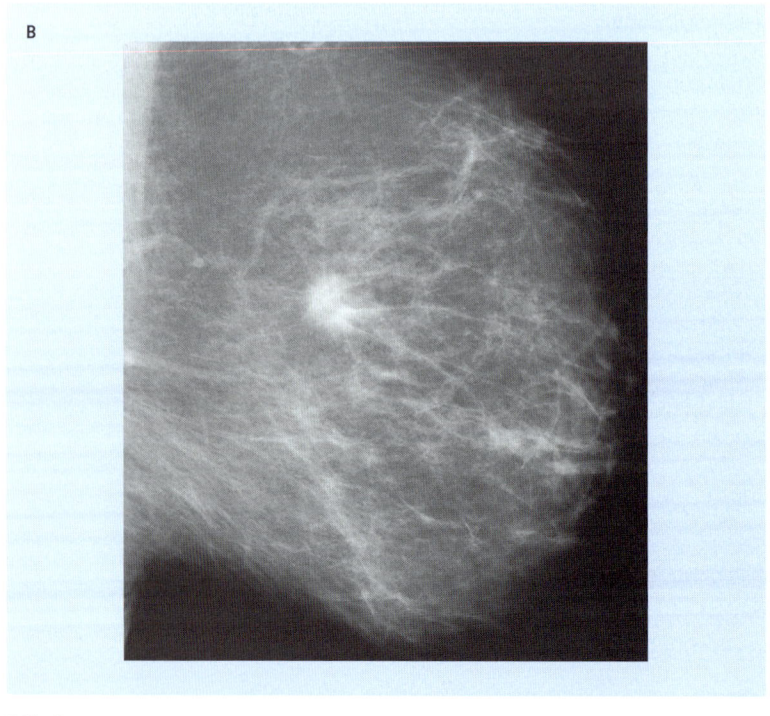

Ultrasonography

Ultrasonography uses the reflection of harmless, high frequency sound waves to determine the composition of various tissues throughout the body. The reflected sound waves are processed to produce an image in which fluid appears black and denser structures appear white.

Ultrasonography may be used to evaluate the size and composition of masses, determining if they are cystic or solid. Masses are commonly examined by ultrasound in the breast, ovaries, testes, thyroid gland and adrenal glands. An example is shown in Fig. 17.11.

Ultrasound imaging is routinely used to examine the developing fetus and to investigate any abnormalities that may be suspected clinically. An example is shown in Fig. 17.12.

Computed tomography

Computed tomography (CT) produces cross-sectional images using X-rays, typically in the axial or horizontal plane. The X-ray emitter rotates about the patient and the computer reconstructs an image by combining views from the multiple X-ray detectors. The computer can differentiate over 2000 densities; this is significantly more than conventional X-ray films. Bone appears white and other soft tissues are grey to black. An example is shown in Fig. 17.13.

CT scanning has applications in almost all disease processes, particularly oncology and neurology. It can also be used for planning accurate biopsy and interventions.

In a similar manner to conventional radiography, contrast media are routinely used in CT to enhance imaging of tissues and vasculature. The same contrast media can be used as for conventional radiography.

Magnetic resonance imaging

Magnetic resonance imaging (MRI) produces cross-sectional images without using ionizing radiation. MRI uses strong external magnetic fields formed by magnetic coils around the patient to manipulate the protons that form the nucleus of hydrogen atoms.

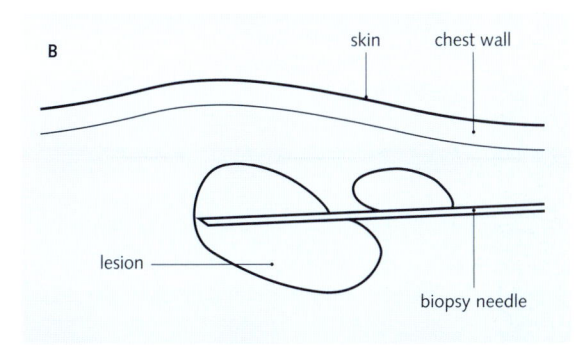

Fig. 17.11 (A) Ultrasound-guided needle biopsy of a breast mass. (B) Line drawing of A. Note that the needle is introduced parallel to the chest wall. (A, from Sutton, 6th edn.)

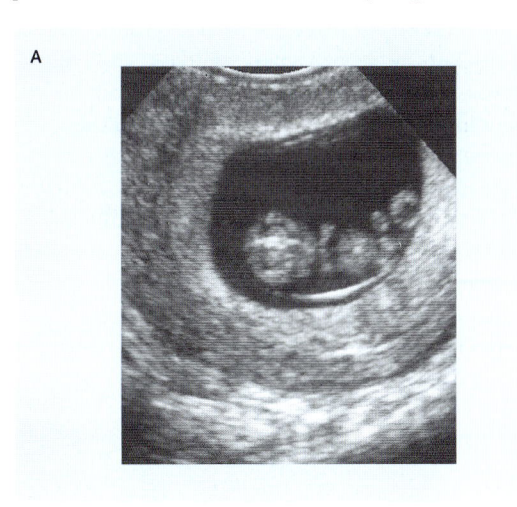

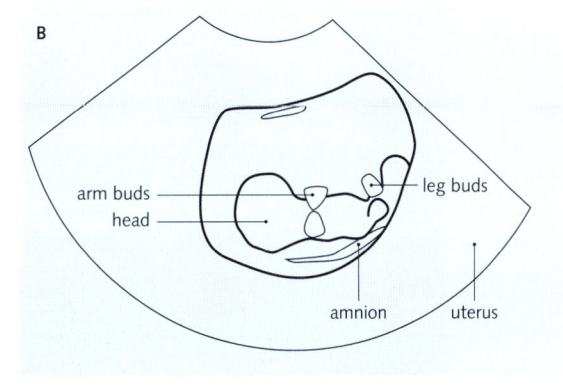

Fig. 17.12 (A) Transvaginal ultrasound scan of a normal 8–9 week fetus. (B) Line drawing of A. (A from Grainger & Allison, 4th edn.)

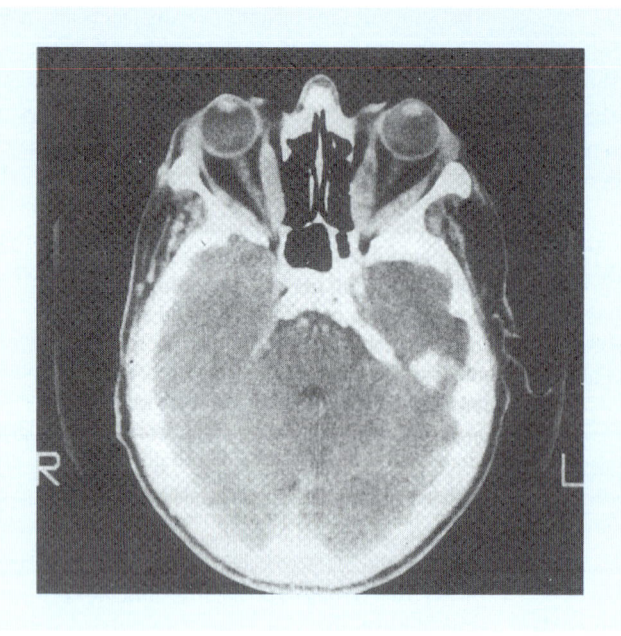

Fig. 17.13 CT scan showing exophthalmos in a patient with Graves' disease. (From Edwards et al. *Davidson's Principles and Practice of Medicine*, 17th edn.)

The protons behave like miniature magnets which line up to the strong magnetic field and gain energy in the process. Once the magnetic field is turned off the protons release the energy they gained by inducing a current in the magnetic coils that produced the magnetic field. This current is detected and processed into a computerized image.

The hydrogen atoms detected are generally in water (H_2O). There are two main methods of processing the image:
- T1 weighted images: good anatomical detail; water is black.
- T2 weighted images: good detail for pathological changes; water is white.

An example is shown in Fig. 17.14.

Radioisotope scans

Certain chemicals are absorbed more rapidly by different tissues, for example iodine is actively absorbed by the thyroid gland. By tagging a chemical with a radioactive element (i.e. a radioisotope), uptake in the target tissue can be monitored using a gamma camera (this is called scintigraphy). This

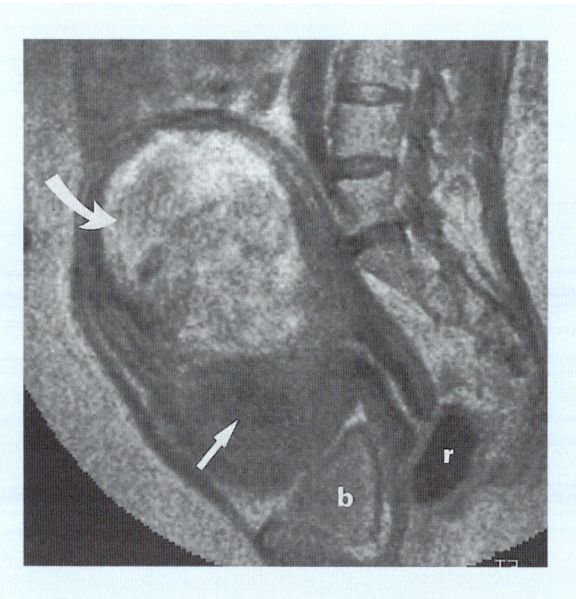

Fig. 17.14 MRI scan (T2-weighted spin echo image) of large uterine leiomyomas, showing areas of high signal (curved arrow) and low signal (straight arrow). b, bladder; r, rectum. (From Sutton, 6th edn.)

pattern of isotope uptake within the gland can expose abnormal areas.

This method is especially suitable for the thyroid gland because of the highly selective uptake of iodine. By administering an oral dose of ^{123}I (a radioactive isotope of iodine) the activity of different areas within the gland can be detected between 5 and 24 hours later. It allows measurement of:
- Hyperactivity.
- Abnormal anatomy.
- Tumours or nodules, including size and location.

Most tumours show up as inactive 'cold' spots, however hypersecretory tumours show up as active 'hot' spots.

A better resolution can be achieved using intravenous technetium-99m pertechnetate. This substance is also less toxic and allows scanning just 20 minutes after injection. Examples are shown in Fig. 17.15.

Different substances can be used to scan the parathyroid and adrenal glands:
- Parathyroid glands—technetium-99m Sestamibi.
- Adrenal cortex—^{131}I-iodonorcholesterol (NP-59) or ^{75}Se-selenomethylnorcholesterol.
- Adrenal medulla—^{131}I-metaiodobenzylguanidine (mIBG).

Radioactive iodine is potentially harmful to the thyroid gland. Lugol's solution, which contains non-radioactive iodine, is given both the day before and on the test day to reduce the uptake of radioactive iodine by the thyroid gland.

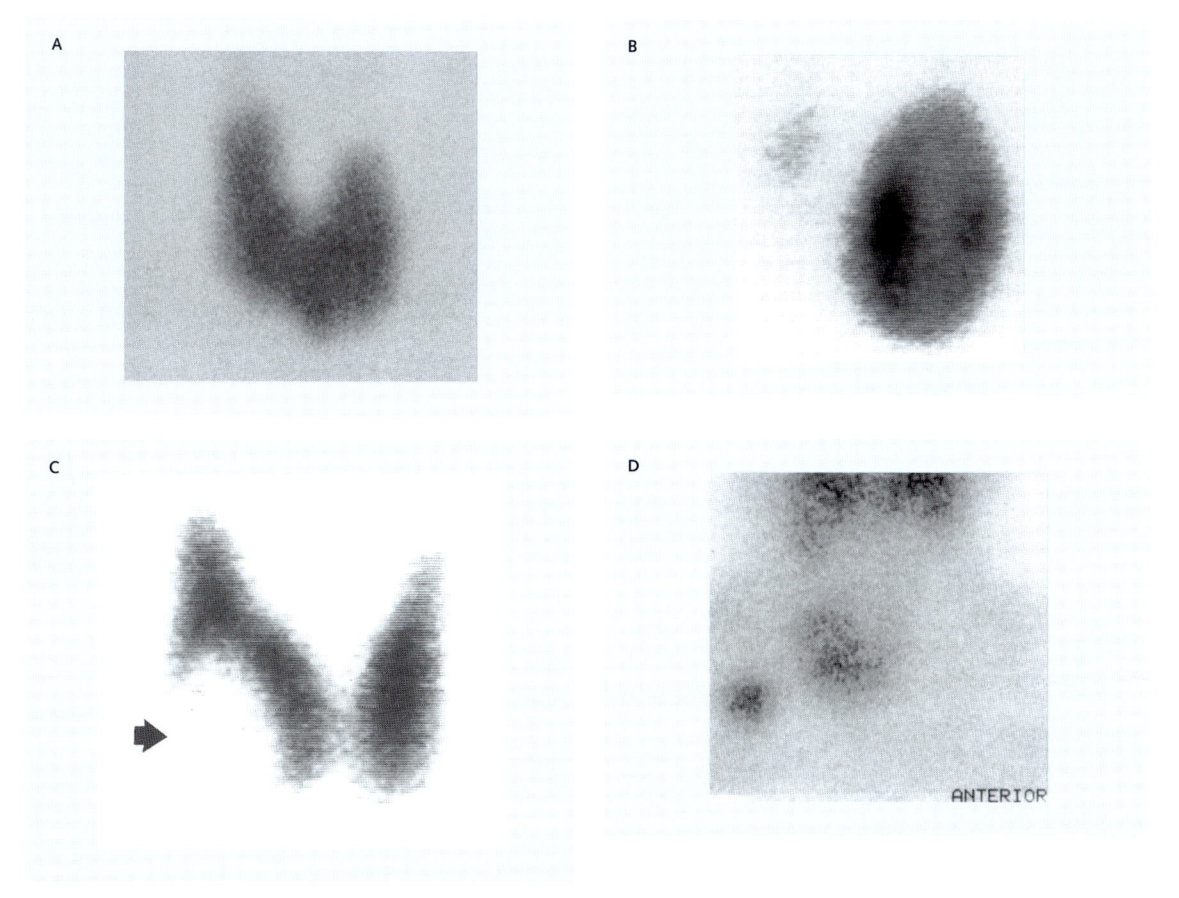

Fig. 17.15 Radioisotope thyroid images using technetium-99m pertechnetate. (A) Scan showing increased uptake throughout the gland in a patient with Graves' disease. (B) A patient with a single toxic nodule; the remainder of the gland is suppressed. (C) Increased uptake in a patient with Graves' disease but decreased uptake in a non-toxic (cold) lump (arrowed) consistent with a thyroid cancer or cyst. (D) ^{99m}Tc MIBI scan of recurrent papillary thyroid carcinoma not seen on ^{131}I scanning. (A, from Grainger & Allison, 4th edn; B–D, from Murray & Ell, 2nd edn.)

Investigations and Imaging

- Describe the ELISA technique and its use in endocrinology.
- Which hormones are measured indirectly and why?
- With an example, describe the situations in which a suppression test is used.
- With an example, describe the situations in which a stimulation test is used.
- Describe the triple stimulation test of pituitary function. Why are all these hormones tested together?
- With examples, describe how laboratory-based investigations other than for hormone levels can aid diagnosis.
- Describe the use of plain and contrast enhanced radiography in diagnosis of endocrine and reproductive disorders.
- Describe the use of ultrasound in diagnosis of endocrine and reproductive disorders.
- Describe the use of CT and MRI scans in diagnosis of endocrine and reproductive disorders.
- Describe the use of radioisotope scans in diagnosis of endocrine and reproductive disorders.

SELF-ASSESSMENT

Multiple-choice Questions	243
Short-answer Questions	253
Essay Questions	254
MCQ Answers	255
SAQ Answers	264

Multiple-choice Questions

Indicate whether each answer is true or false.

Chapter 1—Overview of the Endocrine System

1. Concerning types of hormones:

(a) Steroid hormones act through intracellular receptors.
(b) Polypeptide hormones readily cross plasma membranes.
(c) Many polypeptide hormones are synthesized by the cleavage of larger polypeptides.
(d) Several major hormones are synthesized by modification of the amino acid tyrosine.
(e) Steroid hormones are stored in secretory vesicles ready for release.

2. Insulin receptors:

(a) Insulin acts through tyrosine kinase receptors.
(b) Tyrosine kinase receptors are located in the surface of cells.
(c) This type of receptor is also used by adrenaline.
(d) The serine side chains of amino acid residues are phosphorylated.
(e) Their main effect is to stimulate protein synthesis directly.

3. G-protein receptors:

(a) Are located on the cell membrane.
(b) Often use cATP as a second messenger.
(c) G-proteins bind GDP in their resting state.
(d) When the receptor is active following stimulation the G-protein is attached to the receptor protein.
(e) The receptor protein is a type of glycoprotein.

4. Concerning eicosanoids:

(a) Prostaglandins are usually transported in the blood to act on distant cells.
(b) Prostaglandins are synthesized from a phospholipid found in the cell membrane.
(c) Leukotrienes are synthesized by the cyclooxygenase pathway.
(d) Prostaglandins are only synthesized by specialized cells.
(e) Eicosanoids are stored in secretory granules.

5. Hormonal feedback:

(a) Polypeptide hormones can readily cross the blood–brain barrier.
(b) Feedback to the hypothalamus is by intracellular receptors only.
(c) If a variable altered by a hormone (e.g. plasma osmolarity) affects the regulation of the same hormone this is considered to be feedback.
(d) The hypothalamus and pituitary gland are the only sites of hormonal feedback.
(e) In a healthy person thyroid hormones inhibit the synthesis of TSH at all times.

Chapter 2—The Hypothalamus and the Pituitary Gland

6. The hypothalamus:

(a) Is located at the base of the brain either side of the 3rd ventricle.
(b) Secretes hormones that act directly on many endocrine organs.
(c) Synthesizes mainly steroid hormones.
(d) Stimulates ACTH secretion by direct neural stimulation of the pituitary gland.
(e) Contains nerve cells derived from the fetal ectoderm layer.

7. The anterior pituitary gland:

(a) Secretes the gonadotrophins LH and FSH.
(b) Is a direct extension of the hypothalamus.
(c) Is a relation of the optic chiasma.
(d) Secretes only polypeptide and glycoprotein hormones.
(e) Secretes a hormone that is the main regulator of aldosterone secretion.

8. The posterior pituitary gland:

(a) Secretes aldosterone, which regulates fluid balance.
(b) Is essential for breastfeeding.
(c) Synthesizes two small polypeptide hormones.
(d) Is covered superiorly by a layer of dura mater.
(e) Contains glial support cells called pituicytes.

9. A 48-year-old woman presents with loss of her central vision:

(a) This is a well recognized symptom of pituitary adenomas.
(b) Pituitary adenomas are most commonly prolactinomas.
(c) Pituitary adenomas often present with disorders of several pituitary hormones.
(d) The diagnosis could be confirmed by ultrasound.
(e) Pituitary adenomas are often treated by surgical removal.

Chapter 3—The Thyroid Gland

10. Concerning thyroid hormones:

(a) They regulate the rate of metabolism in cells.
(b) They are transported in the blood bound to a plasma protein called thyroglobulin.
(c) Most tissues in the body can add an iodine molecule to T_3 to form the more active molecule called T_4.
(d) They act through intracellular receptors.
(e) They are synthesized from the same amino acid as adrenaline.

243

Multiple-choice Questions

11. The thyroid gland:

(a) Responds to TSH secreted by the anterior pituitary gland.
(b) Is drained by veins that are closely related to nerves to the larynx.
(c) Is involved in the regulation of calcium.
(d) Is bound to the trachea by the pre-tracheal fascia.
(e) Usually has a pyramidal lobe, which joins the isthmus.

12. A 43-year-old woman presents with weight gain and lethargy. A blood test shows a raised level of TSH:

(a) Hypothyroidism is a common disease, especially in women.
(b) Hypothyroidism is responsible for 10% of obesity.
(c) In children hypothyroidism causes dwarfism.
(d) Hypothyroidism is often a long-term complication of the treatment of hyperthyroidism.
(e) Hypothyroidism should be treated with thyrotrophin releasing hormone (TRH) to return blood TSH levels to normal.

13. A 45-year-old woman presents with weight loss, irritability and sweating that is diagnosed as hyperthyroidism:

(a) Graves' disease is an autoimmune disease that stimulates thyroid cells causing hyperthyroidism.
(b) A history about what clothes she wears is of clinical interest.
(c) The diagnosis of hyperthyroidism can be made by measuring TSH, T_3 and T_4 only.
(d) The history of weight loss and excess thyroid hormone secretion makes carcinoma a likely diagnosis.
(e) It could be treated with the drug bromocriptine.

14. A 79-year-old woman presents with a lump in her throat that has developed over the last few months. Thyroid function tests and fine-needle aspiration are performed:

(a) A papillary carcinoma develops from the follicle cells.
(b) If she also has features of Cushing's syndrome then it is more likely to be a medullary carcinoma.
(c) If it is a carcinoma then her thyroid hormone levels are likely to be high.
(d) Anaplastic carcinoma has a particularly poor prognosis.
(e) Anaplastic carcinoma of the thyroid invades mainly by vascular spread.

15. A 53-year-old woman presents with a lump in the front of her throat:

(a) If it rises when she protrudes her tongue, it is almost certainly part of the thyroid gland.
(b) If the lump is a toxic adenoma her TSH levels will be raised.
(c) Cardiac arrhythmia is a sign of hyperthyroidism.
(d) A fine-needle aspiration should almost always be performed for cytology (having taken into account other factors).
(e) Surgery could potentially result in chronic hypercalcaemia.

Chapter 4—The Adrenal Glands

16. The cortex of the adrenal gland:

(a) Secretes hormones derived from cholesterol.
(b) Develops from neural crest tissue.
(c) Is arranged in follicles that contain stored hormone.
(d) Responds directly to corticotrophin releasing hormone (CRH).
(e) The zona fasciculata secretes cortisol.

17. Adrenocorticotrophic hormone (ACTH):

(a) Acts mainly on the medulla of the adrenal gland.
(b) Secretion is stimulated by CRH and inhibited by cortisol.
(c) Is secreted by corticotrophs in the hypothalamus.
(d) The peak of secretion occurs early in the morning before waking.
(e) Inhibits the release of adrenal androgens.

18. Stress:

(a) Stimulates the release of growth hormone.
(b) Stimulates the release of cortisol.
(c) Causes a rise in blood glucose levels.
(d) Can be tolerated better in the absence of cortisol.
(e) Activates the parasympathetic nervous system.

19. Concerning the hormonal release of catecholamines:

(a) These hormones prepare the body for fight or flight.
(b) More adrenaline is secreted than noradrenaline.
(c) They increase the heart rate and stroke volume.
(d) They lower blood glucose.
(e) They are secreted by cells that are equivalent to preganglionic sympathetic neurons.

20. A 59-year-old man has been taking corticosteroids for 7 years for an inflammatory condition, but he is starting to notice side effects:

(a) He is suffering from Cushing's disease.
(b) His blood glucose levels will probably be high.
(c) He will probably have lost weight.
(d) His skin will be thick and prone to bruising.
(e) The steroids should be stopped immediately to prevent further complications.

21. A 41-year-old man with vomiting, hypotension and confusion is diagnosed with Addison's disease.

(a) This is an excess of aldosterone secretion.
(b) If the cause was primary adrenal insufficiency the patient is likely to have increased skin pigmentation.
(c) It can be diagnosed using the Synacthen® test.
(d) A history of growing beetroot is of clinical interest.
(e) The treatment must often be continued for life.

22. **A 43-year-old man is investigated for hypertension; this reveals that he has Conn's syndrome.**

(a) Conn's syndrome is a type of primary hyperaldosteronism.
(b) The plasma potassium levels will probably be low.
(c) Conn's syndrome is an adenoma of the adrenal medulla.
(d) About 27% of hypertension is caused by raised aldosterone levels.
(e) Testing the blood levels of aldosterone alone can show hyperaldosteronism.

23. **A 25-year-old man was admitted to hospital with a severe headache and found to have a very high blood pressure. This resolved spontaneously before treatment could be given:**

(a) You should measure the quantity of vanillylmandelic acid in his urine to exclude phaeochromocytoma.
(b) Phaeochromocytoma is a tumour of the adrenal cortex.
(c) Phaeochromocytoma can be treated with diuretics and ACE inhibitors and repeated scans to monitor growth.
(d) The hypertensive episodes are caused by fluctuations in blood volume.
(e) During the hypertensive attack his pupils were probably dilated.

Chapter 5—The Pancreas

24. **Insulin:**

(a) Deficiency causes diabetes mellitus.
(b) Is normally secreted after a meal.
(c) Stimulates the uptake and use of glucose in muscle cells.
(d) Is secreted by the α-cells of the islets of Langerhans in the pancreas.
(e) Secretion ceases entirely during prolonged starvation.

25. **The pancreas:**

(a) Has a retroperitoneal position.
(b) Secretes hormones into the duodenum via the pancreatic duct.
(c) Is a relation of the left kidney, stomach and spleen.
(d) Develops from two buds that grow out of the endodermal digestive tract.
(e) The hormone secreting cells are richly innervated.

26. **Concerning hypoglycaemia:**

(a) Is present if blood glucose levels are 3.7 mmol/L.
(b) Stimulates the release of glucagon.
(c) This glucagon is released by the exocrine pancreas.
(d) Inhibits the release of adrenaline.
(e) Can be detected directly by the islet cells of the pancreas.

27. **Glucagon:**

(a) Consists of multiple glucose molecules.
(b) Inhibits the secretion of insulin.
(c) Inhibits the actions of insulin.
(d) Secretion is inhibited by insulin.
(e) Has a catabolic and antianabolic effect.

28. **A 10-year-old boy presents with tiredness for the last 3 months, with malaise and vomiting in the last month:**

(a) This could be caused by diabetes mellitus (DM).
(b) DM could be excluded if there is no glucose in his urine.
(c) A history of weight gain and polyuria would increase the probability of this diagnosis.
(d) His diabetes could be treated by carefully regulating his diet for several months until the type of diabetes is determined.
(e) If diabetes mellitus is confirmed he will need an urgent referral to an ophthalmologist to treat retinopathy as this can cause irreversible blindness.

29. **An elderly woman is referred from an ophthalmologist due to concerns about her vision:**

(a) This could be the presentation of a type of diabetes mellitus treatable with oral medications.
(b) Retinopathy is a common microvascular complication of all types of diabetes mellitus.
(c) Blindness caused by retinopathy can be prevented by frequent screening and treatment.
(d) Microvascular complications of diabetes mellitus are associated with hyaline arteriolosclerosis.
(e) Diabetes mellitus could be excluded by a urine dipstick test.

30. **A 76-year-old man is found to have glycosuria:**

(a) His blood glucose levels must be above 10 mmol/L.
(b) This is a common presentation of NIDDM.
(c) If this is NIDDM he can be reassured that he will not need to inject insulin.
(d) NIDDM is often associated with obesity and hypotension called syndrome X.
(e) NIDDM is caused by autoantibodies against the β-cells of the islets of Langerhans.

Chapter 6—Up and Coming Hormones

31. **Concerning leptin:**

(a) It is a hormone secreted by muscle cells.
(b) It is essential for fertility and puberty.
(c) It helps to regulate food intake.
(d) Oral leptin will reduce food intake in the majority of patients.
(e) Inherited leptin deficiency is a common cause of obesity.

Multiple-choice Questions

32. Melatonin:

(a) Is a polypeptide hormone.
(b) Is secreted mainly by the pineal gland.
(c) Increased secretion causes pigmentation of the skin.
(d) Acts to reset the preoptic nucleus of the hypothalamus.
(e) Is secreted in response to stimulation of the retina.

33. A 53-year-old man has suffered from recurrent peptic ulcers that are difficult to treat medically:

(a) In Zollinger–Ellison syndrome there is a tumour in the stomach.
(b) This condition affects about 1 in 1000 people.
(c) The tumour secretes gastrin.
(d) Gastrin stimulates the chief cells of the stomach to secrete acid.
(e) This tumour is fairly often associated with MEN I.

Chapter 7—Endocrine Control of Fluid Balance

34. Renin:

(a) Is an enzyme secreted by the juxtaglomerular complexes in the kidney.
(b) Is secreted in response to low sodium or low blood pressure.
(c) Acts on angiotensinogen to form angiotensin I.
(d) Release is inhibited by angiotensin converting enzyme (ACE) inhibitors.
(e) Indirectly stimulates aldosterone release.

35. Antidiuretic hormone (ADH):

(a) Increases water excretion.
(b) Is secreted from specialized neurons.
(c) Acts on the proximal tubule of the kidney nephrons to raise permeability to water.
(d) Secretion is stimulated by raised plasma osmolarity.
(e) Causes vasodilation of arteries to control blood pressure.

36. The following factors cause a rise in blood pressure:

(a) An increase in blood volume.
(b) Reduced peripheral vascular resistance.
(c) High renin levels.
(d) High angiotensinogen levels.
(e) High adrenaline levels.

37. A 27-year-old man has recently started to excrete large volumes of urine:

(a) His fasting blood sugar should be measured in case he has diabetes mellitus.
(b) It could be caused by hypocalcaemia.
(c) He is likely to have reduced skin turgor.
(d) It could be caused by a urinary tract infection.
(e) The water stimulation test is used to confirm diabetes insipidus.

Chapter 8—Endocrine Control of Calcium Homeostasis

38. The parathyroid glands:

(a) In most people there are three pairs of parathyroid glands, which lie posterior to the thyroid gland.
(b) Secrete parathyroid hormone and calcitonin to regulate calcium levels.
(c) Are derived from the mesenchyme surrounding the thyroid gland.
(d) Secrete parathyroid hormone in response to low blood calcium.
(e) Are supplied by branches of the lingual artery.

39. Concerning calcium:

(a) The majority of calcium within the body is stored in bone.
(b) A rise in intracellular calcium levels stimulates muscle contraction.
(c) A fall in intracellular calcium levels stimulates exocytosis (e.g. hormone secretion).
(d) Osteoclasts release calcium by eroding bone.
(e) Low calcium stimulates the activation of vitamin D by acting directly on the kidney.

40. Vitamin D:

(a) Is derived from cholesterol.
(b) Deficiency results in poor bone formation.
(c) Acts mainly on the kidneys to increase the reabsorption of calcium.
(d) Can be synthesized in the skin by the action of infra-red light.
(e) Must be activated by conversion to 24,25-dihydroxy-vitamin D_3 in the kidney.

41. Parathyroid hormone:

(a) Is usually secreted by four small glands located anteriorly to the thyroid gland in the neck.
(b) Is secreted in response to low blood calcium levels.
(c) Is a modified amino acid hormone.
(d) Acts chiefly on the intestines to increase the absorption of calcium.
(e) Stimulates the activation of 25-hydroxyvitamin D_3 by acting on 1α-hydroxylase in the kidney.

42. Calcitonin:

(a) This hormone is secreted by parafollicular cells in the thyroid gland.
(b) The main stimulus for calcitonin release is high blood calcium.
(c) Is essential for normal calcium regulation.
(d) Is a polypeptide hormone.
(e) The secretory cells are types of APUD cells.

43. A 52-year-old man presents with a fractured head of femur. A history is difficult to obtain as the patient seems confused but he does complain of abdominal pain:

(a) Hypercalcaemia would be consistent with this history.
(b) Hypercalcaemia can be caused by secondary hyperparathyroidism.
(c) Hypercalcaemia can be caused by a parathyroid adenoma.
(d) Failure to treat parathyroid adenomas increases the risk of renal failure.
(e) If a parathyroid adenoma was found in a very young patient it could be a component of MEN I.

Chapter 9—Endocrine Control of Growth

44. Growth hormone:

(a) Is released from the posterior pituitary gland.
(b) Is a polypeptide hormone.
(c) Stimulates growth in many tissues by direct stimulation.
(d) Secretion usually increases at night.
(e) Secretion normally ceases after puberty.

45. Insulin-like growth factors:

(a) Are peptide hormones secreted mainly by the kidney.
(b) Blood levels are higher during sleep.
(c) Stimulate the uptake of glucose into cells.
(d) Stimulate the release of growth hormone.
(e) Act via tyrosine kinase receptors on the cell surface.

46. A 32-year-old man is referred by his dentist with change in his jaw that the dentist cannot explain:

(a) This could be because the jaw has enlarged due to an excess of growth hormone.
(b) If there is an excess of growth hormone he is likely to be taller than he was 5 years ago.
(c) This excess of growth hormones is called gigantism.
(d) He has an increased risk of insulinomas.
(e) The most common cause is a somatotroph adenoma of the anterior pituitary gland.

Chapter 10—Endocrine Disorders of Neoplastic Origin

47. A 79-year-old man has been diagnosed with lung cancer. This is may be associated with secretion of:

(a) Parathyroid hormone causing hypercalcaemia.
(b) Cortisol causing hypoglycaemia.
(c) Prolactin causing gynaecomastia.
(d) Testosterone causing aggression.
(e) ADH causing diabetes insipidus.

48. The man in question 23 is investigated further:

(a) The adrenal tumour will probably be palpable on abdominal examination.
(b) He also has a lump in his neck: this suggests MEN II.
(c) You should order a CT scan to exclude pancreatic tumours found in MEN II.
(d) The thyroid lump probably has a papillary pattern
(e) Other members of his family may have similar tumours.

Chapter 11—Development of the Reproductive System

49. Regarding development of the male reproductive system:

(a) The ductus deferens develops from the mesonephric (Wolffian) ducts.
(b) Early differentiation from the indifferent stage is caused by expression of the sex-determining region (SRY) on the X chromosome.
(c) The scrotum develops from the same structures that form the labia majora in the female.
(d) The testes have a blood supply directly from the aorta.
(e) The primordial germ cells originate on the dorsal body wall of fetal abdomen.

50. Regarding development of the female reproductive system:

(a) The ovaries are derived from the same structure that forms the testes in the male.
(b) Müllerian inhibiting substance (MIS) is secreted to cause the male reproductive system to regress.
(c) The primordial oocytes begin a meiotic division before birth that is not completed until after puberty.
(d) Female and male development is the same until the sixth week.
(e) Ovarian stroma develops as an outgrowth of the paramesonephric ducts.

51. Concerning male puberty:

(a) Puberty occurs at an earlier age in boys than girls.
(b) Testosterone stimulates growth of the long bones.
(c) Hypothalamic GnRH secretion is inhibited prior to puberty.
(d) Puberty is said to have started when the testes reach 12 mL.
(e) When the gonads begin to secrete sex steroids at puberty it is called adrenarche.

52. Concerning female puberty:

(a) Body weight is a better predictor of when menarche will occur than age.
(b) Menarche is the first sign of female puberty.
(c) Pubic and axillary hair grow as a result of adrenal androgens.
(d) The first ovulation usually occurs at menarche.
(e) The ovarian follicles mature into secondary follicles during puberty.

Chapter 12—The Female Reproductive System

53. Concerning the ovaries:

(a) They are located on the posterior of the broad ligament of the uterus.
(b) New follicles can be formed until puberty.
(c) Each month several follicles develop in response to FSH.
(d) At ovulation, the oocyte leaves the ovary and enters the peritoneal cavity.
(e) All blood vessels, lymphatics and nerves reach the ovary via the hilum.

54. Concerning the fallopian tubes:

(a) They form the superior border of the broad ligament.
(b) Fertilization usually occurs in their uterine section.
(c) Their blood supply is derived from the ovarian and uterine arteries.
(d) They open into the peritoneal cavity forming a connection to the outside world via the vagina.
(e) They develop from the paramesonephric (Müllerian) ducts.

55. Concerning the cervix:

(a) The cervix is considered to be a section of the vagina.
(b) The opening of the cervix into the vagina is called the internal os.
(c) The endocervix is usually lined by columnar epithelium.
(d) During childbirth the cervix dilates to 5 cm diameter.
(e) The action of progesterone allows the cervix to dilate at childbirth.

56. Concerning the vagina:

(a) The acidic environment is maintained by commensal *Candida* yeast.
(b) The epithelial lining of the vagina is shed during menstruation.
(c) It is lined by simple columnar epithelium throughout.
(d) It is a posterior relation of the bladder.
(e) It develops as an outgrowth from the urethra.

57. Concerning the breast:

(a) The glandular tissue of the breast is derived from specialized hair follicles.
(b) Until puberty the male and female breasts are structurally indistinguishable.
(c) A single lactiferous duct opens at each nipple.
(d) The breast tissue can move freely over the underlying muscle.
(e) Progesterone stimulates the development of the secretory tissue during puberty.

58. Concerning oestrogen:

(a) Blood levels are usually higher in females than males.
(b) It acts via cell-surface receptors.
(c) It is essential for the normal development of female external genitalia.
(d) It is not secreted in the luteal phase of the menstrual cycle.
(e) It acts on the anterior pituitary gland where it has a variable effect of LH and FSH secretion.

59. Concerning progesterone:

(a) Blood levels are low between menstruation and ovulation.
(b) Progesterone causes smooth muscles to relax during pregnancy.
(c) In the non-pregnant female, the main source of progesterone is the endometrium.
(d) Progesterone is a complex steroid hormone that is synthesized in many steps from cholesterol.
(e) Progesterone's main role is to promote growth of the reproductive organs.

60. Hormones during the menstrual cycle:

(a) LH levels are highest during menstruation.
(b) Oestrogen and progesterone stimulate the release of gonadotrophins around the middle of the cycle.
(c) Human chorionic gonadotrophin (hCG) rises in the follicular phase of the menstrual cycle.
(d) LH is a glycoprotein hormone released by the anterior pituitary gland.
(e) Oestrogen stimulates the synthesis of progesterone.

61. Concerning ovarian follicles:

(a) Each follicle contains a single oocyte.
(b) They develop into primary follicles from primordial follicles during puberty.
(c) The thecal cells are the major site of oestrogen secretion.
(d) Just before ovulation the follicles contain a fluid filled cavity called the antrum.
(e) Several follicles begin to develop at the start of each menstrual cycle.

62. The corpus luteum:

(a) Is the main source of hCG for the first 6 weeks after fertilization.
(b) Is formed by the remaining granulosa and thecal cells of the mature follicle after ovulation.
(c) Regresses in the absence of fertilization due to falling LH levels.
(d) Has a characteristic blue colour due to the rapid metabolism required for hormone synthesis.
(e) Is usually found in both ovaries during early pregnancy.

63. Concerning cyclical changes in the endometrium:

(a) The endometrium begins to proliferate after ovulation.
(b) The uterine glands secrete a fluid that is rich in glycogen in response to oestrogens.
(c) Approximately 30 mL blood is lost during menstruation.
(d) The entire endometrium is shed during menstruation.
(e) Menstruation is caused by contraction of the uterine arteries causing ischaemia of the uterus.

64. A woman presents with infertility and amenorrhoea. On investigation this is diagnosed as polycystic ovarian syndrome:

(a) She is likely to have the features of Cushing's syndrome.
(b) Ultrasound is very useful for diagnosis.
(c) This diagnosis would have been suspected if she had galactorrhoea.
(d) This diagnosis would have been suspected if she had hirsutism.
(e) Her blood levels of LH and FSH are likely to be raised.

65. A 32-year-old woman presents with menorrhagia and pre-menstrual abdominal pain that is diagnosed as endometriosis:

(a) Endometriosis can be caused by bacterial infections following trauma.
(b) She has a higher risk of infertility.
(c) Endometriosis is usually diagnosed through ultrasound.
(d) Endometriosis is often associated with hirsutism.
(e) Adhesions can develop causing constant pain.

66. A 46-year-old woman presents with abdominal pain and tenderness a month after having an IUD inserted:

(a) She probably has chronic pelvic inflammatory disease.
(b) The most common cause of pelvic inflammatory disease is from STDs.
(c) In this woman's disease, the infectious organism is probably *Chlamydia*.
(d) Pelvic inflammatory disease sometimes causes damage to the fallopian tubes.
(e) The woman will probably need a laparoscopy.

67. A 37-year-old woman presents with menorrhagia and deep dyspareunia. On bimanual vaginal examination, a lump is felt outside the uterus on the left:

(a) The lump is likely to be palpable on careful abdominal examination.
(b) A fluid-filled cyst will have a dark centre on an ultrasound scan.
(c) If the lump is found to contain tissues that resemble teeth and skin it is probably malignant.
(d) A yellow coloured cyst would suggest the woman has endometriosis.
(e) Most ovarian neoplasia originates from the epithelial component of the ovary.

68. A 27-year-old woman presents with an intensely itchy vulva and vaginal discharge:

(a) This is a common presentation of herpes simplex infection.
(b) A vaginal swab showing the presence of lactobacilli would prompt antibiotic treatment with metronidazole.
(c) A thick, white discharge is commonly caused by the fungal infection often called thrush.
(d) Thrush is normally a sexually transmitted infection.
(e) If *Chlamydia* infection is not treated it can cause pelvic inflammatory disease.

69. A 51-year-old woman presents with menorrhagia that is diagnosed as leiomyoma:

(a) This is a common, benign tumour often called a fibroid.
(b) Fibroids form in the myometrium.
(c) Fibroids grow mainly in the presence of progesterone.
(d) They can progress into endometrial carcinoma.
(e) After the menopause it will probably get better.

70. A 49-year-old woman presents with a lump in her left breast:

(a) This is likely to be a cyst caused by fibrocystic change.
(b) It is likely to be a fibroadenoma.
(c) The presence of skin dimpling on the breast when the woman leans forwards suggests breast cancer.
(d) The presence of a red eczema-like rash around the nipple suggests an infection.
(e) The possibility of breast cancer can often be excluded by careful examination.

71. A 50-year-old woman presents with nipple discharge in her right breast:

(a) If a lump is palpable near the nipple this is likely to be mammary duct ectasia.
(b) If the lump is malignant the breast may have the appearance of an orange skin.
(c) The appearance in 'b' is caused by local invasion of the sweat glands.
(d) A strong family history of ovarian cancer would prompt investigation for a *BRAC* mutation.
(e) Early menarche is a risk factor for breast cancer.

72. A 51-year-old woman is suffering from oligomenorrhoea and hot flushes that are diagnosed as the early stages of the menopause. She has previously had two children and no gynaecological surgery:

(a) This is quite early to start the menopause.
(b) If she desires HRT oestrogen should be given alone.
(c) If she took HRT it would lower her risk of fracturing the neck of her femur.
(d) HRT is associated with an increased risk of cardiovascular disease.
(e) Her gonadotrophin levels are probably reduced.

Chapter 13—The Male Reproductive System

73. Concerning the testes:

(a) After puberty they continuously produce sperm.
(b) The seminiferous tubules are lined by testosterone-secreting Leydig cells.
(c) They are surrounded by the tunica vaginalis that is derived from the peritoneum.
(d) They are divided into lobes, each of which contains 20 seminiferous tubules.
(e) Blood from the right testis usually drains into the right renal vein.

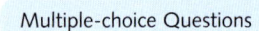

74. The epididymis:

(a) Is 'Z' shaped.
(b) Is a major site of spermatogenesis.
(c) Is formed from a single tube.
(d) Is found within the testis.
(e) Transmits sperm to the seminal vesicles where they are stored in preparation for ejaculation.

75. Concerning the penis:

(a) It is composed of two corpora cavernosa and one corpus spongiosum.
(b) The body of the penis is derived from the genital tubercle.
(c) Erection is maintained by the parasympathetic innervation.
(d) The structures that transmit sperm and urine are on opposing sides of the penis.
(e) The penis enters the cervix during sexual intercourse.

76. Concerning spermatogenesis:

(a) Spermatocytes are stem cells that develop into spermatozoa.
(b) As they develop spermatids migrate between the Sertoli cells.
(c) FSH stimulates the Sertoli cells to synthesize testosterone receptors.
(d) Spermatogenesis can only take place at temperatures of 37°C or above.
(e) Spermatozoa only complete their second meiotic division when they reach the epididymis.

77. Concerning spermatozoa:

(a) The front of the nucleus is surrounded by a giant lysosome.
(b) Mitochondria are found mainly in the head of the sperm.
(c) The tail contains microtubules derived from a centriole.
(d) The spermatozoon is only motile after maturation in the epididymis.
(e) The end-piece forms the majority of the spermatozoon's length.

78. An elderly man presents with recurrent urinary tract infections and on questioning he admits he has difficulty urinating:

(a) He probably has carcinoma of the prostate gland.
(b) Prostatic cancer usually develops in the larger peripheral zone glands.
(c) hCG is a useful tumour marker for prostatic cancer.
(d) Prostatic cancer can be treated by transurethral resection of the prostate.
(e) Prostate cancer is known to metastasize to bones, where it forms characteristic osteosclerotic lesions.

79. A 20-year-old man presents with a lump on the surface of his right testicle:

(a) If the lump is malignant it is probably a seminoma.
(b) If the lump is a teratoma it is probably benign.
(c) Teratomas commonly spread to the lungs.
(d) Seminomas are very sensitive to radiotherapy.
(e) 50% of testicular tumours are derived from the germ cells of the testes.

80. A 16-year-old boy presents with a swollen, tender, and painful right testicle:

(a) This is consistent with epididymo-orchitis.
(b) A history of recent ballet lessons is relevant.
(c) This condition can be diagnosed from the history and treated with antibiotics.
(d) This history is consistent with a varicocoele.
(e) Varicocoeles can reduce fertility.

81. On a well baby check a transluminable lump is found in the scrotum:

(a) This condition is much more common on the left of the scrotum.
(b) If the testis is not distinguishable from the lump, it is probably a hydrocoele.
(c) Congenital hydrocoeles are caused by absorption of amniotic fluid.
(d) The majority of congenital hydrocoeles will resolve without treatment.
(e) If a hydrocoele develops in an adult, the entire testicle should be removed.

82. A 7-year-old boy presents with a non-retractile foreskin:

(a) This is called paraphimosis.
(b) The boy will need a circumcision.
(c) The foreskin usually becomes retractile by 5 years of age.
(d) Before performing a circumcision it is important to check for hypospadias.
(e) Infection of the glans with *Candida albicans* is called balanitis xerotica obliterans (BXO).

Chapter 14—The Process of Reproduction

83. Concerning fertilization:

(a) It usually occurs in the uterus.
(b) The sperm are capable of fertilization as soon as they enter the uterus.
(c) The oocyte only completes the first meiotic division when the sperm penetrates the plasma membrane.
(d) The sperm penetrates the zona pellucida by the secretion of acrosin from the acrosome.
(e) Polyspermy (repeated fertilizations) is prevented by depolarization of the oocyte membrane and changes in the zona pellucida.

84. Hormonal contraception:

(a) Oestrogen is used as a hormonal contraceptive agent on its own.
(b) Progesterone can prevent implantation of the blastocyst.

(c) The main contraceptive action of oestrogen is to inhibit follicle development.
(d) Both oestrogen and progesterone inhibit LH to prevent ovulation.
(e) Progesterone directly inactivates sperm.

85. Concerning the contraceptive pill:

(a) The combined contraceptive pill must be taken without breaks to be effective.
(b) The combined pill contains oestrogens and progesterone, both of which suppress ovulation.
(c) The combined pill causes fewer adverse side effects than the progesterone-only pill.
(d) Using the combined pill reduces the risk of ovarian cancer.
(e) Small amounts of irregular bleeding in the first month requires urgent referral to a specialist.

86. Early fetal development:

(a) At ovulation, the oocyte has completed its meiotic divisions.
(b) For the first 4 days after conception the zona pellucida prevents an increase in embryo size.
(c) After the first mitotic division the cells are called blastomeres.
(d) Compaction is the process by which an inner and outer layer of cells is formed.
(e) The zona pellucida is shed from the morula when it enters the uterus.

87. Implantation:

(a) At the point of implantation the embryo is called a blastocyst.
(b) The blastocyst is formed from two layers and a fluid filled cavity.
(c) The embryoblast splits into two layers, one of which has many nuclei but no dividing cell membranes.
(d) The cytotrophoblast initially invades the functional endometrium.
(e) The blastocyst normally implants with the embryoblast closest to the uterine cavity.

88. Development of the placenta:

(a) When the first finger-like projections of the blastocyst develop spaces filled with maternal blood they are called primary chorionic villi.
(b) The secondary chorionic villi invade the myometrium.
(c) The lacunar spaces are eventually supplied by the spiral arteries.
(d) The tertiary villi contain fetal blood vessels.
(e) Before birth, the maternal and fetal blood are separated by just three layers of cells.

89. Concerning the placenta:

(a) It is derived mainly from maternal tissue.
(b) The chorionic villi secrete hCG.
(c) It is expelled from the uterus during the second stage of labour.
(d) It contains a large space in which maternal and fetal blood can mix.
(e) It is divided into cotyledons by fibrous septa from the fetal side.

90. Regarding maternal adaptations to pregnancy:

(a) The uterine myometrial cells hypertrophy.
(b) Initial adaptations by the circulatory system are mostly due to an increase in blood volume.
(c) Blood pressure rises throughout pregnancy.
(d) The maternal metabolism uses more glucose, to conserve fat for storage.
(e) The immune system becomes more active to prevent the risk of infection.

91. Labour and birth:

(a) Progesterone inhibits contractions of the uterus.
(b) When the baby's head is first seen, it usually faces the mother's thigh.
(c) The pain of early labour is caused by stretching of the cervix.
(d) Stimulation of the cervix causes oxytocin release.
(e) Oxytocin stimulates the release of locally acting thromboxanes that cause contractions.

92. Regarding lactation:

(a) Suckling during breastfeeding stimulates the secretion of oxytocin but inhibits prolactin secretion.
(b) Colostrum has a lower fat content than normal maternal milk.
(c) Oxytocin initiates the production of milk.
(d) During pregnancy, oestrogen inhibits the production of milk.
(e) Prolactin inhibits LH and FSH release, lowering fertility.

93. A young woman had her last period 7 weeks ago; she now has abdominal pain; an ectopic pregnancy is suspected:

(a) Raised hPL would suggest that this could be an ectopic pregnancy.
(b) In ectopic pregnancy the embryo almost always implants in a fallopian tube.
(c) Vaginal bleeding would alter this diagnosis.
(d) Pelvic inflammatory disease predisposes to ectopic pregnancy because the zygote travels more slowly in the damaged fallopian tube.
(e) Rupture almost always causes a life-threatening intra-pelvic haemorrhage.

94. A pregnant woman of 34 weeks gestation presents with a headache and a blood pressure of 156/104:

(a) She probably has eclampsia.
(b) Her urine probably has more than 300 mg/L protein.
(c) Antihypertensive treatment could cure this condition.
(d) She should be treated as an emergency.
(e) Magnesium sulfate would be an appropriate treatment.

Multiple-choice Questions

95. **A 6 week pregnant woman presents with vaginal bleeding, a blood pressure of 142/98 and severe vomiting most mornings:**

   (a) A likely diagnosis is a hydatidiform mole.
   (b) This diagnosis could be largely excluded if she has high levels of hCG.
   (c) This diagnosis would be corroborated by an abnormally large uterus.
   (d) An ultrasound scan could confirm this diagnosis.
   (e) The woman has an increased risk of developing choriocarcinoma.

96. **A young couple present in clinic with a failure to conceive after a year of unprotected sexual intercourse:**

   (a) The cause of infertility will almost certainly be an abnormality in the woman.
   (b) Female infertility is usually caused by amenorrhoea or abnormalities of the fallopian tubes.
   (c) The blood levels of several hormones are routinely investigated for infertility.
   (d) In-vitro fertilization (IVF) has an 80% success rate per cycle.
   (e) IVF is the main fertility treatment for women with obstructed fallopian tubes.

97. **Concerning male and female sterilization:**

   (a) In a vasectomy the epididymis is separated from the testis.
   (b) Reliable contraception occurs more rapidly with a vasectomy than tubal ligation.
   (c) Vasectomies are more effective at preventing pregnancy than tubal ligation.
   (d) Tubal ligation is normally performed by laparotomy.
   (e) Both procedures are irreversible.

98. **A 28-year-old woman is 8 weeks pregnant when she experiences a small amount of bleeding from her vagina. She presents in floods of tears saying she has had a miscarriage:**

   (a) She is probably right.
   (b) An amniocentesis should be performed to assess the state of the baby.
   (c) If her cervix is dilated then urgent medical treatment is needed to prevent spontaneous abortion.
   (d) The spontaneous abortion probably occurred because of a fetal abnormality.
   (e) Therapeutic abortion can legally be performed up to 24 weeks gestation.

Chapter 15—Common Presentations of Endocrine and Reproductive Disease

99. **The following conditions can cause weight loss:**

   (a) Cushing's syndrome.
   (b) Diabetes mellitus.
   (c) Addison's disease.
   (d) Amenorrhoea.
   (e) Hypothyroidism.

Chapter 17—Investigations and Imaging

100. **Concerning the investigation of endocrine pathology:**

   (a) Thyroid hormone levels can be measured directly using ELISA.
   (b) A stimulation test is used to investigate a deficiency of hormone levels.
   (c) The triple stimulation test is used to investigate the function of the adrenal cortex.
   (d) The water deprivation test is used to investigate diabetes mellitus.
   (e) An MRI scan involves the use of X-rays.

Short-answer Questions

1. Describe the mechanism of action of a G-protein linked receptor.

2. Compare the development of the anterior and posterior pituitary gland.

3. What is negative feedback in respect of control of hormone secretion? Draw a flow diagram to show the negative feedback inhibition involved in cortisol release.

4. Describe the synthesis of thyroid hormones.

5. Explain the actions of parathyroid hormone, vitamin D, and calcitonin.

6. Briefly describe the divisions of the adrenal glands and the hormones they secrete.

7. What is the role of progesterone in pregnancy?

8. Draw and label a diagram of a mature spermatozoon.

9. Discuss the neuroendocrine control of lactation.

10. Describe the endometrial changes that occur over a 28 day menstrual cycle if fertilization does not occur.

11. A 15-year-old boy is concerned about his short height. Describe how growth hormone regulates growth and determines final height.

12. A 49-year-old woman has gone through the menopause and wants to know about hormone replacement therapy (HRT). List reasons for prescribing HRT to women during and after the menopause and the type of HRT used.

13. A 19-year-old woman has recently had unprotected sex with a man she met on holiday. She has developed lower abdominal pain and tenderness. How is pelvic inflammatory disease (PID) caused, diagnosed and treated?

14. A 74-year-old man with a fractured neck of femur is investigated for secondary hyperparathyroidism. What is this condition and how is it caused?

15. A 30-year-old man presents with a scrotal lump. List five conditions that this could be and discuss how they can be differentiated on examination.

16. A 25-year-old woman wants to go on the 'Pill' but doesn't know which type to use. How does the progestogen-only pill provide contraceptive protection? List its advantages and disadvantages compared with the combined oral contraceptive pill.

17. An abnormality of the adrenal gland is seen on a CT scan of a 65-year-old man. In preparation for the radiologist's report, what are: Addison's disease, Cushing's syndrome, Cushing's disease and Conn's syndrome?

18. A thyroid function test of a 54-year-old woman gives the following results:

Thyroid function test results

Hormone	Result	Normal levels
TSH	7.8 mU/L	0.5–5.7 mU/L
T_4	55 nmol/L	70–140 nmol/L
T_3	0.8 nmol/L	1.2–3 nmol/L

What type of thyroid disease is this and what symptoms would you expect?

19. A 78-year-old woman presents with tiredness, polyuria and weight loss. State what disease this is likely to be. How is it diagnosed and what are the treatment options?

20. It is time to witness your first delivery, but the midwife expects you to know what is going on. Describe the three stages of labour.

Essay Questions

1. Discuss the causes of thyroid disease.

2. Outline the embryological development of the reproductive tract in both males and females, paying particular attention to the hormones involved.

3. Describe the regulation of blood glucose.

4. Describe the hormones secreted by the different regions of the adrenal gland. Explain the effects of deficiency or excess of adrenal hormones.

5. Draw a diagram to show the changes in blood levels of oestrogen, progesterone, LH and FSH during the menstrual cycle. Discuss how these changes are brought about.

6. Describe how the combined pituitary test is used to evaluate pituitary hormone secretion. Draw a graph (hormones measured in arbitrary units) showing the expected result if the patient has a prolactinoma.

7. List the functions of the placenta and the features that allow it to perform these functions.

8. Why is calcium homeostasis important and how are its plasma concentrations controlled?

9. Outline the sequence of events in spermatogenesis and explain its hormonal control.

10. Using diagrams, compare the secretion and action of polypeptide and steroid hormones.

MCQ Answers

1. (a) True—they can cross cell membranes.
   (b) False—they are generally too large.
   (c) True—the larger proteins are called prohormones.
   (d) True—for example, thyroid hormones and adrenaline.
   (e) False—steroid hormones are secreted immediately, but polypeptide hormones are stored in vesicles.

2. (a) True—as do insulin-like growth factors.
   (b) True—insulin is a polypeptide, so it cannot cross the cell membrane.
   (c) False—adrenaline acts via G-protein receptors.
   (d) False—as the name suggests, it is the tyrosine side chains that are phosphorylated.
   (e) False—that is the action of an intracellular receptor, instead it activates preformed proteins.

3. (a) True—they bind many polypeptides that cannot cross this membrane.
   (b) False—it is cAMP that is the common second messenger.
   (c) True—hence the name; GDP is released when the receptor is active.
   (d) False—the G-protein leaves the receptor protein to split and activate effectors.
   (e) True—it is a cell-surface receptor with glycoprotein binding sites.

4. (a) False—prostaglandins act locally, usually by diffusion through extracellular fluid.
   (b) True—arachidonic acid is a phospholipid.
   (c) False—they are synthesized by the lipoxygenase pathway.
   (d) False—most cells in the body secrete prostaglandins.
   (e) False—they are released immediately after synthesis.

5. (a) False—they often need special transporter proteins.
   (b) False—individual hormones use similar receptors all over the body.
   (c) True—negative feedback does not have to be directly hormonal.
   (d) False—feedback is used in many different cells throughout the body.
   (e) True—it is only the feedback to oestrogens and progesterones that changes.

6. (a) True—it forms a 'V' shape beneath the thalamus.
   (b) False—the hormones secreted by the hypothalamus act on the anterior pituitary gland.
   (c) False—they are polypeptides and modified amino acids.
   (d) False—ACTH is an anterior pituitary hormone, so the stimulation is hormonal via the portal veins.
   (e) True—nerve cells are derived from the neuroectoderm.

7. (a) True—in response to GnRH.
   (b) False—that is the posterior pituitary; the anterior pituitary comes from the endodermal Rathke's pouch.
   (c) True—it is directly below the optic chiasma.
   (d) True—it does.
   (e) False—ACTH regulates aldosterone, but angiotensin II is the main regulator.

8. (a) False—aldosterone is from the adrenal cortex; the posterior pituitary secretes ADH, which regulates fluid balance.
   (b) True—it secretes oxytocin to stimulate milk ejection.
   (c) True—ADH and oxytocin are both made of 9 amino acid residues; they are secreted in the hypothalamus.
   (d) True—this layer is called the diaphragma sellae.
   (e) True—it is made of neural tissue with these glial cells.

9. (a) False—it is the peripheral vision that is lost: a bitemporal hemianopia.
   (b) True—functioning prolactinomas are the most common pituitary adenomas.
   (c) True—they grow to a large size causing hypopituitarism.
   (d) False—a CT scan of the pituitary fossa is needed.
   (e) True—the pituitary gland can be accessed through the nose leaving no visible scar.

10. (a) True—this is their primary function.
    (b) False—thyroglobulin is the stored form found in the follicles; thyroid-binding globulin is the relevant plasma protein.
    (c) False—most tissues can *remove* an iodine from T_4 to make T_3.
    (d) True—thyroid hormones readily cross the cell membrane.
    (e) True—these hormones are made from tyrosine.

MCQ Answers

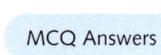

11. (a) True—as its name suggests, TSH stimulates the thyroid gland.
    (b) False—it is the arteries that are closely related to these nerves.
    (c) True—calcitonin is released from the parafollicular cells.
    (d) True—this is why it moves with the larynx on swallowing.
    (e) False—it is only found in about 10% of people.

12. (a) True—it is about three times more common in women than men.
    (b) False—most obesity has no known cause.
    (c) False—it causes cretinism.
    (d) True—ablation of the thyroid gland by surgery or radioactive iodine often results in hypothyroidism with time.
    (e) False—it should be treated with thyroxine (T_4) to return TSH levels to normal.

13. (a) True—the autoantibodies stimulate the TSH receptors.
    (b) True—she may suffer from heat intolerance causing her to wear few clothes despite cold weather.
    (c) True—TSH will be low despite raised T_3 and T_4.
    (d) False—thyroid carcinoma rarely causes hyperthyroidism.
    (e) False—carbimazole is used to treat hyperthyroidism.

14. (a) True—these cells would be seen on fine-needle aspiration.
    (b) True—these can secrete ectopic ACTH.
    (c) False—thyroid carcinoma usually causes hypothyroidism.
    (d) True—the 5-year survival is very low.
    (e) False—it invades locally and rarely causes distant metastases.

15. (a) False—this is the description of a thyroglossal cyst.
    (b) False—they will be lowered due to the excess thyroid hormones.
    (c) True—especially atrial fibrillation.
    (d) True—thyroid carcinoma cannot be excluded by history, examination or blood tests.
    (e) False—loss of the parathyroid glands can cause chronic hypocalcaemia.

16. (a) True—all steroid hormones are derived from cholesterol.
    (b) False—that is the adrenal medulla, the cortex is from surrounding mesenchyme.
    (c) False—that is the thyroid gland; the adrenal cortex is arranged in cords.
    (d) False—CRH stimulates the anterior pituitary to release ACTH, which stimulates the adrenal cortex.
    (e) True—remember GFR—the middle layer secretes cortisol.

17. (a) False—it acts mainly on the adrenal cortex as suggested by its name.
    (b) True—CRH is a hypothalamic stimulant, and cortisol has a negative feedback effect.
    (c) False—it is secreted by the corticotrophs, which are found in the anterior pituitary gland.
    (d) True—ACTH has a very marked circadian rhythm that peaks in the early morning.
    (e) False—it stimulates adrenal androgen release (e.g. in congenital adrenal hyperplasia).

18. (a) False—it inhibits growth hormone release.
    (b) True—to cope with the stress.
    (c) True—adrenaline and cortisol inhibit the actions of insulin to raise blood glucose
    (d) False—cortisol allows the body to cope with stresses.
    (e) False—it is the sympathetic nervous system that is activated.

19. (a) True—they have similar actions to the sympathetic nervous system.
    (b) True—the sympathetic neurons secrete mainly noradrenaline while the adrenal gland secretes mainly adrenaline.
    (c) True—so that cardiac output increases.
    (d) False—they raise blood glucose; only insulin can lower blood glucose.
    (e) False—they are secreted by cells that are equivalent to postganglionic sympathetic neurons.

20. (a) False—Cushing's disease is caused by a pituitary tumour. Exogenous corticosteroids may cause a presentation similar to Cushing's syndrome.
    (b) True—corticosteroids raise blood glucose.
    (c) False—he will probably have gained weight.
    (d) False—his skin will probably be thin and prone to bruising.
    (e) False—long-term steroid therapy must never be stopped suddenly as ACTH would be suppressed.

21. (a) False—it is a deficiency of cortisol and aldosterone.
    (b) True—due to the excess ACTH secretion.
    (c) True—this is a stimulation test using synthetic ACTH.
    (d) False—it is completely irrelevant.
    (e) True—he will probably need life long cortisol replacement.

22. (a) True—it is a disease of the adrenal gland, so it is a primary disease.
    (b) True—aldosterone increases potassium excretion but retains sodium.

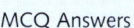

(c) False—it is an adenoma of the adrenal cortex.
(d) False—it is a rare cause of hypertension.
(e) False—renin and potassium levels should also be measured.

23. (a) True—severe hypertensive episodes like this need investigating.
(b) False—it is found in the adrenal medulla and secretes catecholamines.
(c) False—the phaeochromocytoma should be removed as soon as possible, though this is a dangerous operation.
(d) False—adrenaline causes vasoconstriction which raises blood pressure.
(e) True—adrenaline causes pupil dilation.

24. (a) True—especially IDDM.
(b) True—in response to the high blood glucose.
(c) True—this acts to lower blood glucose.
(d) False—these secrete glucagon; the β-cells secrete insulin.
(e) False—in most people, insulin secretion is always maintained at basal levels.

25. (a) True—it is located on the back wall of the abdomen with the duodenum behind the peritoneum.
(b) False—it secretes digestive enzymes via the pancreatic ducts, but pancreatic hormones enter the blood directly.
(c) True—it passes over the left kidney, under the stomach, to touch the spleen with its tail.
(d) True—the ventral and dorsal buds.
(e) False—the hormone-secreting cells are sparsely innervated since their activity is mainly regulated by blood glucose levels.

26. (a) False—this is the lower end of the normal glucose range.
(b) True—to raise blood glucose.
(c) False—it is released by the endocrine pancreas.
(d) False—it stimulates adrenaline release, which also raises blood glucose.
(e) True—glucose is the main regulator of insulin and glucagon release.

27. (a) False—it is a polypeptide; it is glycogen that is made from glucose molecules.
(b) False—it stimulates the secretion of insulin.
(c) True—it prevents glucose uptake and use.
(d) True—this allows insulin to act unopposed after a meal.
(e) True—it causes large molecules (e.g. glycogen) to be broken down, whilst preventing them from being synthesized.

28. (a) True—this is a common presentation.
(b) False—glycosuria is a poor diagnostic measure; his fasting blood glucose should be measured on two occasions.
(c) False—weight loss and polyuria would be expected.
(d) False—his young age suggests that he has IDDM and will almost certainly need insulin injections.
(e) False—in IDDM diabetic complications take several years to develop so the appointment does not need to be urgent.

29. (a) True—this is a presentation of NIDDM.
(b) True—though it can only be a presenting feature in NIDDM.
(c) True—good glucose control also helps prevent retinopathy.
(d) True—this is the hyalinization of arterioles.
(e) False—glycosuria is not a good diagnostic measure for diabetes.

30. (a) False—the threshold for glycosuria is very variable.
(b) True—though NIDDM can be present without glycosuria.
(c) False—he may well need insulin for good glucose control.
(d) False—syndrome X is hypertension, obesity and insulin resistance.
(e) False—these are found in IDDM, but NIDDM is usually due to insulin resistance.

31. (a) False—it is a hormone secreted by fat cells.
(b) True—it permits the synthesis of GnRH in the hypothalamus.
(c) True—it causes satiety (fullness).
(d) False—leptin must be injected as it is a protein.
(e) False—it is incredibly rare.

32. (a) False—it is a modified amino acid made from tryptophan.
(b) True—found at the back of the 3rd ventricle.
(c) False—melatonin is not the same as melanin.
(d) False—it acts to reset the suprachiasmatic nucleus of the hypothalamus.
(e) False—melatonin is secreted in response to the dark when the retina is not stimulated.

33. (a) False—the tumour is in an islet of Langerhans within the pancreas.
(b) False—it is a very rare condition.
(c) True—it is called a gastrinoma.
(d) False—it stimulates the parietal cells to secrete acid.
(e) True—about 30% of gastrinomas are caused by MEN I.

MCQ Answers

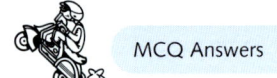

34. (a) True—it is also considered to be a hormone.
    (b) True—it acts to raise blood volume.
    (c) True—it forms angiotensin I from angiotensinogen.
    (d) False—renin secretion increases due to the loss of negative feedback.
    (e) True—via angiotensin II.

35. (a) False—increases water reabsorption; its name gives the answer
    (b) True—the neurons in the hypothalamus and posterior pituitary.
    (c) False—it has this action, but on the collecting duct.
    (d) True—it lowers blood osmolarity by retaining water.
    (e) False—it can cause vasoconstriction following severe hypotension.

36. (a) True—conversely, severe haemorrhage causes hypotension.
    (b) False—increased peripheral resistance can raise blood pressure.
    (c) True—causes vasoconstriction via angiotensin II.
    (d) False—has no effect without the presence of renin.
    (e) True—causes vasoconstriction.

37. (a) True—this is a good diagnostic indicator of diabetes mellitus.
    (b) False—but hypercalcaemia is associated with polyuria.
    (c) True—this is a sign of the resulting dehydration.
    (d) False—this causes urinary frequency, but the volume of urine remains relatively constant.
    (e) False—it is the water deprivation test that is used.

38. (a) False—there are two pairs of parathyroid glands.
    (b) False—they only secrete parathyroid hormone.
    (c) False—they are derived from APUD cells.
    (d) True—it raises calcium levels.
    (e) False—their blood supply is mainly from the arteries supplying the thyroid.

39. (a) True—over 99%, in fact.
    (b) True—this is true in all types of muscle throughout the body.
    (c) False—it is a rise in calcium that stimulates this process.
    (d) True—so excess osteoclast activity can weaken bones.
    (e) False—it acts via parathyroid hormone.

40. (a) True—it has the same characteristics as a steroid hormone.
    (b) True—e.g. the disease rickets.
    (c) False—its main site of action is the gastrointestinal tract, where it stimulates calcium absorption.
    (d) False—it is ultraviolet light.
    (e) False—it is activated from 25-hydroxyvitamin D_3.

41. (a) False—the four small parathyroid glands are located posterior to the thyroid gland.
    (b) True—it raises blood calcium in response.
    (c) False—it is a polypeptide.
    (d) False—that is the action of vitamin D, instead it acts on the kidneys and bones.
    (e) True—PTH stimulates the activation of vitamin D.

42. (a) True—as their name suggests, these are found next to the follicles.
    (b) True—it has the opposite role to that of parathyroid hormone.
    (c) False—excess or absence have no clinical effects on calcium regulation.
    (d) True—it is.
    (e) True—they secrete the characteristic polypeptides.

43. (a) True—remember 'bones, stones, abdominal groans and psychic moans'.
    (b) False—secondary hyperparathyroidism is a result of hypocalcaemia.
    (c) True—this is a relatively common cause.
    (d) True—chronic hypercalcaemia is associated with renal failure.
    (e) True—though parathyroid hyperplasia is more common.

44. (a) False—it is from the anterior pituitary gland.
    (b) True—all anterior pituitary hormones are polypeptides or glycoproteins.
    (c) False—it acts mostly via insulin-like growth factors.
    (d) True—it follows a circadian rhythm and is stimulated by sleep.
    (e) False—growth hormone secretion continues throughout life to maintain tissues.

45. (a) False—the liver is the main site of synthesis, though other tissues are important too.
    (b) True—growth hormone is stimulated during sleep.
    (c) False—they have the opposite action on glucose metabolism to that of insulin.
    (d) False—they inhibit growth hormone release by negative feedback.
    (e) True—they act on similar receptors to insulin.

46. (a) True—the jaw is not a long bone, so it can continue to grow with excess growth hormone.
    (b) False—the growth plates of his long bones will have fused.
    (c) False—it is called acromegaly in adulthood.
    (d) False—he has an increased risk of insulin resistance.
    (e) True—this can cause visual disturbances.

MCQ Answers

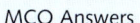

47. (a) True—this is a polypeptide and it is also secreted from APUD cells, making it more common as an ectopic hormone.
    (b) False—ectopic hormones are usually polypeptides as these only require the activation of a single gene. Cortisol would cause hyperglycaemia anyway.
    (c) True—this is a polypeptide.
    (d) False—this is a steroid.
    (e) False—it is the deficiency of ADH that causes diabetes insipidus.

48. (a) False—the adrenal glands are posterior in the abdomen and endocrine tumours do not need to grow large to cause symptoms.
    (b) True—these are both features of MEN II and he has multiple tumours at a young age.
    (c) False—these are a feature of MEN I.
    (d) False—it will almost certainly have a medullary pattern.
    (e) True—MEN syndromes often have an autosomal dominant inheritance.

49. (a) True—as do most of the internal genitalia.
    (b) False—this gene is found on the Y chromosome.
    (c) True—the labioscrotal folds.
    (d) True—the testes retain their abdominal blood supply when they pass through the inguinal canal.
    (e) False—they originate outside the embryo in the yolk sac and migrate into the fetal gonads.

50. (a) True—the indifferent gonads.
    (b) False—MIS causes regression of female ducts.
    (c) True—this division is completed just before ovulation; the second division is only completed on fertilization.
    (d) True—this is called the indifferent stage of development.
    (e) False—it develops from the mesenchyme.

51. (a) False—girls typically go through puberty earlier than boys.
    (b) True—it also stimulates fusion of the growth plates.
    (c) True—puberty is the reactivation of GnRH secretion.
    (d) False—it has started at 4 mL.
    (e) False—this is called gonadarche; adrenarche is when the adrenal glands start secreting androgens.

52. (a) True—47 kg is the average weight at menarche, and this is relatively constant.
    (b) False—it is a relatively late sign; breast budding or the growth spurt are noticed first.
    (c) True—in females, adrenal androgens are an important androgen source.
    (d) False—it occurs 10 months later, on average.
    (e) False—they mature from primordial follicles into primary follicles.

53. (a) True—they are held by the mesovarium and ligaments.
    (b) False—no new follicles are formed after birth.
    (c) True—but only one reaches full maturity.
    (d) True—the oocyte bursts out of the ovary into the peritoneal cavity. It remains close to the ovarian surface.
    (e) True—this is the site of the mesovarium.

54. (a) True—they are surrounded by the fold of peritoneum that makes up the broad ligament.
    (b) False—fertilization normally occurs in the ampulla of the fallopian tubes.
    (c) True—there are anastomoses between these vessels.
    (d) True—though the cervix prevents easy access.
    (e) True—these are mesenchymal structures.

55. (a) False—it is a section of the uterus.
    (b) False—it is the external os.
    (c) True—though squamous metaplasia can occur.
    (d) False—it dilates to 10 cm.
    (e) False—it is the action of oestrogens and relaxin that do this.

56. (a) False—it is the commensal lactobacilli bacteria that maintain the acidic environment.
    (b) False—it is the uterine lining (the functional endometrium) that is shed during menstruation.
    (c) False—the lining is a stratified squamous epithelium.
    (d) True—it is located between the bladder and rectum.
    (e) True—it is an endodermal structure.

57. (a) False—the glandular tissue is derived from apocrine sweat glands.
    (b) True—it is only at puberty that the female breast changes.
    (c) False—there about 15–20 lactiferous ducts opening onto each nipple.
    (d) True—if it becomes tethered it is a sign of neoplasia.
    (e) False—oestrogen stimulates the growth of the lobules and acini.

58. (a) True—it is a female sex steroid.
    (b) False—it crosses cell membranes to act on intracellular receptors.
    (c) True—it is secreted during development and puberty.
    (d) False—it is secreted along with progesterone.
    (e) True—feedback is normally negative, but it becomes positive before ovulation.

MCQ Answers

59. (a) True—progesterone is mainly secreted in the second half of the menstrual cycle.
    (b) True—it inhibits contractions in the uterus, but it also causes oesophageal reflux and urinary incontinence.
    (c) False—it is the corpus luteum.
    (d) False—it only requires two steps from cholesterol, via pregnenolone.
    (e) False—that is oestrogen's main role; progesterone maintains the endometrium and promotes secretion.

60. (a) False—LH peaks just before ovulation.
    (b) True—this positive feedback causes the LH and FSH surge.
    (c) False—this hormone is only found during pregnancy.
    (d) True—it is secreted in response to GnRH.
    (e) False—it is LH that stimulates progesterone release.

61. (a) True—though several follicles and oocytes begin to develop each month.
    (b) True—the primordial follicles are present at birth.
    (c) False—thecal cells secrete androgens that granulosa cells convert to oestrogens.
    (d) True—the fluid is under pressure to allow the oocyte to be ejected.
    (e) True—but only one reaches full maturity.

62. (a) False—during early pregnancy hCG is secreted by the chorionic villi, as its name suggests.
    (b) True—they secrete progesterone.
    (c) True—after fertilization hCG replaces the falling LH.
    (d) False—lutein means yellow in Greek and this is its characteristic colour.
    (e) False—only one follicle ovulates, so only one corpus luteum is formed.

63. (a) False—it begins to proliferate after menstruation under the influence of oestrogen.
    (b) False—they secrete this fluid in response to progesterone.
    (c) True—up to 80 mL is considered normal.
    (d) False—only the functional layer is shed.
    (e) False—it only causes ischaemia in the functional endometrium.

64. (a) False—cortisol levels are unaffected by POS.
    (b) True—it shows the multiple cysts.
    (c) False—that would suggest a prolactinoma.
    (d) True—this is a recognized symptom of the high androgen levels.
    (e) False—LH will be raised, but FSH should be low.

65. (a) False—that is endometritis.
    (b) True—this is an association of endometriosis.
    (c) False—laparoscopy is the best means of diagnosis.
    (d) False—polycystic ovarian syndrome is associated with hirsutism.
    (e) True—caused by rupture of the ectopic endometrial lining.

66. (a) False—but this could be acute pelvic inflammatory disease.
    (b) True—especially *Chlamydia* and *Gonococcus*.
    (c) False—it is more likely to be *Staphylococcus* or *Streptococcus*.
    (d) True—in about 10% of infections.
    (e) False—the IUD's location should be checked by feeling the threads and the PID should be treated with antibiotics.

67. (a) False—only very large pelvic masses are palpable on abdominal examination.
    (b) True—fluid appears dark on ultrasound scans.
    (c) False—this is the presentation of a benign cystic teratoma.
    (d) False—it is a brown (chocolate) cyst that suggests endometriosis leading to endometrioid tumours.
    (e) True—including cystadenomas and endometrioid tumours.

68. (a) False—herpes forms itchy, red blisters; it does not cause a vaginal discharge.
    (b) False—lactobacilli are normal residents of the vagina and they prevent infection.
    (c) True—this is the classic presentation of *Candida albicans* infection.
    (d) False—*Candida albicans* is commonly found in the vagina; it has purely overgrown.
    (e) True—it is the most common organism causing PID.

69. (a) True—they are the most common neoplasia of the female reproductive tract.
    (b) True—they are tumours of the smooth muscle cells.
    (c) False—it is oestrogen that mainly stimulates their growth.
    (d) False—these two conditions occur in different layers of the uterus.
    (e) True—in the absence of oestrogen they begin to regress and menorrhagia will no longer be a problem.

70. (a) True—this is the most probable diagnosis, but it cannot be assumed.
    (b) False—these mainly affect younger women.
    (c) True—this is caused by interference with the connective tissue.

(d) False—it suggests Paget's disease; it is another sign of malignancy.
(e) False—breast cancer can only be excluded by ultrasound scan with fine-needle aspiration.

71. (a) True—these are most common just before the menopause.
(b) True—peau d'orange is a classic sign of breast cancer.
(c) False—it is caused by interference with lymphatic drainage.
(d) True—*BRAC* mutations increase the risk of both breast and ovarian cancer.
(e) True—it increases the lifetime oestrogen exposure.

72. (a) False—51 years is the average age of menopause.
(b) False—she must use oestrogen and progestogen to protect her endometrium from carcinoma.
(c) True—oestrogen reduces the risk of osteoporosis.
(d) False—HRT reduces the risk of cardiovascular disease.
(e) False—they will be increased in the early stages of menopause due to reduced feedback of oestrogen.

73. (a) True—this is continuous until death.
(b) False—they are lined by Sertoli cells; the Leydig cells lie between the seminiferous tubules.
(c) True—this is the tissue cavity in which hydrocoeles develop.
(d) False—each lobe contains 1–4 seminiferous tubules.
(e) False—it usually drains into the inferior vena cava. It is the left testis that drains to the left renal vein.

74. (a) False—they have a comma shape.
(b) False—spermatogenesis only occurs in the testes.
(c) True—this tube is about 5 m long.
(d) False—it is within the scrotum next to the testes.
(e) False—sperm are stored in the epididymis.

75. (a) True—these are the erectile components.
(b) False—it is derived from the urogenital folds.
(c) True—remember 'point and shoot'.
(d) False—both fluids are transmitted by the urethra.
(e) False—hopefully not; it should only enter the vagina.

76. (a) False—the spermatogonia are the stem cells.
(b) True—they reach the lumen surface as they develop.
(c) True—this allows them to respond to testosterone that stimulates spermatogenesis.
(d) False—spermatogenesis requires temperatures below 37°C, this is why the testes are outside of the abdominal cavity.
(e) False—they complete meiosis as they become spermatids.

77. (a) True—this is called the acrosome.
(b) False—they are arranged in spirals in the middle piece.
(c) True—they are arranged in the characteristic 9+2 pattern.
(d) True—the membrane phospholipids are rearranged allowing motility.
(e) False—it is the principal-piece that is the longest.

78. (a) False—benign prostatic hypertrophy is much more common.
(b) True—while benign prostatic hypertrophy affects the periurethral glands.
(c) False—it is prostate-specific antigen (PSA) that is a useful marker.
(d) True—though the staging determines treatment.
(e) True—most bony metastases are osteolytic; osteosclerotic lesions suggest prostatic carcinoma.

79. (a) False—due to his young age it is likely to be a teratoma.
(b) False—male teratomas are almost always malignant.
(c) True—by vascular spread.
(d) True—while teratomas respond well to chemotherapy.
(e) False—97% are derived from the germ cells, including seminomas and teratomas.

80. (a) True—this is an infection of the testis and epididymis.
(b) True—any intensive exercise can predispose to testicular torsion.
(c) False—testicular torsion must be excluded surgically.
(d) False—varicocoeles are non-tender and they are felt in the spermatic cord.
(e) True—they raise the temperature of the testes.

81. (a) False—it is varicocoeles that are more common on the left.
(b) True—this is the most common testicular lump at all ages.
(c) False—they are caused by patency of the processus vaginalis.
(d) True—though they can need surgery if they do not.
(e) False—though it suggests there is an infection or malignancy.

82. (a) False—it is phimosis; paraphimosis is when the foreskin becomes stuck behind the glans.
(b) False—this is only needed if BXO is present.
(c) True—but it is a common disorder after 5 years of age.

261

(d) True—the foreskin is essential for the repair of hypospadias.
(e) False—it is called balanitis; the cause of BXO is not known.

83. (a) False—the ampulla of the fallopian tubes is the most common site.
(b) False—they must undergo capacitation.
(c) False—it completes the second meiotic division at this point.
(d) True—this is stimulated by binding to ZP3 receptors.
(e) True—the fast and slow block.

84. (a) False—oestrogen is only used with progesterone, but progesterone can be used alone. Oestrogen alone would cause endometrial hyperplasia.
(b) True—this is one of its main contraceptive effects.
(c) True—oestrogen inhibits FSH, which is required for follicle development.
(d) True—this is a contraceptive action that both hormones can cause.
(e) False—this is the action of a copper IUD.

85. (a) False—it is only taken for 21 days with a 7 day break; progesterone-only pills are taken continuously.
(b) True—both hormones inhibit LH secretion.
(c) False—it is more effective, but it causes more side effects.
(d) True—it also reduces the risk of endometrial carcinoma.
(e) False—this is a common side effect.

86. (a) False—the second meiotic division is only completed if the oocyte is fertilised.
(b) True—only after the zona pellucida is shed can the blastocyst begin to grow.
(c) True—the zygote splits into two blastomeres.
(d) True—the cells clump together and arrange themselves in these layers.
(e) False—the zona pellucida is shed from the blastocyst in the uterus.

87. (a) True—it has a blastocyst cavity, which distinguishes it from the morula.
(b) True—the trophoblast, embryoblast and blastocyst cavity.
(c) False—the trophoblast splits into the syncytiotrophoblast and cytotrophoblast.
(d) False—it is the syncytiotrophoblast that invades the functional endometrium.
(e) False—the embryoblast usually faces the endometrium.

88. (a) False—this is the lacunar phase; the primary villi are formed when the cytotrophoblast invades the syncytiotrophoblast.
(b) False—further invasion by the placenta is prevented by the decidua basalis. It does not invade past the functional endometrium.
(c) True—failure of this invasion predisposes to pre-eclampsia.
(d) True—the chorionic villi enter their tertiary phase when the fetal mesenchyme forms blood vessels.
(e) False—there are only two layers because the cytotrophoblast regresses.

89. (a) False—the chorionic villi and umbilical cord are fetal tissues.
(b) True—this is essential for maintaining the corpus luteum and progesterone in early pregnancy.
(c) False—expulsion of the placenta and membranes is the third stage of labour.
(d) False—the fetal and maternal blood are separated by the syncytiotrophoblast and fetal blood vessel walls.
(e) False—these septa come from the maternal tissue and are incomplete on the fetal side.

90. (a) True—this allows the uterus to enlarge and prepare for birth.
(b) False—blood volume rises mostly in the second half of pregnancy; stroke volume increases initially.
(c) False—blood pressure initially decreases, then rises back to normal levels.
(d) False—the metabolism relies on fat more than normal to spare glucose for the fetus.
(e) False—the immune system is less active.

91. (a) True—it prevents early expulsion.
(b) False—the head usually faces backwards, but then it turns to the thigh.
(c) False—the pain of early labour is caused by hypoxia of the uterus during contractions; stretching of the cervix usually causes pain in the second stage.
(d) True—this is called the Ferguson reflex.
(e) False—it is prostaglandins, especially PGE_2, that are stimulated by oxytocin.

92. (a) False—both prolactin and oxytocin are stimulated.
(b) True—this is why it doesn't appear white.
(c) False—it is prolactin that does this; oxytocin stimulates the ejection of milk.
(d) True—though many women express small amounts towards the end of pregnancy.
(e) True—breastfeeding has a contraceptive effect.

MCQ Answers

93. (a) False—it is a raised hCG that is used to diagnose pregnancy.
    (b) True—99% of ectopic pregnancies are in the fallopian tubes.
    (c) False—this would suggest that a tubular abortion of an ectopic pregnancy has occurred.
    (d) True—any disease process that slows the transport along the fallopian tube predisposes to ectopic pregnancies.
    (e) False—this is a rare complication.

94. (a) False—eclampsia is seizures following pregnancy induced hypertension or pre-eclampsia; it is pre-eclampsia that she probably has.
    (b) True—she has severe hypertension, so she will probably have proteinuria.
    (c) False—it may help prevent eclampsia, but the only cure is birth.
    (d) True—this is a very serious condition.
    (e) True—this should help bring her blood pressure down and reduce the risk of a seizure.

95. (a) True—she has early pre-eclampsia and severe morning sickness, so this is a likely diagnosis.
    (b) False—hCG levels are massively raised in the presence of hydatidiform moles.
    (c) True—another symptom of 'excessive pregnancy'.
    (d) True—it will show a snow-storm-like uterine cavity.
    (e) True—recurrent moles often become malignant.

96. (a) False—both men and women can be responsible for infertility.
    (b) True—these are the two most common causes.
    (c) True—including prolactin, oestrogen, progesterone, testosterone, LH and FSH.
    (d) False—it has a 20–30% success rate.
    (e) True—it is the only means of implanting an embryo in the uterus.

97. (a) False—it is the ductus deferens that is cut.
    (b) False—other contraceptive measures should be used for a couple of months after a vasectomy.
    (c) True—the failure rate for vasectomies is 10 times lower.
    (d) False—it is usually a laparoscopic procedure.
    (e) False—there is a 50% chance of being able to restore fertility.

98. (a) False—usually the bleeding is from other causes, but miscarriage is high on the list of differential diagnoses.
    (b) False—an ultrasound scan will be less harmful and more useful.
    (c) False—miscarriage is inevitable at this point; the fetus and afterbirth should be expelled and checked.
    (d) True—60% of miscarriages are due to fetal abnormalities, especially chromosome defects.
    (e) True—although the majority are performed before 14 weeks.

99. (a) False—associated with weight gain.
    (b) True—due to the inhibition of anabolism and unopposed catabolism.
    (c) True—due to the deficiency of cortisol.
    (d) False—this can be caused by weight loss.
    (e) False—it is hyperthyroidism that causes weight loss.

100. (a) True—ELISA is the main technique for measuring many hormone levels.
     (b) True—while suppression tests are used for excess secretion.
     (c) False—it is used to investigate hypopituitarism.
     (d) False—it is used to investigate diabetes insipidus.
     (e) False—CT scans use X-rays; MRI uses magnets and radio waves.

SAQ Answers

1. G-protein coupled receptors consist of two main elements: a glycoprotein receptor and an associated protein bound to GDP. The receptor spans the membrane with a hormone binding site on the extracellular surface and a G-protein binding site on the intracellular surface. Binding of the hormones stimulates a change in shape that affects the attached G-protein. The G-protein has two components: the α-subunit that binds GDP in the resting state and the βγ-complex that is bound to the α-subunit in the resting state. The change in shape of the receptor causes the α-subunit to exchange GDP for GTP. The G-protein then leaves the receptor and splits into the two subunits described above, both of which bind to effector proteins also found on the inside of the cell membrane. The effector proteins stimulate other molecules (e.g. ATP is converted to cAMP) that act as second messengers (see Fig. 1.9).

2. The anterior pituitary gland develops as an endodermal outgrowth of the primitive mouth called Rathke's pouch. This grows upwards until it meets, and fuses with, the down-growing infundibulum. The anterior pituitary is, therefore, composed of non-neural secretory epithelial tissue and it is not directly connected to the hypothalamus. It communicates with the hypothalamus through the portal veins, which replace the blood supply from the mouth. The posterior pituitary gland is derived from the neuroectoderm of the primitive brain tissue. It develops as an outgrowth from the hypothalamus called the infundibulum. Axons from the hypothalamus pass through the posterior pituitary to create a direct neuronal connection which is the main communication from the hypothalamus to the posterior pituitary (see Fig. 2.4).

3. Negative feedback is when the presence, or effects, of a hormone act on the cells in its regulatory pathway to suppress further secretion. In the endocrine system this negative feedback often acts on the hypothalamus and pituitary gland. For example, see Fig. 4.1.

4. T_3 and T_4 are derived from the amino acid tyrosine and iodine. Tyrosine is converted into the glycoprotein thyroglobulin, which is secreted into the follicle lumen. Iodine ions are actively transported into the follicular cells where they are oxidized into reactive iodine atoms that are also secreted into the follicle lumen. The iodine binds to thyroglobulin by the action of thyroperoxidase resulting in monoiodothyrosine and diiodotyrosine. These molecules are coupled together to form tri-iodothyronine and thyroxine. These are secreted by reabsorbing the thyroglobulin and breaking down this large molecule in lysosomes. The thyroid hormones are released. See Fig. 3.6.

5.

Actions of parathyroid hormone, vitamin D, and calcitonin

Hormone	Parathyroid hormone (PTH)	Vitamin D	Calcitonin
Secreted/activated in response to:	Low blood calcium	PTH	High blood calcium
Kidneys	Calcium reabsorbed; vitamin D activated	Calcium reabsorbed	Calcium excreted
Bones	Calcium released	Calcium trapped	Calcium trapped
Intestines	Negligible	Calcium absorbed	Negligible

6. The adrenal gland is divided into two regions: the adrenal cortex and the adrenal medulla. The adrenal cortex is further divided into three layers:
   - Zona glomerulosa—secretes mineralocorticoids (e.g. aldosterone).
   - Zona fasciculata—secretes glucocorticoids (e.g. cortisol).
   - Zona reticularis—secretes glucocorticoids and androgens (e.g. testosterone).

   The adrenal medulla secretes catecholamines (e.g. noradrenaline and adrenaline).

7. Progesterone induces the uterine glands to secrete nutrient-rich 'milk' for the developing embryo. It also prepares and maintains the endometrium for implantation. Progesterone secretion increases continually as pregnancy proceeds and is essential for its maintenance. After implantation its effects include:
   - The prevention of premature labour by inhibiting prostaglandin secretion, which stimulates contractions in the myometrium.
   - Promotes the storage of body fat.
   - Maintenance of the functional endometrium.
   - Physiological adaptation to pregnancy, e.g. changes in cardiovascular, renal and respiratory system.
   - Relaxation of smooth muscle throughout the body.

8. See Fig. 13.7.

9. The nipples are stimulated during suckling and this stimulus is conveyed to the hypothalamus via neural pathways. It causes the secretion of oxytocin from the posterior pituitary gland and prolactin from the anterior pituitary gland. Prolactin initiates the synthesis of milk in the acini once the plasma oestrogen levels

decline after pregnancy; it also maintains the secretory ability of the breast. Oxytoxin stimulates milk ejection by the contraction of the myoepithelial cells. Lactation is maintained by frequent suckling via a positive feedback loop. See Fig. 14.19.

10. There are three phases to the menstrual cycle in the endometrium:
    1. Menstrual phase (days 1–4)—the absence of progesterone causes spiral arteries to constrict and coil causing the functional endometrium to become ischaemic and necrotic. It is shed through the vagina, along with blood from the damaged blood vessels, in a process called menstruation.
    2. Proliferative phase (days 4–13)—cells in the basal epithelium proliferate in response to oestrogen to form a new functional endometrium with new spiral arteries. Endometrial glands are formed within this layer, but they do not secrete.
    3. Secretory phase (days 14–28)—progesterone causes the endometrial glands to enlarge and develop a corkscrew shape. They begin to secrete a glycogen-rich fluid in preparation for implantation. The endometrium continues to thicken and becomes oedematous. Towards the end of this phase, low progesterone levels cause the spiral arteries to contract causing ischaemia.

11. Growth hormone stimulates the production of insulin-like growth factors (IGFs) in many tissues, but especially the liver. These polypeptide hormones stimulate the growth of soft tissues and bones. They stimulate the uptake and anabolism of amino acids causing cell growth that also stimulates cell division. In the bones IGFs stimulate growth of the chondrocytes at the epiphyseal growth plates causing the bone to lengthen. The growth factors also stimulate fusion of this growth plate preventing further growth. Final height is determined by the rate of growth and the time at which these growth plates fuse.

12. Hormone replacement therapy (HRT) preparations are prescribed during and after the menopause to:
    - Treat menopausal symptoms such as hot flushes and sweating.
    - Protect against osteoporosis caused by chronic oestrogen deficiency.
    - Protect against cardiovascular disease caused by chronic oestrogen deficiency.
    - Protect against collagen loss, which can cause uterovaginal prolapse, immobility, muscle weakness, and skin wrinkling.

    In most women oestrogen can only be given with progesterone to prevent the risk of endometrial carcinoma. This causes withdrawal bleeding similar to that when using the pill. If the woman has had a hysterectomy then oestrogen can be given alone.

13. Pelvic inflammatory disease (PID) is an infection of the endometrium, fallopian tubes or ovaries. It is most commonly caused by sexually transmitted disease including *Chlamydia trachomatis* (60% of PID) or *Neisseria gonorrhoeae* (30% of PID). The remaining 10% comes from other routes:
    - Direct infection following trauma due to childbirth, surgical abortion or insertion of a coil.
    - Blood-borne infection (e.g. tuberculosis).
    - Transperitoneal infection (e.g. from appendicitis).

    It is diagnosed from the history and examination followed by screening the urine for signs of the infectious agent and taking vaginal and cervical swabs. It is treated with doxycycline and metronidazole.

14. Secondary hyperparathyroidism is a disorder where parathyroid hormone (PTH) secretion is elevated in response to persistent hypocalcaemia. Persistent hypocalcaemia and secondary hyperparathyroidism can be caused by:
    - Calcium malabsorption (e.g. vitamin D deficiency or coeliac disease).
    - Renal failure causing uncontrolled calcium excretion and a failure to activate vitamin D (called renal osteodystrophy).

15. Indirect hernia, testicular tumour (seminoma or teratoma), hydrocoele, varicocoele, epididymal cyst:
    - Indirect hernia arises in the abdomen so you cannot feel above the mass in the scrotum; it may be reducible or tender.
    - A mass that is solid and feels to be part of the testis is likely to be a testicular tumour.
    - A hydrocoele is cystic (i.e. translucent when a light is shone through it) and surrounds the testis (i.e. the testis is indistinguishable).
    - A mass caused by a varicocoele lies above the testis, feels like a 'bag of worms' and often reduces when the patient lies flat.
    - A mass caused by an epididymal cyst is small, firm, cystic and lies within the epididymis (i.e. it feels separate from the testis).

16. The progestogen-only pill does not reliably suppress ovulation; it provides contraceptive protection by:
    - Inhibiting the changes in the cervical mucus that normally occur around ovulation so that the passage of sperm is reduced.
    - Increasing the rate of ovum transport so it reaches the endometrium before implantation can take place.
    - Inhibiting endometrial proliferation so that implantation does not occur.

The progestogen-only pill does not cause an increased risk of cardiovascular disease, breast cancer or other side effects associated with oestrogen. It must be taken at the same time every day ±3 hours and it is still less effective than the combined pill, especially in younger women. It often causes irregular bleeding and may cause symptoms of premenstrual syndrome (PMS).

The combined pill carries the risks and side effects associated with oestrogen, but it is more effective, does not need to be taken at the same time each day, and reduces the risk of ovarian and endometrial cancer.

17. These four conditions are disorders of the adrenal cortex hormones:.
    - Addison's disease—a deficiency of glucocorticoids (cortisol) and mineralocorticoids (aldosterone) caused by destruction of the adrenal gland, often by an autoimmune process.
    - Cushing's syndrome—an excess of glucocorticoids (cortisol) caused by any disease causing a characteristic set of symptoms.
    - Cushing's disease—an ACTH-secreting pituitary adenoma that causes Cushing's syndrome along with hyperpigmentation.
    - Conn's disease—an adenoma of the zona glomerulosa causing primary hyperaldosteronism.

18. Raised TSH and low thyroid hormones indicates primary disease of the thyroid gland causing hypothyroidism. The symptoms are shown in Fig. 3.10.

19. It is probably non-insulin dependent diabetes mellitus (NIDDM). It is diagnosed by blood glucose levels above 7.8 mmol/L after an overnight fast on two occasions. There are three treatment options:
    - Diet alone.
    - Diet and oral hypoglycaemic agents.
    - Diet and injected insulin.

20. **First stage**—from the onset of labour until full cervical dilatation (10 cm). The uterine contractions become stronger and more frequent and they push the fetal head into the pelvis towards the cervix. The woman feels pain due to hypoxia of the uterus caused by occlusion of the blood vessels following the muscular contractions. The amniotic membrane often ruptures during this stage.

    Second stage—from full cervical dilatation until the birth of the baby. The fetal head descends sideways through the pelvis and it rotates 90° so that it faces the sacrum. Contractions continue and are assisted by voluntary pushing by the mother. Once the head is born it rotates back 90° to face the mother's leg and the rest of the baby is born shortly afterwards. The pain is most severe during this stage; it is caused by stretching of the cervix, vagina and perineum.

    Third stage—from birth of the baby until the delivery of the placenta and membranes. The entire placenta and decidua basalis detach and are expelled, causing haemorrhage from the ruptured blood vessels. The haemorrhage is stopped by the muscle fibre arrangements within the uterus. The contractions slowly subside once the afterbirth has been expelled.

Index

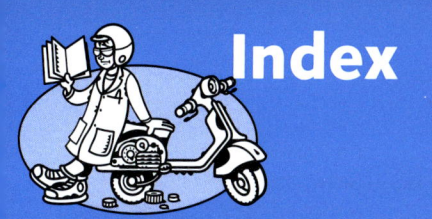

Page numbers in *italics* refer to figures and tables.

A

abacterial prostatitis 161
abdomen
  examination 219–20
  palpation and percussion *219*
abortion
  spontaneous 193
  therapeutic 178
  tubal 192
acarbose 68
ACE *see* angiotensin converting enzyme
acromegaly 100, *101*
  as cause of diabetes mellitus 64
ACTH *see* adrenocorticotrophic hormone
activin 124
Addison's disease 51–2
  ACTH stimulation test 51, 231
adenocarcinoma, fallopian tubes 131
adenohypophysis *see* anterior pituitary
adenomas
  adrenal cortex 51
  anterior pituitary 23–6
    functioning 23
    non-functioning 23, 25–6
  fallopian tubes 131
  ovary 129
adenomyosis, uterus 131
adenosarcoma, uterus 134
adenylate cyclase 10, 11
adipose tissue, endocrine role of 78
adrenal cortex
  development 42
  disorders 48–53
  hormones 8, 45–8
  microstructure 42–4, *44*
adrenal cortex insufficiency 51–3
adrenal cortical failure, acute 52–3
adrenal glands 41–55
  anatomy 42, *43*
  blood supply, nerves and lymphatics 42, *43*
  function tests 230–1
  in pregnancy 185–6
adrenal medulla
  development 42

disorders 54–5
  hormones 9, 53–4
  microstructure 44
adrenaline 12, 62
  ectopic secretion *104*
  secreting cells 44
  see also catecholamines
adrenarche 47–8, 112
adrenocorticotrophic hormone (ACTH) 19, 21, 22, *41*, 45, 46
  in adrenal cortex disorders 50, 50, 52, 53
  assays 228, 230
  ectopic production *104*, 105
  normal range 229
  in pregnancy 185, *186*
  stimulation test, Synacthen® 51, 231
  suppression test, dexamethasone 49, 230–1
adrenogenital syndrome 51
AFP (α-fetoprotein) 159
albumin
  and calcium 88
  and oestrogen 122
  and progesterone 123
  and testosterone 156
aldosterone 45, *46*, 81, 82, *83*, 85
  assays 230
  deficiency 51–3
  excess 49, 51
  normal range 229
  in pregnancy 186
  see also steroid hormones
α-fetoprotein (AFP) 159
amenorrhoea 135, 199–200
  primary and secondary 135
amine precursor uptake and decarboxylation (APUD) cells 73
amino acid metabolism, in pregnancy 186
amniotomy 189; 190
amplification 6–7
androblastoma 129
androgens
  adrenal 47
  adrenarche 47–8, 112

conversion to oestrogens 122, *123*
  embryological secretion 110
  excess 51
  in females 122, 123
  testicular 156–7
  see also testosterone
androstenedione 47
ANF (atrial natriuretic factor) 85
angiography 234
angiotensin converting enzyme (ACE) 84
  inhibitors 84, 85
angiotensin II 45, 81, 82, *83*, 84–5
  renin–angiotensin II system *83*, 84–5
anorgasmia 169–70, 203
anovulatory cycles 132
anterior pituitary
  development 17–18
  disorders 23–7
  hormones
    hypothalamic regulation 20, 21–2
    secreted by 20–2
    target organs 22
    see also specific hormones
  investigations 227–9
  microstructure 18–19
antidiuretic hormone (ADH) 20, 22, 81, 83–4
  deficiency 27, 84
  excess 27, 84, *104*
  investigations 229–30
anti-thyroglobulin (anti-TgAb) assay 230
anti-thyroid-peroxidase (anti-TPO) assay 230
apareunia 170
APUD cells 73
arginine vasopressin (AVP) *see* antidiuretic hormone (ADH)
assisted delivery 190
atherosclerosis 66–7
atrial natriuretic factor (ANF) 85
atrophic hypothyroidism, primary 37
autoantibody detection *see* thyroid autoantibody assays

267

Index

autocrine 3, *4*
autoimmune granulomatous orchiditis 159
autonomic (sympathetic/parasympathetic) nervous system 12, 42, *43*, 168

B

bacterial prostatitis 160–1
　chronic 161
bacterial vaginosis 140
balanitis 164
balanitis xerotica obliterans (BXO) 163
Bartholin's cyst 142
Bartholin's glands 120
　in sexual arousal 168
benign neoplasias *see* neoplasias (benign and malignant)
benign prostatic hypertrophy 161
biguanides 68
bimanual examination, vagina and uterus 220–1
biopsy 233–4
　endometrial 132
　vulval 142, 143
bitemporal hemianopia 26
blastocyte 172
　see also trophoblast
blood glucose *see* glucose
blood osmolarity 83–4
　normal range *233*
　and urine osmolarity 229, 230
blood pressure
　control 81–2, 84–5
　in pregnancy 184
　see also hypertension
blood vessels, ADH effects 83, 84
blood volume
　control 82, 84–5
　in pregnancy 184
body mass index (BMI) 206
body posture *215*
body weight *see* weight gain; weight loss
bone demineralization *see* osteomalacia; osteoporosis; renal osteodystrophy
bone effects
　calcitonin 92
　PTH 90
　vitamin D 91
Braxton Hicks contractions (false labour) 188

breast
　embryological development 112
　female 120–1
　　development in pregnancy 190–1
　　development in puberty 113
　　disorders 144–7
　　microstructure 121
　male, disorders 165, 200, *201*
breast abscess 144
breast carcinomas
　female 146–7, 175
　male 165
breast examination 221, *222*, *223*
breast lumps 144–7, 200
　types and features *144*, *223*
　see also breast carcinomas; breast examination
breast milk 192
　abnormal production *see* galactorrhoea
　hormonal control of secretion 191
breast tissue, accessory 144
breastfeeding 190–1, 192
breech presentation of fetus 187
Brenner tumour 129
broad ligament 118, 131
bromocriptine 25, 177
BXO (balanitis xerotica obliterans) 163

C

C peptide 60
caesarean section, indications for 190, 194
calcitonin 91–2
　actions, synthesis and receptors 91
　ectopic secretion *104*
　normal range *229*
　treatments 94, 95
calcium
　disorders of regulation 92–5, *104*
　hormones involved in homeostasis 89–92
　mechanisms involved in homeostasis 87–8
　normal distribution and movements 89
　normal range *233*
　role 87
Candida albicans
　female 141
　male 164

capillary blood spot testing 69
carbimazole 35
carbohydrate metabolism *see* glucose
carcinomas
　adrenal cortex 51
　female
　　breast 146–7, *147*, 175
　　cervix 139–40
　　fallopian tubes 131
　　ovaries 130–1
　　trophoblastic origin 195
　　uterus 134
　　vagina 142
　　vulva 142–3
　lung 51, *104*
　male
　　breast 165
　　penis 164–5
　　prostate 161–2
　　testes 159
　thyroid gland 39
cardiovascular system
　following menopause *137*
　in pregnancy 183–5
　review *212*
catecholamines 53
　actions 53
　assays 230
　breakdown 53–4
　physiological effects *54*
　-producing tumours 54
　receptors 10
　regulation 53
　synthesis 53
　see also adrenaline; noradrenaline
CCK (cholecystokinin) *74*, 75
cell maturation 97
cell-surface receptors 10–11
cephalic presentations of fetus 187, *188*
cephalopelvic disproportion 190
cervical carcinoma 139–40
　dysplastic changes leading to 138
cervical dilatation 187–8
　disorders 189, 190
cervical ectopy 138
　in puberty 119–20
cervical intraepithelial neoplasia (CIN) 138–9
cervical polyps 140
cervical smears 138–9
　positive abnormal (Pap) 138–9, 142

268

Index

cervical swabs 220
cervicitis 138
cervix 118
  disorders 138–40
  in labour *see* cervical dilatation
  microstructure 119–20
  in pregnancy 183
  speculum examination 220
Chlamydia trachomatis
  cervical swabs for 220
  female 127, 128, 138
  male 158, 160, 161
cholecalciferol *see* vitamin D
cholecystokinin (CCK) 74, 75
cholesterol 8, 122
choriocarcinoma 195
  ovarian 130
chorionic villi
  abnormal development 195
  normal development 178–80
  see also human chorionic gonadotrophin (hCG)
chromophobe adenomas *see* adenomas, anterior pituitary, non-functioning
chromophobes 19
CIN (cervical intraepithelial neoplasia) 138–9
clear cell tumours 130
clinical chemistry 232–4
clomiphene 26, 177
coitus interruptus 173
collagen breakdown 137
colostrum 190, 192
colposcopy 139
coma 200, *202*
  differential diagnosis *202*
combined oral contraceptive pill (COCP) 175
combined pituitary test (CPT) 228
communication 211–14
  non-verbal skills 213
  obstacles to 212–13
  verbal skills 213
compression of the pituitary gland 26
computed tomography (CT) 237, *238*
condoms 174
cone biopsy 139, 140
congenital abnormalities
  female 144
  male 157, 162–3
congenital adrenal hyperplasia 51

Conn's syndrome 49, 51
  aldosterone assays 230
consultation, objectives of 213–14
contraception 172–6
  barrier 174
  hormonal 174–5
    complications 175
    types 175
    see also oral contraceptives
  irreversible 176
  natural 173–4
  prevention of implantation 176
contrast media 234–5
corpus luteum phase of pregnancy 181–2
corticosteroid-binding globulin (CSG) 123
corticotrophin releasing hormone (CRH) 20, 46
corticotrophs 19, *20*
cortisol 44, 45–8
  actions 47, 62
  assays 230–1
  deficiency 51–3
  excess 49–50
    see also Cushing's disease; Cushing's syndrome
  intracellular actions 47
  normal range *229*
  regulation *41*, 46–7
  see also steroid hormones
cranial nerve palsies 26
craniopharyngiomas 26, 27
cretinism 37, 99
CRH (corticotrophin releasing hormone) 20, 46
cryotherapy 139, 164
cryptorchidism 157
CSG (corticosteroid-binding globulin) 123
Cushing's disease 50, 51
Cushing's syndrome *48*, 49–50, 51
  as cause of diabetes mellitus 64
  dexamethasone suppression test 49, 230–1
cyclic AMP 10–11, 33, 61–2, 83, 89
cyclooxygenase pathway 10
cyproterone acetate 162
cystadenocarcinomas 130
cystadenomas 129
cystic teratomas, benign 129
cysts
  Bartholin's 142
  epididymal 159

  ovarian 128–9
  thyroglossal 31, 38
cytology 233–4

D

danazol 132
De Quervain's thyroiditis 37
dehydration 85, *86*
dehydroepiandrosterone (DHEA) 47
deiodination 34
depolarization 13, 60
depot progesterone 175
dermoid cysts 129
desmopressin
  test 230
  treatment 27
dexamethasone suppression test 49, 230–1
  high-dose 231
  low-dose 231
DHEA (dehydroepiandrosterone) 47
diabetes insipidus (DI) 27
  investigations 229–30
diabetes mellitus 63–70
  complications 66–7
    screening for 69
  diagnosis 67, 321
  hypoglycaemic attacks 69
  monitoring glucose control 69
  symptoms and presentation 64–6
  treatment 67–9
  types
    gestational 186
    IDDM/type 1 63, 64, 65, 66, 68
    NIDDM/type 2 63, 64, 65–6, 67, 68
diabetic coma 65
diaphragm, barrier contraception 174
dietary intake of calcium 87–8
dietary treatment
  of diabetes mellitus 68
  of osteoporosis 95
dihydrotestosterone (DHT) 156
disseminated intravascular coagulation (DIC) 52, 194
diuretic hormones 85
dopamine 9, *20*
  deficiency 23
  as natriuretic factor 85
drug history 210–11
duct papilloma, female breast 145

269

Index

ductus (vas) deferens 152
  vasectomy 176
dwarfism 23, 100
dysgerminomas 130
dysmenorrhoea 136
dyspareunia
  female 170
  male 163

E

eclampsia 193, 194–5
ectopic hormone syndromes 103–4
ectopic pregnancies 192–3
eicosanoids 9–10
ejaculation 153, 168
  premature 169
  retrograde 161, 162
ejaculatory ducts 152
ELISA 227, *228*, 229, 230
embryological development
  of gender 107–12
  see also development *under specific glands*; fertilization
empty sella syndrome 26
endocrine organs 4, *5*
endocrine secretion 3–4
endocrine system
  main components 4–6
  organization 4–7
  in pregnancy 185–6
  relationship to nervous system 12–13
  role 3
endocrine tissue
  arrangement 4
  definition 3, 4
  neurosecretory cells of hypothalamus 13, 18
  peripheral 5
endometrial carcinoma 134
endometrial cysts 129
endometrial hyperplasia 132
endometrial polyps, benign 133
endometrial stromal sarcoma 134
endometrioid tumours 130
endometriosis 131–2
endometritis 131
endometrium 119
  blastocyte implantation 172
  changes in menstrual cycle 126–7
  disorders 131–4
enteroendocrine (APUD) cells 73

enzyme linked immunosorbent assays (ELISA) 227, *228*, 229, 230
epididymal cysts 159
epididymis 150–2
epididymitis 158
epiphyseal growth plates, fusion 100
epispadias 162–3
epithelial hyperplasia, female breast 145
epithelial masses, ovarian 129, 130
erection
  impotence 169, 203
  mechanism 153, 168
erogenous zones 167
examination 214–24
exophthalmos 36, *218*
external genitalia
  embryological development 109–10, 111–12
  examination
    female 220, *221*
    male 220–2, *223*
eyes
  examination *217–8*
  visual disturbances 26, 36
  see also retinopathy, diabetic

F

face, examination *214*
fallopian (uterine) tubes 117–18
  ectopic pregnancy 192, 193
  fertilization site 171
  ligation 176
  PID 127, 128
  tumours 131
family history 211
fasting blood glucose 67, 231
  normal range *233*
fat metabolism, in pregnancy 186
fat necrosis, female breast 145
feedback 6
  hypothalamic–pituitary 21–2
  negative 6, 122, 123
feet
  general examination 215–17
  screening for peripheral neuropathy 69
female condom 174
female embryological development 111–12
female fetus, meiosis 121
female infertility *see* infertility
female orgasm and resolution 168
female puberty 113–14

female reproductive system 115–47
  development 111–12
  disorders 127–47
  examination 220–1
  hormones 122–4
  menstrual cycle 124–7
  oogenesis 121–2
  organization 116–21
female sexual dysfunction 169–70
femoral hernia 157
fertility
  male 155
  see also infertility
fertilization 170–2
  meiosis at 122
fetal growth 99
fetus
  female, meiosis 121
  position, onset of labour 187
fibroadenomas, female breast 145
fibrocystic change, female breast 145
fibroids *see* leiomyomas, benign (fibroids)
fibromas, ovarian 129
fluid balance 81–6
  disorders 85–6
  hormones involved in 83–5
  natriuretic factors 85
  regulation 82–3
fluid retention 85, *86*
fluorescein angiography 234
follicle stimulating hormone (FSH) 19, 21, *24*, 122, 123
  assays 227, 232
  deficiency, in PCOS 128
  GnRH stimulation test 232
  hormonal contraceptive effects 174
  infertility investigation and treatment 177, 178
  in male 157
  puberty 113
  in menopause 136–7
  in menstrual cycle 126
  oogenesis 121–2
  in pregnancy 185, *186*
  replacement 27
follicular cysts 129
follicular phase of menstrual cycle 124–7
forceps (assisted) delivery 190

Index

fructosamine, blood glucose testing 69
FSH *see* follicle stimulating hormone

G

G-cells 75
G-protein coupled receptors 10–11
galactorrhoea *104*, 200, *201*
gamete intrafallopian transfer (GIFT) 178
Gardnerella vaginalis 140
gastrin 74, 75
  Zollinger–Ellison syndrome 70
gastrointestinal system
  polypeptide hormones 7, 73–6
  review *212*
  see also intestine
gender, embryological development of 107–12
general inspection 214, *214–16*
genital ducts, embryological development 109
genital warts 142
  male 164
germ cell tumours
  female 129, *130*
  male 159–60
gestational diabetes mellitus 186
GI tract *see* gastrointestinal system; intestine
GIFT (gamete intrafallopian transfer) 178
gigantism 100
GIP (glucose-dependent insulinotrophic peptide) 74, 76
gliomas
  anterior pituitary 26, 27
  hypothalamus 23
glucagon 57, 61–2, *63*, 64
  gastrin-stimulated secretion 75
glucagon pens 69
glucagonomas 70
glucocorticoids *see* cortisol
glucose
  control of homeostasis 62–3, *64*
  control and monitoring in diabetes mellitus 69
  investigations 67, 69, 231
  metabolism in pregnancy 186
glucose tolerance test (OGTT) 25, 67, 231
glucose-dependent insulinotrophic peptide (GIP) 74, 76
glycosuria 67

goitres 37–9, 203, *206*
gonadarche 112
gonadotrophin releasing hormone (GnRH)
  analogue, treatments 132, 133
  male 162
  congenital deficiency 23
  and leptin 78
  in puberty 112
  stimulation test 232
  triple stimulation (CPT) test 228
gonadotrophs 19, *20*
gonads
  embryological development 107–9, *108*
  function tests 232
gonorrhoea *see Neisseria gonorrhoeae*
Graafian follicle 122, 126
granulosa cell tumours 129
Graves' disease 36
  goitre 38
  thyroid-stimulating antibodies (TsAb) assay 230
growth 97–101
  direct control 97–8
  disorders 100–1
  indirect control 98–100
growth factors 99
  see also insulin-like growth factors (IGFs)
growth hormone (GH) 19, 21, *24*, 97–8
  assay 228
  deficiency 100
  effects 98, 100
  excess 100
  see also acromegaly
  normal range *229*
  in pregnancy *186*
  regulation of secretion 98
  stimulation test 228–9
  suppression test 25
  treatment 27, 100
growth hormone inhibiting hormone (GHIH/somatostatin) *20*, 74, 76, 98
growth hormone releasing hormone (GHRH) *20*, 98
  congenital deficiency 23
growth spurt, pubertal 113
gumma 159
gummata 164
gynaecomastia *104*, 165, 200, *201*

H

haematocoele, testicular 157–8
haematology 233
haematoma, testicular 157
hands, examination 215
Hashimoto's thyroiditis 37
  autoantibody assays 230
HbA_{1C}, blood glucose testing 69
head, examination 217–19
height, determinants of 100
hernias, male 157
herpes simplex virus (HSV)
  female 141–2
  male 163–4
hirsutism 200, *202*
histopathology 233–4
history taking 209–11
  preparations 209
  structure 209–11
  summary 211
HONK coma 65
hormonal control of neurons 12
hormone measurement methods 227
hormone replacement therapy (HRT) 27, 137
hormones
  control of secretion 6
  definition 3
  types 3–4
  and secretion 7–9
HPV *see* human papilloma virus
human chorionic gonadotrophin (hCG) 182
  neoplasias of trophoblastic origin 195
  pregnancy test 232
    ectopic pregnancy 192
  secretion following implantation 181, 182
  and TSH 185
  tumour marker 159
human papilloma virus (HPV) 138, 139
  genital warts 142, 163
  penile neoplasias 164, 165
human placental lactogen (hPL) 182–3, 186, 190
HVS *see* herpes simplex virus
hydatidiform mole 195
hydrocoeles, congenital 157
hydrosalpinx 128
5-hydroxytryptamine (5-HT)
  ectopic secretion 39
  possible role in PMS 136
  respiratory tract secretion 73

271

Index

hyperaldosteronism 49, 51
hypercalcaemia 92, *93*, 104
hyperglycaemia 63
　symptoms 65
hyperosmolar non-ketotic (HONK) coma 65
hyperparathyroidism
　ectopic hormone syndromes *104*
　primary 92–4
　secondary 94–5
　tertiary 95
hyperpituitarism 23–5
hyperprolactinaemia 23–4, *25*, 26, 199–200, *201*
hypertension
　paroxysmal 54
　pregnancy-induced (PIH) 193–5
hyperthyroidism 35–6
　see also thyrotoxicosis
hypertrichosis 200
hypocalcaemia 94, 95
hypoglycaemia 63
　caused by diabetes treatment 69
　in non-diabetic patients 69–70, *104*
　symptoms *70*
hypoglycaemic agents, oral 68
hypokalaemia *104*
hyponatraemia *104*
hypoparathyroidism 95
hypopituitarism 25–7
hypospadias 162, *163*
hypothalamic–pituitary dysfunction 232
hypothalamus 4–5, 12, 15
　anatomy 15, *16*
　blood supply 17
　development 17
　disorders 23
　hormones
　　modified amino acid 9
　　polypeptide 7
　　regulating anterior pituitary 20, 21–2
　　released from posterior pituitary 20, 22, *23*
　investigations 227, 228–9, 232
　microstructure 18
　negative feedback of ovarian sex steroids 122, 123
　neurosecretory cells 13, 18
　pituitary gland development 17–18
　suprachiasmatic nucleus (SCN) 77, 78

hypothyroidism 37
hypovolaemia 81
hysterosalpingography 235

I

ICSI (intracytoplasmic sperm injection) 178
IGFs *see* insulin-like growth factors
imaging 234–9
implantation 171–2
　prevention of 176
impotence 169, 203
in vitro fertilization (IVF) 178
indirect inguinal hernia 157, *158*
induction of labour 189
infarction of the pituitary gland 26
infections
　breast 144–5
　cervix (HPV) 138, 139
　endometrium 131
　penis 163–4
　PID 127, 128
　prostate gland 160–1
　testes and epididymis 158–9
　vagina 140
infertility 176–8
　female causes 177
　　chronic PID 128
　　leiomyomas 133
　　role of leptin 78
　investigations 177, 232
　male causes 176
　　cryptorchidism 157
　treatment 177–8
　see also fertility
inguinal canal 149
inguinal hernia, indirect 157, *158*
inhibin 124, 183
insulin 59–61, 62, 63, *64*
　actions 12, 60–1
　control of secretion 60
　gastrin-stimulated secretion 70, 75
　in PCOS 128
　receptors 11, 60
　subcutaneous injections 68–9
　synthesis 60
　triple (CPT) stimulation test 228
insulin dependent diabetes mellitus (IDDM/type 1) 63, 64, 65, 66, 68
　see also diabetes mellitus

insulin tolerance test 26
insulin-like growth factors (IGFs) 98, 100
　assay 228
　receptors 11
insulinomas 70
intermenstrual bleeding 200, *203*
　differential diagnosis *203*
internal genitalia
　female
　　embryological development 111
　　examination 220–1
　male, embryological development 110–11
　see also specific organs
intestine
　calcium homeostasis 90, 91
　vitamin D effects 91
　see also gastrointestinal system
intra-uterine devices (IUD) 176
　progesterone-releasing (Mirena®) 175
intracellular receptors 10, 11, *12*
intracytoplasmic sperm injection (ICSI) 178
introducing yourself to patient 209, 214
invasive mole 195
investigations 227–32
iodide (I_2) oxidation 32
iodination of thyroglobulin 32
iodine deficiency 38
iodine (I^-) trapping 32
iodine metabolism 33
iodine, radioisotope 238–9
IUD *see* intra-uterine devices
IVF (in vitro fertilization) 178

J

jet-lag 77–8

K

Kallmann's syndrome 23
ketoacidosis
　absence of, in NIDDM 65–6
　symptoms in IDDM 65
kidney *see* nephropathy, diabetic; *entries beginning* renal
kinins 85
Kjelland's forceps 190
Klinefelter syndrome 111, *201*
Krukenberg tumour 130

L

labour 187–90
  disorders 189–90
  hormonal control 189
  induction 189
  initiation 189
  position of fetus 187
  sequence 187–8
lactation 190–2
Lactobacillus vaginalis 120, 138
lactotrophs 19, *20*
laparoscopy, investigative
  endometriosis 132
  ovarian masses 130–1
laser therapy 139
leiomyomas, benign (fibroids) 133, *134*
  cervical 140
leiomyosarcomas 134
leptin 12, 78
  in puberty 112–3
leucoplakia, vulval 142, *143*
leukotrienes 10
Leydig cell(s) 150, 156, 157
  fetal 110, 111
  tumours 160
LH *see* luteinizing hormone
libido, reduced 169, 203
lichen sclerosus 142, *143*
limbs, examination 215, *216*
lipoxygenase pathway 10
loss of consciousness 200, *202*
  differential diagnosis *202*
lump, examination 220
lung carcinoma 51, *104*, 105
luteal cysts 129
luteal phase of menstrual cycle 127
  inadequate 132
luteinizing hormone (LH) 19, 21, 24, 122, 123
  assays 227, 232
  excess, in PCOS 128
  in follicular phase of menstrual cycle 126
  GnRH stimulation test 232
  and hCG 182
  hormonal contraceptive effects 174
  infertility investigation and treatment 177, 178
  in male 157
    puberty 113
  in menopause 136–7
  oogenesis 122
  in pregnancy 185, *186*
  replacement 27
lymphoma, primary 160

M

macroadenomas *see* adenomas, anterior pituitary, non-functioning
magnesium sulphate 194–5
magnetic resonance imaging (MRI) 237–8
male condom 174
male embryological development 110–11
male infertility *see* infertility
male orgasm and resolution 168
male puberty 113
male reproductive system 149–66
  development 110–11
  disorders 157–65
  hormones 156–7
  organization 149–54
  spermatogenesis 154–6
male sexual dysfunction 169
mammary duct ectasia 145
mammography 236
MAO (monoamine oxidase) 53
mass, examination 220
mastitis, acute 144
maternal adaptations to pregnancy 183–6
mature-onset diabetes (MODY) 64
mean parental height (MPH) 100
medical termination of pregnancy 178
mefenamic acid 136
meiosis
  in female 121–2
  following fertilization 171
  in male 154
melanocyte-stimulating hormone (MSH) 185
melatonin 9, 12, 77
  treatment for jet-lag 77–8
menarche 113–4
menopause 136–7
  endometrial changes 133
menorrhagia 135–6, 200, *203*
  differential diagnosis *203*
menstrual cycle 121–2, 124–7
  disorders 135–6
  endometrial changes 126–7
  follicular (first) phase 124–7
  luteal (second) phase 127
  onset in puberty 114
  ovarian changes 126, 127

menstruation 124–6
  retrograde 131
metaplasia
  of endocervix 138
  of peritoneal epithelium 131
metastatic endometrial tissue 131
metastatic ovarian tumours 130
metastatic prostate carcinoma 162
methotrexate 193
microadenomas 23
microbiology 233
mifepristone 178
mineralocorticoids 45
  see also aldosterone
Mirena® (IUD) 175
MIS (Müllerian inhibiting substance) 111, 157
miscarriage 193
mitosis, following fertilization 171
mnemonics
  adrenal cortex, order of layers 44
  adrenal hormone diseases 48
  anterior pituitary hormones 21
  consultation objectives 214
  hypercalcaemia 92
  hypoglycaemia in non-diabetic patients 69–70
  presenting history of pain 210
  spermatic cord contents 152
  symptoms of hyperthyroidism 35
modified amino acid(s) 9
  secreting cells 9
MODY (mature-onset diabetes in the young) 64
monoamine oxidase (MAO) 53
'morning-after' pill 176
morula 171–2
MPF (mean parental height) 100
MRI (magnetic resonance imaging) 237–8
MSH (melanocyte-stimulating hormone) 185
Müllerian inhibiting substance (MIS) 111, 157
multinodular goitre 39
multiple endocrine neoplasia (MEN) syndromes 103, *104*
mumps 159
musculoskeletal system review *212*
myometrial disorders
  leiomyomas 133, *134*
  leiomyosarcomas 134
myometrium 119
myxoedema (hypothyroidism) 37
myxoedema coma 37

273

N

nails, examination 216
natriuretic factors 85
neck, examination 217–19
negative feedback 6
  ovarian sex steroids 122, 123
Neisseria gonorrhoeae
  female 127, 138, *141*
  male 158, 160, 164
neoplasias (benign and malignant)
  ectopic hormone syndromes 103–5
  endocrine organs
    adrenal cortex 51
    adrenal medulla 54–5
    anterior pituitary 17, 23–6, 27, 50
    hypothalamus 23
    pancreas, islet-cell 70
    parathyroid glands 92–3
    thyroid gland 39
  female
    cervix 138–40
    fallopian tubes 131
    ovarian 128–31
    trophoblastic origin 195
    uterus 133–4
    vagina 142
    vulva 142–3
  male
    breast 165
    penis 164–5
    prostate 161–2
    testes 159–60
  MEN syndromes 103
nephropathy, diabetic 68
  screening for 69
nervous system
  autonomic (sympathetic/parasympathetic) 12, 42, *43*, 168
  relationship to endocrine system 12–13
  review *212*
  sexual arousal 167, 168
  see also peripheral neuropathy
neural control of secretion 6, 12
neurocrine 3, *4*
neuroendocrine cells 73
neurohypophysis *see* posterior pituitary
neurosecretory cells/granules, hypothalamus 13, 18, 19
Neville–Barnes forceps 190
nipples
  inversion 144
  supernumerary 144
non-gonococcal urethritis 164
non-insulin dependent diabetes mellitus (NIDDM/type 2) 63, 64, 65–6, 67, 68
  see also diabetes mellitus
non-verbal communication skills 213
noradrenaline 12
  secreting cells 42, 44
  see also catecholamines

O

obesity 205
  and diabetes 64
  differential diagnosis *207*
  leptin as treatment for 78
octreotide 25
oestradiol-17β assay 232
oestrogen(s) 122, *123*, *124*
  in anovulatory cycles 132
  assays 227
  contraception 174, 175
  excess exposure
    breast cancer 146
    cryptorchidism 157
  following implantation 182
  in males 156, 157
  in menopause 136–7
  in menstrual cycle 126–7
  in PMS 136
  in pregnancy 182
    breast development 190, 191
    in puberty 112, 113, 119–20
  replacement 27, 137
  see also steroid hormones
OGTT (oral glucose tolerance test) 25, 67, 231
oocyte 170
  fertilization 171
  primary 121–2
  secondary 122
oogenesis 121–2
oral contraceptives 175
  drug history 211
  effects 126, 127, 174–5
  endometrial changes 133
  treatment
    of endometriosis 132
    of menorrhagia 136
oral glucose tolerance test (OGTT) 25, 67, 231
oral hypoglycaemic agents 68
oral mifepristone 178
orchiditis
  acute 158
  autoimmune granulomatous 159
  viral 159
orgasm and resolution
  female 168
  male 168
osmolarity *see* blood osmolarity
osteomalacia 92, *93*, 94
osteoporosis 95, *137*
ovarian changes, in menstrual cycle 126
ovarian cysts 128–9
ovaries 116–17
  disorders 127–31
  embryological development 111
  hormones 8, 122–4
    see also oestrogen(s); progesterone
  microstructure 117, *118*
  oogenesis 121–2
over-the-counter (OTC) preparations 211
ovulation 127
oxytocin 20, 22
  control of parturition 189
  deficiency 27
  and ergometrine (Syntometrine®) 188
  induction of labour 189, 190
  milk ejection 191

P

palpation
  breast *223*
  male external genitalia *223*
  and percussion of abdomen 219
pancreas 57–71
  blood supply, lymphatics, and nerves 59
  development 59, *60*
  disorders 63–70
  function tests 231
  hormones 7, 59–62, 76
  location and anatomy 58–9
  microstructure 59
pancreatic duct 58–9
pancreatic polypeptide 76
papillary hidradenoma 143
paracrine hormones 3, *4*, 9–10
parathyroid glands
  anatomy and blood supply 30
  development 31
  microstructure 31
parathyroid hormone (PTH) 89–90
  actions 89–90
  deficiency 95

Index

excess 92–5, *104*
  synthesis 89
parathyroid receptors 89
parturition *see* labour
past medical history 210
patient details, history taking 209–10
PC *see* presenting complaint
PCOS (polycystic ovarian syndrome) 128
pelvic inflammatory disease (PID) 127
  acute and chronic 128
  and endometritis 131
'penile intraepithelial neoplasia' (PIN) 164–5
penis 153–4
  development in puberty 113
  disorders 162–5
  microstructure 153
  see also erection
peripheral endocrine tissues 5–6
peripheral neuropathy 68
  screening for 69
peripheral resistance 81
permission
  to examine 214
  to take history 209
phaeochromocytomas 54–5
  VMA levels 54, 230
phimosis 163
phyllodes tumours 145
PID *see* pelvic inflammatory disease
PIN ('penile intraepithelial neoplasia') 164–5
pineal gland 9, 76–8
  function and regulation 77
  melatonin 9, 12, 77–8
  structure 76–7
pituitary gland 5, 7
  anatomy 15–17
  blood supply 17
  see also anterior pituitary; posterior pituitary
pituitary insufficiency *see* hypopituitarism
placenta 178–81
  development 178–80
    see also trophoblast
  functions 180–1
  hormonal secretion 181, 182
  structure 180
placenta previa 193
placental abruption 193
placental lactogen 99, *104*
placental phase of pregnancy 182

plain X-ray radiography 234
plasma osmolarity *see* blood osmolarity
PMS (premenstrual syndrome) 136
polycystic ovarian syndrome (PCOS) 128
polydipsia 203
polymastia 144
polypeptide hormones 7–8
  GI tract 7, 73–6
  receptors 10
polypeptide secreting cells 8
polyuria 203, *204*
  differential diagnosis *204*
postnatal development 112
postcoital douche 173
postcoital medication 176
posterior pituitary
  development 18
  disorders 27
  hypothalamic hormones released from 20, 22, *23*
  investigations 229–30
  microstructure 19
pre-eclampsia 193–5
pregnancy
  breast development 190–1
  disorders 192–5
  maternal adaptations 183–6
  reproductive hormones in 181–3
  signs and symptoms 183, *185*
  thyroid hormones in 39
pregnancy test 232
  ovarian masses 130
premature ejaculation 169
premenstrual syndrome (PMS) 136
presenting complaint 210
  history of 210
primary atrophic hypothyroidism 37
primary dysfunctional labour 189–90
primary lymphoma 160
primordial follicles 121–2
progesterone 122–3, *124*
  assays 227, 232
  contraception 174, 175
  following implantation 181–2
  inadequate luteal phase 132
  in males 156
  in menstrual cycle 127
  neoplasia of trophoblastic origin 195
  in pregnancy 182
    breast development 190, 191
  replacement 27, 137
  see also steroid hormones

progesterone-only pill (POP) 175
progestogens *see* progesterone
prolactin 19, *20*, 21, 22
  assay 232
  ectopic secretion *104*
  excess 23–4, *25*, 26, 199–200, *201*
  normal range *229*
  in pregnancy 183, 185, *186*
    breast development and milk secretion 190–1
prolactin releasing hormone *20*
prolactinomas 23–4, *25*, 26
prolonged labour 189–90
prostaglandins 10, 47
  control of parturition 189
  induction of labour (at term) 189
  medical termination of pregnancy 178
  renal 85
prostate 152–3
  disorders 160–2
  examination 222, *224*
  microstructure 153
'prostate-specific antigen' 161
'prostatism' 161
PTH *see* parathyroid hormone
pubertal growth spurt 113
puberty 112–14
  female 113–14
    cervical ectopy 119–20
  male 113
  sex hormones in 99–100
  thyroid hormones in 39
pyosalpinx 128

R

radioactive iodine therapy 35
radioisotope scans 238–9
radiotherapy 23, 25
receptor mediated control 11
receptors 3, 10–11
relaxin 124, 183, 189
'releasing' hormones 5
renal effects
  ADH 83–4
  aldosterone 85
  angiotensin II 84
  calcitonin 92
  natriuretic factors 85
  PTH 89–90
  vitamin D 91
renal failure, chronic 94–5
renal osteodystrophy 94–5
renal prostaglandins 85
renal system, in pregnancy 184–5

275

Index

renin 81, 82, 84
  normal range 229
renin–angiotensin II system 83, 84–5
reproductive system
  embryological development 107–12
  review 212
  see also female reproductive system; male reproductive system
respiratory system
  in pregnancy 184
  review 212
retinopathy, diabetic 67, 68
  screening 69
rhythm method 173–4
rickets 92, 93
round ligament 118

S

sclerosing adenosis, female breast 145
SCN (suprachiasmatic nucleus) 77, 78
scrotal lumps 203, 204
  examination 222, 224
scrotum 152
SDY (sex-determining region) 107, 110, 111
second messenger, definition 3
secondary adrenocortical insufficiency 53
secondary arrest of labour 190
secondary hyperaldosteronism 49
secretin 74, 75
seminal vesicles 152
seminoma 159
  mixed 160
Sertoli cell(s) 150, 154, 155, 157
  in autoimmune granulomatous orchiditis 159
  fetal 111
  tumours 160
sex steroids see androgens; oestrogen(s); progesterone; testosterone
sex-cord stromal tumours 160
sex-determining region (SDY) 107, 110, 111
sex-hormone-binding globulin (SHBG) 122, 156
sexual arousal 167–8
sexual dysfunction 169–70, 203, 205

sexual intercourse, physiology 167–8
sexually transmitted diseases (STDs) 127, 140–2, 163–4
Sheehan's syndrome 26
shoulder presentation of fetus 187
SIADH (syndrome of inappropriate secretion of ADH) 27, 84
Sipple's syndrome (MEN-IIa) 103
skin, examination 214
social history 211
sodium
  excess/deficiency 86, 104
  importance of 82
  natriuretic factors 85
  normal range 233
  renin–angiotensin II system effects 84–5
  see also aldosterone
solid teratoma 130
somatomedins see insulin-like growth factors (IGFs)
somatostatin (GHIH) 20, 74, 76, 98
somatotrophin see growth hormone (GH)
somatotrophs 19, 20
speculum, internal examination with 220
spermatic cord 152
spermatid 149
spermatocoele 159
spermatogenesis 149, 154–6
  in puberty 113
spermatozoa 149, 170–1
  abnormalities 176, 177
  mature structure 155–6
spontaneous abortion 193
squamous cell hyperplasia, vulval 142, 143
staging, of carcinomas
  breast 146, 147
  cervical 140
  endometrial 134
  ovarian 131
  prostatic 162
  vulval 143
starvation, symptoms in IDDM 65
STDs (sexually transmitted diseases) 127, 140–2, 163–4
sterilization 176
steroid hormones
  receptors 11
  synthesis 8–9, 45
  see also specific hormones
steroid secreting cells 8, 9

stimulation tests 26, 227
  ACTH (Synacthen®) 51, 231
  GH 228–9
  GnRH 232
  TRH 230
  triple (CPT) 228
STOP (surgical termination of pregnancy) 178
stress response 47, 53, 62
stromal tumours
  female 129, 134
  male 160
struma ovarii 129
subdermal progesterone implants 175
sulphonylureas 68
supernumerary nipples 144
suppression tests 24–5, 227
  desmopressin 230
  dexamethasone 49, 230–1
  GH 229
suprachiasmatic nucleus (SCN) 77, 78
surgical termination of pregnancy (STOP) 178
Synacthen® (ACTH) stimulation test 51, 231
  prolonged 231
  short 231
syndrome of inappropriate secretion of ADH (SIADH) 27, 84
'Syndrome X' 64
syphilis 159, 164
systems review 211, 212

T

T_3 (tri-iodothyronine) see thyroid hormones
T_4 (thyroxine) see thyroid hormones
target cells 3, 5–6, 10
TB (tuberculosis) 159
technetium–99m pertechnetate 239
teratomas
  female 129, 130
  male 160
testes 150
  development in puberty 113
  disorders 157–60
  embryological development 110
  hormones 156–7
    see also androgens; testosterone
  microstructure 150, 151

testicular feminization syndrome 135
testicular haemorrhage 157–8
testosterone 156–7
  assay 232
  control of testicular production 157
  deficiency 157
  normal range 229
  in puberty 112, 113
  replacement 27
  see also androgens; steroid hormones
testosterone-secreting interstitial cells see Leydig cells
thecomas 129
therapeutic abortion 178
thiazolidinediones 68
thorax, examination 219
thromboembolism 175
thromboxanes 10
thrush see Candida albicans
thyroglobulin 29, 31
  iodination 32
  synthesis 32
thyroglossal cyst 31, 38
thyroid autoantibody assays 230
  Graves' disease 35, 230
  Hashimoto's thyroiditis 37, 230
thyroid function tests 230
thyroid gland 29–40
  anatomy 30
  blood supply, nerves, and lymphatics 30
  development 31
  disorders 34–9
  microstructure 31, 32
  in pregnancy 185
thyroid hormones 9, 29, 32–4, 35
  actions 34
  assays 227, 230
  deficiency 12
  diagnosis and treatment of thyroid disorders 35, 37
  feedback 34
  influence on growth 99, 100
  normal ranges 229
  in puberty and pregnancy 39
  regulation 33
  synthesis 32–3
  transport 34
thyroid lumps 203, 206
  differential diagnosis 206
thyroid stimulating hormone (TSH) 5, 19, 21, 24, 29, 33–4
  in hyperthyroidism 35, 36
  in hypothyroidism 37, 38
  normal range 229
  in pregnancy 185, 186
thyroid-stimulating antibodies (TsAb) assay 230
thyroiditis 39
thyrotoxic crisis 35
thyrotoxicosis 35, 36
  as cause of diabetes mellitus 64
  see also hyperthyroidism
thyrotrophin see thyroid stimulating hormone (TSH)
thyrotrophin-releasing hormone (TRH) 5, 20, 33
  stimulation test 230
  triple stimulation (CPT) test 228
thyrotrophs 19, 20
thyroxine (T_4) see thyroid hormones
torsion of the testes 158
toxic goitre 37
tranexamic acid 136
transformation zone (cervix) 120, 138, 139
  large loop excision (LLETZ) 139
transurethral resection of the prostate (TURP) 161, 162
Treponema pallidum 159, 164
TRH see thyrotrophin-releasing hormone
tri-iodothyronine (T_3) see thyroid hormones
Trichomonas vaginalis 138, 141
triple (CPT) stimulation test 228
trophoblast
  abnormal development 194, 195
  differentiation 172
  hCG secretion 181, 182
  neoplasias 195
tryptophan 9
TsAb assay 230
TSH see thyroid stimulating hormone
tubal abortion 192
tubal ligation 176
tubal rupture 192
tuberculosis (TB) 159
tumours see carcinomas; neoplasias
Turner syndrome 111, 135
TURP (transurethral resection of the prostate) 161, 162
tyrosine 9, 32, 33
tyrosine kinase receptors 11, 60

U

ultrasonography 237
urethritis, non-gonococcal 164
urinary system, review 212
urinary tract infections (UTI) 158, 160
urine
  and blood osmolarity 229, 230
  24-hour urinary free cortisol assay 230
  monitoring
    in diabetes mellitus 69
    in pre-eclampsia and eclampsia 193, 194
uterine contractions
  Braxton Hicks (false labour) 188
  in labour 188, 189
  inefficient 189, 190
uterine ligaments 118–19
uterine neoplasias 133–4
uterine tubes see fallopian (uterine) tubes
uterosacral ligament 118–19
uterus 118–19
  bimanual examination 220–1
  microstructure 119
  in pregnancy 183

V

V receptors 83
vagina 120
  bimanual examination 220–1
  disorders 140–2
  microstructure 120
  in pregnancy 183
vaginal speculum, examination with 220
vaginismus 170
vanillyl mandelic acid (VMA) 53, 54, 230
varicocoele 158
vascular disorders
  diabetes mellitus 66–7
  testes 157–8
vas (ductus) deferens 152
vasectomy 176
vasopressin see anti-diuretic hormone (ADH)
ventouse extraction 190
verbal communication skills 213
VIN (vulval intraepithelial neoplasia) 142–3
virilism 200

virology 233
vitamin D 90–1
  deficiency 94–5
VMA (vanillyl mandelic acid) 53, 54, 230
vulva 120, *121*
  disorders 141–3
  embryological development 111–12
  examination 220, *221*
vulval carcinoma 143
vulval dystrophies 142, *143*
vulval intraepithelial neoplasia (VIN) 142–3

W

water
  excess and deficiency 85, *86*
  excretion 82
  importance of 82
  intake 82
water deprivation test 229–30
Waterhouse–Friderichsen syndrome 52–3
weight gain 205, *207*
  differential diagnosis *207*
  in puberty 112–3
weight loss 206
  differential diagnosis *207*

Wermer's syndrome (MEN-I) 103
Wrigley's forceps 190

X

X chromosome 107, 111
X-ray radiography 234

Y

Y chromosome 107
yolk sac tumours 130

Z

Zollinger–Ellison syndrome 70
zygote 171, *172*